W0115562

"As Medical Director of the National Rehabilitation Hospital, I was impressed that most patients with Post Polio Syndrome were highly motivated and persevered through the adversity of childhood and young adulthood muscle weakness. For some, the resulting weakness made even the simplest tasks very difficult, requiring creative solutions, physical compensations, and reliance on technology and assistive devices. A number of these individuals—corporate executives, physicians, lawyers, competitive college athletes—maximized their functional abilities and accomplished impressive achievements, and a few even had successful short careers in professional sports. I have been just as impressed with individuals with greater involvement with weakness who raised families and found a way to work full- or part-time. However, I do not recall any other patient who maintained such an active, stressful, and dangerous lifestyle for so long a time as General William Matz. He was fortunate in that the distribution and extent of his weakness primarily involved a foot drop in one of his lower limbs for which a military boot could adequately serve as a splint, reducing functional impairment resulting from the weakness.

When I asked him which physical compensatory methods or techniques he relied on over the many years, his response was always, 'It was natural … I just knew what I had to do.' Listening to him then, I sensed his story could be inspirational to others, especially those with Post Polio Syndrome. Having now read his manuscript, I am pleased to have recommended that he tell his unique story. *My Toughest Battle* is a book for our times."

—John N. Aseff, MD, former Medical Director, Post Polio Program,
MedStar National Rehabilitation Hospital

"The extraordinary life story of my good friend and fellow infantryman, Major General Bill Matz. As his commander, I served alongside him for six years in some of the Army's toughest infantry units, and saw firsthand the true grit and warrior spirit that allowed him to persevere and triumph over the most difficult circumstances. No task was too tough, and few knew he was battling the effects of polio as he "gutted out" every morning PT run and road march. Soldiers held him in high esteem, and servicemen and women today will benefit by reading his story. This heartfelt, lively story of courage and faithful service to family, country and the traditional American values of liberty, patriotism and religion is sure to inspire, encourage and reassure readers of all ages and persuasions. Americans searching for an "uplifting" story of selfless service and overcoming physical adversity will find it in this book."

—Lieutenant General Carmen J. Cavezza, U.S. Army (Ret.),
Former Commanding General I Corps and Fort Lewis, Chairman &
CEO Emerita, National Infantry Museum Foundation

"An incredible story of courage, perseverance, and commitment. Bill Matz is extraordinary, a leader of character with engrossing perspectives on behind-the-scenes military circumstances. His account of his time as a rifle company commander with the Mobile Riverine Force in Vietnam and his description of the Al-Qaeda attack on his project management compound in Saudi Arabia are fascinating. An illuminating, thought-provoking, inspirational narrative and a great read for anyone who desires to learn more about crisis management."

—Lieutenant General Robert F. Foley, U.S. Army (Ret.),
Vietnam Medal of Honor Recipient

"More than ever, America today needs real heroes whom average citizens can identify with, look up to, and emulate. General Bill Matz provides just such an inspirational story, starting from humble beginnings, struggling with polio, leading courageously in war and peace, and accomplishing heroic missions for our nation. ... Reading this book reinforces the greatness of America and those who serve it so effectively. It will inspire future generations to follow General Matz's example of perseverance and service."

—Brigadier General Michael J. Meese, PhD, U.S. Army (Ret.)

"In this fine, carefully documented book, you meet William Matz—Billy, Bill, General, Military Contractor, Association Chair, Commission Executive—who vividly details the many phases, adventures and challenges of his life. You experience through his eyes the triumphs and travails of youth, the frustrations of the rice-paddy war in Vietnam, the horror and carnage of terrorism, walking and working the halls of Congress, and caring for the overseas graves of our fallen heroes. ... And in the Afterword, you will read the honest, forthright expressions of concern and frustration, from the personal view of one from the post-WWII generation, over the direction he sees his nation is now heading if left unchecked, veering away from the traditional values he grew up enjoying and that shaped his life. I commend this book to all. It will be exciting to some, nostalgic to some, instructive to some, but a worthy read to all."

—Fred F. Fielding, Attorney and Counselor, Former Counsel to the
President (Presidents Reagan and George W. Bush),
Former Commissioner, 9-11 Commission

"Inspirational people worthy of emulation are in short supply. This book was written by one of them. Bill Matz is a fighter. All his life he has battled one thing or another, most notably polio and America's enemies. He is the embodiment of a phrase we don't hear much anymore: 'The American way of life.'

In 1959, when I met the kid who would become a general, Bill was no hero to me. It was enough that he would become my fraternity brother and friend. Never did I suspect that mornings in the bathroom of the Phi Gamma Delta house at Gettysburg College, I was brushing my teeth next to a hero. I had no inkling of the battle he was waging with our generation's signature disease. Or that he would soon be parachuting from combat planes to help protect and preserve my right to go to grad school, dream my dreams, and down an occasional beer at Meadowbrook Tavern.

Bill Matz would never call himself a hero. I would ... and am happy to endorse his story."

—Jerry Spinelli, author and Newbery Medal Winner

MY TOUGHEST BATTLE

A Soldier's Lifelong Struggle with Polio

MAJOR GENERAL WILLIAM M. MATZ, JR.,
U.S. ARMY (RET.)

CASEMATE
Pennsylvania & Yorkshire

Published in the United States of America and Great Britain in 2025 by
CASEMATE PUBLISHERS
1950 Lawrence Road, Havertown, PA 19083, USA
and
47 Church Street, Barnsley, S70 2AS, UK

Copyright © 2025 Major General William M. Matz, Jr., U.S. Army (Ret.)

Hardcover Edition: ISBN 978-1-63624-498-3
Digital Edition: ISBN 978-1-63624-499-0

A CIP record for this book is available from the British Library

All rights reserved. No part of this book may be reproduced or transmitted in any form or by any means, electronic or mechanical including photocopying, recording or by any information storage and retrieval system, without permission from the publisher in writing.

Printed and bound in the United States of America by Integrated Books International
Typeset in India by DiTech Publishing Services

For a complete list of Casemate titles, please contact:

CASEMATE PUBLISHERS (US)
Telephone (610) 853-9131
Fax (610) 853-9146
Email: casemate@casematepublishers.com
www.casematepublishers.com

CASEMATE PUBLISHERS (UK)
Telephone (0)1226 734350
Email: casemate@casemateuk.com
www.casemateuk.com

Cover images: (left to right) Being presented the Distinguished Service Cross by General Creighton Abrams, September 1968. (US Army photograph, 9th Infantry Division); Major General Bill Matz, 1995. (Private collection); Receiving Sister Kenny moist hot pack therapy for polio at Home of the Merciful Savior for Crippled Children Hospital in 1944. (Courtesy of Home of the Merciful Savior for Crippled Children)

Cover concept by Julie Brown
Cover design by Casemate Publishers

Contents

Foreword

During a long military career, one gets to work with and meet many different types of people across the various branches of service, often never having the opportunity to learn their personal stories. Despite having known Bill for a number of years, due to his humility I wasn't totally familiar with his story of overcoming childhood adversity and how the effects of polio later presented extreme challenges during his Army service and afterward. I must say that I am even more impressed now by his story of courage and perseverance during his outstanding military career.

As a fellow Vietnam veteran, I understood his story of infantry combat and leadership challenges; many of us faced similar experiences to some degree, a few of us, perhaps, more successfully than others. Bill's demonstrable leadership and commitment to accomplish every mission and secure every objective, without exception, shine through. This is where character comes into play, resulting in the inner strength to do whatever is required to get the job done. Bill Matz served above and beyond what was expected in both the Army and in the federal government at the highest levels, and his record of performance speaks for itself. He placed duty to our country above all else and the needs of others above his own. Bill is the essence of an American fighting man!

By reading this book, I think that many people in general, but younger Americans in particular, will learn all facets of love of country, ethics, perseverance in the face of incredible adversity, but most of all, loyalty and honor from a man who led by example. This book is more than an autobiography; it is truly a walk through time, taking the reader through decades of our country's history, illustrating the true goodness of our nation and our men and women who wear the uniform. Our nation was well served by his exemplary service, and I am proud to call Bill a fellow warrior and friend.

Major General James E. Livingston USMC (Ret.)
Medal of Honor, Vietnam War

Acknowledgements

Writing the story of my life has been no small task, but more rewarding than I could have ever imagined. I have a legacy to pass on to my friends and family, particularly my seven grandsons, where no recorded legacy existed before. Stricken with infantile paralysis (polio) in my childhood years, I was blessed with many who helped me overcome the effects of the life-threatening disease.

To my mother and father who raised·me and my sisters, Becky and Diane, in a God-fearing home and taught us right from wrong, I can never thank them enough for their sacrifice, dedicated care, and optimism during those years recovering from polio.

I am indebted to Sister Elizabeth Kenny whose innovative treatment facilitated my ability to regain the use of my paralyzed leg, and to the staff at the Home of the Merciful Savior for Crippled Children for researching and providing me with my medical records and photographs from 75 years ago. Without their care, there would not be this story to tell.

I am grateful to Doctor John Aseff for his encouragement to "tell my story now." He convinced me to begin writing.

I am thankful to my friend, Major General Jim Livingston, U.S. Marine Corps (Ret.), author and Medal of Honor recipient, for his encouragement and for his suggestion that I use his author to help me navigate the process of writing a book. Colin Heaton, a veteran of both the Army and the Marine Corps, has been invaluable throughout this project. I couldn't have had a better editor and writing mentor. He guided me through the complicated process of authoring and launching a book. His skill, patience, and familiarity with the military have been valuable beyond estimation, including introducing me to his principal editor, Marilyn Walton.

Marilyn, the daughter of a World War II POW, is an award-winning author and writer of five World War II books. Her meticulous editing and organizational suggestions enhanced the story immeasurably. She accomplished this despite the tragedy of unexpectedly losing her pilot son in an airplane crash two-thirds of the way through the project. Both she and Colin displayed inexhaustible patience and were a joy to work with.

To my friends Lieutenant Generals Carmen Cavezza, Medal of Honor recipient Bob Foley, and Keith Kellogg, as well as Brigadier General Mike Meese, and

Colonel Mike Conley, I am grateful for their constructive suggestions. Special thanks to Steve Hein and Rick Jones, at the National Association for Uniformed Services (NAUS), for their assistance in helping me remember the events of those busy six and a half years fighting for veterans and their families.

Colonel (Ret.) Dan Brownlee's and Glen Edmonson's "eyewitness" accounts of the Al-Qaeda attack in Saudi Arabia were extremely helpful. Their vivid memories helped me fill in the gaps in my narrative of that horrific, never-forgotten event in 2003. My cousins, Tom Bauer and Nancy Hill, provided insights into my boyhood that confirmed my memory of things, and I am grateful for their reviewing several parts of the book.

My thanks to my computer-savvy neighbors, Julie Brown, Visagar Shyamsundar, and Alex Suhenko, whom I often called to assist me with a computer problem. They were indispensable.

I will always be especially grateful to the thousands of soldiers whom I had the privilege and honor to serve with. To my fellow infantrymen, who fell in battle and didn't come home, you are always on my mind. You hold a special place in my heart as you were with me when times were toughest.

A special thank you to my wife, Linda, for her support and patience throughout this long process and for encouraging me from the beginning to tell my story. Her forbearance during these months of writing and editing has been amazing! To our children, Bill III, Heather, and Rebecca—and grandsons Will, Ben, Max, Luke, Alexander, Josh, and Ian—you make us proud! We are further blessed by our supportive daughter-in-law, Helen, and son-in-law Pete, and Ben's wife, Alyssa.

They all join me in dedicating this book to my wonderful wife, Linda, our family's constant source of support and inspiration and the quintessential example of caring and courage.

Preface

"Our greatest glory is not in never falling,
but rising every time we fall."
CONFUCIUS

Family members and close friends, who know me and who are well aware of my life-long battle with polio and its effects, as well as my military service, would often say, "You need to write a book." "You need to share with others your story of struggle, courage, perseverance, and service."

I was flattered that some wanted to learn more about my overcoming childhood paralysis and about my experiences in the Vietnam War. But there are many who overcame adversity whose achievements in life are greater than mine. So, the notion of writing a book about my life never resonated with me.

Upon retirement after 33 years in the Army, I was diagnosed with post-polio syndrome (PPS), a serious disorder that involves the nerves and muscles many years after having polio. I had my first post-polio appointment with John Aseff, MD, medical director of the Post-Polio Clinic at the National Rehabilitation Hospital (NRH) in Washington, DC, on August 19, 2014. A renowned physiatrist, Dr. Aseff ranks among the best post-polio-syndrome specialists in the world today. He worked and studied under Dr. Lauro Halstead, founder of the clinic and a pioneer in the discovery of PPS who was present during my examination.

After his thorough examination and review of my medical records and active-duty history as an Army paratrooper and an infantry combat veteran, he said, "Your life's story, being able to accomplish so much despite your earlier polio and this post-polio diagnosis is quite a story. You really had to be motivated and quite tenacious to accomplish all you have." He suggested that now would be a good time to finally tell my story. I replied, "Thank you … we'll see."

Years passed, but during follow-up visits, Dr. Aseff again mentioned "my story." There are many books about soldiers and their courageous actions on the battlefield. But after thinking about it, I began to realize that there is a uniqueness about my story that makes it different and worthy of sharing.

Mine is a story of how perseverance, tenacity, and faith drove a young boy to overcome a serious crippling disease and fulfill a life of service to our country. If my story can inspire and help others to fight and overcome their challenges, then sharing it with others is worthwhile. I see it as my duty. Without the encouragement of Dr. Aseff and others, I might never have come to that conclusion.

Overcoming the debilitating effects of polio and my strong desire to compete in life while meeting the challenges of a military career in the infantry fills the narrative.

The two counterintuitive pursuits are polar opposites! This struggle is what gives the story its "uniqueness," including service with several presidential administrations beginning in the 1960s and ending with the Trump administration, transcending all politics in service to the nation. In writing, I recognized myself as an unwitting witness, if not a participant, in some of America's now-historic defining moments throughout the last 60 years.

The greater reason for this book is to emphasize the value of national or public service, especially serving in the military. I've always believed that the most noble of professions is that of a soldier. Therefore, I hope there are lessons to be learned in leadership and in overcoming obstacles that will help young aspiring officers and noncommissioned officers as they pursue their careers in our armed services.

I hope that parents with young children can relate to the importance of overcoming physical adversity. I have seven grandsons. I hope that they and their parents derive some benefit and encouragement in their own pursuits from my lifetime experience.

Telling my story has allowed me to explore my life's expected and unexpected purposes and to reflect upon to what degree I met them. I hope it also allows the reader to assess their own.

In summary, it is my earnest desire that the reader, when turning the last page, will have enjoyed, learned from, and been inspired by this story—*My Toughest Battle*!

PART I

My Youth

CHAPTER ONE

The Battle Begins

I was born on October 7, 1938, in Drexel Hill, Pennsylvania, a suburb of Philadelphia. My parents were living in Easton, 75 miles north of Philadelphia, where my father worked for a savings and loan company. When my mother went into labor, they returned to their family doctor in Drexel Hill for the birth. My parents were of German, Scottish, and Irish ancestry, and Matz, McDowell, White, and Gunnis were their four family surnames. Their great grandparents immigrated to America in the early and mid-1800s. We were part of a suburban Philadelphia middle-class family whose American roots extended from Chambersburg, Pennsylvania, east to Philadelphia.

Except for my first two years living in Easton, I spent my childhood in the suburban boroughs of East Lansdowne and Lansdowne. Those neighborhoods were ethnically mixed and consisted mostly of families who had migrated from inner-city Philadelphia to the suburbs. The Russos, Lists, Dixons, Hoffmans, Kearneys, Tasjians, Santellas, DeForges, and Carrs, were among the families on our block. They all had kids, including the borough pharmacist, Irvin Werlinsky, who lived at the corner with his family in an apartment over the pharmacy. They had each other's backs and if anyone fell on tough times they were always there with a helping hand.

About 70 percent of the dads served in World War II. Few were college educated. Except for a few teachers, all the moms were homemakers, most of whom volunteered for community projects that supported "the boys overseas" from 1942 to 1945. They were principled, hard-working, wage-earning American families, who were grateful for what they had!

My sister, Rebecca (Becky), was born a year and a half later. We were a family of four living in a rented house in Lansdowne. Despite the Great Depression, and with the loving support of both sets of grandparents (the Whites and Matzes), my parents were able to scrape out a living, but it wasn't easy. I am told that during my toddler and preschool years, I was a "handful," always doing daredevil, adventurous things and misbehaving and challenging my mother constantly. A prominent elder in our church said that I was "defiant," and told my Aunt Gerry, "That kid will never amount to anything." Those early days of innocence were filled with birthdays, Christmases, tree climbing, sledding, catching lightning bugs, and visits with cousins and friends.

During those years, my dad continued to work for various savings and loan companies and as a foreman in a steel mill in Eddystone, PA, until he joined the Navy in World War II. As the homemaker, my mother managed every aspect of the home while being an active member of the East Lansdowne Trinity United Methodist Church, where she sang in the choir and was superintendent of the Sunday School. She believed the sanctity of the Sabbath should be preserved. My sister and I never missed a day of Sunday school. She was loved and respected by all and one of the kindest and most gracious ladies ever. She had a tender way of caring for people and always "giving of herself" to help others. She had a wonderful sense of humor, laughing at her own shortcomings, and enjoyed each day no matter what it brought.

My sister and I are the beneficiaries of her caring, loving, teaching ways and will forever be grateful for the role she played in our development. My father, on the other hand, was the breadwinner. Providing for the family was a compulsory characteristic consistent with his German upbringing. He was restless and unpredictable, which didn't make it easy for my mother. Often changing jobs, never satisfied with what he was doing and always looking for something better. Soon after World War II started, he joined the Navy.

He had an exempt status from his local draft board and didn't have to serve, because he was married with two young children. But his restless nature, coupled with the guilt he felt with many of his closest friends already serving, prompted him to enlist. That resulted in significant hardship for our family. After his discharge at the end of the war, he finally found his career niche as a salesman, and he was highly successful, eventually going into business for himself. He was also a no-nonsense disciplinarian and he meted out punishment when it was necessary. Together, they set the rules and standards by which my sister and I lived and grew.

Polio Strikes!

Then, on August 7, 1944, at age five and a half, my world changed. Adversity unexpectedly struck hard when my lifetime battle with polio began. I was visiting my paternal grandparents in Philadelphia with my mother that day when I fell walking across their living room and couldn't get up; I had no feeling in my right leg. I tried several times to stand up and still couldn't move my leg. Back then, the polio epidemic was at its height in Philadelphia and other East Coast cities. Families lived in fear of this dreadful, debilitating virus. There was no disease that frightened parents so much as polio as paralysis could set in quickly. As a result, swimming pools and theaters closed. The family wondered if polio would dictate my destiny.

Polio could cripple quickly. Dr. Philip Lewin wrote, "Every family shuddered at the thought that any of its members might become crippled. It is not death, so much as the crippling aftermath of the disease, a physical crippling which may last a lifetime and impair the activities of an otherwise healthy person."[1]

Being not yet six years old, many of my memories of that time are limited and might be colored by family accounts. I remember being covered with a white sheet and being wheeled on a gurney. The ambulance took me to the hospital where I was diagnosed with acute poliomyelitis ("polio" for short). While being treated in the emergency room, my mother wasn't allowed in to accompany me. A spinal tap confirmed the diagnosis.

I remember my Uncle Kreuger Matz (my father's brother) came to the hospital. Polio (or "infantile paralysis" because it mostly struck young children) is a disabling and life-threatening disease caused by the poliovirus. The virus spread from person to person through the digestive tract, and infected and damaged a person's spinal cord, around the anterior horn cells, affecting the nerves to their specific muscles, causing paralysis.

It wasn't known exactly how long I had been infected as I showed no physical symptoms until my fall that day. Only a small number of people with the poliovirus infection develop the most serious symptoms that affect the brain and spinal cord, including meningitis and paralysis. Unfortunately, I was in that very small cohort. I had paralysis, or paralytic polio. The virus had reached my spinal cord anterior horn cells affecting the motor neurons to the muscles, and the damage had already begun.

Whenever the question was asked, "How did Billy contract polio?" the answer always came back to the surgery I had a few days earlier on August 2 at the Delaware County Memorial Hospital to remove my tonsils and adenoids. Although that theory was never actually proven, it was based on the logic that the tonsils are the immune system's first line of defense against bacteria and viruses entering the mouth. Once removed, the victim becomes more vulnerable to viruses.

Parents and doctors were warned that tonsillectomies were risky during the polio season of summer and fall. Inflammation of the mouth and throat because of surgery made patients more susceptible to infection. Despite that, the doctor was insistent that I have the operation, and without my father's wise counsel and support, my mother agreed. One could never prove the cause of my polio, but family and friends would never be convinced otherwise. I have done enough research about the causes of polio onset to also agree that the tonsillectomy was the proximate cause of infection. My father often ruminated over that tonsillectomy, and he never forgave that doctor.

It was a tough time for our family. We were living with my maternal grandparents while my father was serving in the Navy. Because of the contagious nature of the virus, my mother wasn't allowed to visit me in the hospital as the nursing units caring for the children with polio were initially quarantined. Early on, I was moved to several hospitals, including the Philadelphia Municipal Hospital for Contagious

Diseases, before it was determined that my condition would best be treated at the Home of the Merciful Savior (HMS) for Crippled Children on Baltimore Avenue in Philadelphia, later called the HMS School for Children with Cerebral Palsy.

When the polio epidemics were at their worst, HMS was converted solely to treating the most serious cases. Home's polio program was funded by the National Foundation for Infantile Paralysis (NFIP) and provided care for poorer children. My case met the requirements. I was admitted on August 21, 1944. NFIP, with its goal of providing care to all polio victims who needed care, was a financial godsend to our family.

Dr. Burton Chance, Home's executive medical director and orthopedic surgeon, and Dr. Wallace Cleland, one of Home's staff physicians, were my physicians until my final release from outpatient treatment in the early 1950s. Treatment for paralytic polio in the 1940s advocated the use of splints and braces of leather and steel to prevent muscles from tightening, deforming joint contractures, and to provide support for the affected muscles. Many paralyzed patients laid in plaster body casts immobilized for months at a time. But prolonged periods in a cast often resulted in atrophy of both the affected and healthy muscles, resulting in further atrophy and weakness.

My right leg was placed in a brace to prevent deformity to the leg as a result of my weakened and paralyzed muscles. Immobilization of the affected limb with a splint or brace prevented the stronger or unaffected muscles from twisting or bending the body unnaturally. That was the regimen followed at Home of the Merciful Savior treating polio when I was a patient.

The first organized treatment of polio began in the late 1930s by Sister Elizabeth Kenny, a World War I Army nurse from Queensland, Australia.[2] Upon her arrival in 1940 to the United States, she challenged the traditional orthodox treatment of braced immobilization. Her novel and controversial approach to the treatment of polio patients rejected immobilization and instead advocated beginning muscle treatment and reeducation within days of the initial diagnosis.[3] She developed a physical therapeutic regimen using moist hot packs, massage, and gentle, carefully supervised exercise. She believed moist hot packs wrapped around the paralyzed limb would help relax and loosen muscles, reduce spasm, relieve pain, and enable the limb to be stretched and strengthened.

Her theory of treatment was muscle re-education with the immediate retraining of the muscles to regain functionality. One of the principles behind the program's success was that it stimulated the remaining nerve cells that had not been killed by the virus. Although the actual neurologic mechanism of strength recovery following acute poliomyelitis, "re-innervation" in the motor neurons, would not be determined for multiple decades, the Sister Kenny treatment significantly lessened the complications seen in children with polio and all the side effects of forced prolonged inactivity.[4]

I was lucky to be one of the few polio patients at HMS selected for the Kenny treatment. I don't know how or why I was selected. I remember the day they removed my steel leg brace and began the treatment. As the nurse began to move my paralyzed muscles, I laid in bed crying. Stretching and pulling the leg was painful. Days earlier, my mother had been told by the medical staff at Municipal Hospital for Contagious Diseases that my paralysis to the right leg was serious, and I would never have full normal use of my leg or ever be able to walk again without the use of a leg brace or a crutch. This was a devastating blow!

There was concern that the virus might infect my left leg and right arm as those limbs were showing weakness and tightness. My mother was extremely upset, but her strong Christian beliefs got her through what she would later call her "darkest days." My father was notified by the Red Cross that his son Billy was hospitalized with polio. His ship had not departed Long Beach, California, for the western Pacific yet, and he got a pass from the Navy to come home for a visit in September 1944.

He and my mother had to stand quite a distance away from my bed during their visit. There was a persistent fear of the virus being contagious so they couldn't touch me. I was quarantined in one of the isolation wards, a long room filled with two rows of children lying on white wrought-iron beds. Many were too sick to make a sound. I was never allowed to leave my bed or even sit up. I remember trying to push myself up to kneel in bed to show my parents I could move and support myself. But I couldn't. My right leg wouldn't move. It was tougher for my parents than it was for me. My dad wore his Navy uniform that September day. That was the last time I recall seeing him until he was discharged from the Navy on Christmas Eve 1945.

The boy in the bed next to me, Clayton Moore, was paralyzed so badly he had to lie in a curled-up fetal position. He couldn't straighten his back or legs. I laid there and stared at him. One day, the nurses took him away. I never heard about him again.

The Kenny treatments continued through 1944 and 1945. I underwent two to three torturous sessions of moist hot pack treatments each day for several weeks. Each treatment was one to two hours long. I would either be carried or wheeled into the treatment room and lying prone on a rubber sheet on the bed, while the nurses and Kenny clinicians wrapped my paralyzed leg, and sometimes my right arm, with dark-gray moist woolen hot packs. That was sometimes followed with a session of gentle therapy in the hospital pool on the lower floor.

The packs were so hot that I could see the steam evaporating from the wool. When the heat subsided, the clinician removed the pack and replaced it with another from the boiler, a large metal tub placed next to the bed. I remember the smell of hot, steamy wool followed by a gentle massage of the limb. My HMR medical records show the treatment regimen was accompanied with dosages of prostigmin and atropine.[5]

As the limb showed progress, the number and length of treatments were reduced. Fortunately, my respiratory muscles hadn't been affected, so I never had to go into an iron lung for breathing. The iron lung wing of the hospital was close to my ward, and when they wheeled me by in a wheelchair, I could hear the slow, steady whooshing sound of the giant ventilator machines as the respirator expanded and compressed to help the patients breathe. The iron lung room frightened me. I didn't want to go near it.

The greatest effect of the virus was to my right leg, with paralysis from hip to toe. When released from the hospital to go home for short, temporary visits, I had to wear a metal and leather leg brace and use either a wheelchair or forearm crutches for stabilization. I was slowly improving, but the leg had atrophied significantly with little muscle left. The virus also destroyed muscles in my right foot resulting in slow growth and permanent deformity. It would take months of additional treatments and physical therapy before I experienced muscle development and improved functional strength.

As a means of measuring strength and ability to regain use of my leg, the hospital clinicians charted my progress by how far I could walk unassisted without a brace. The procedure became a competitive event for those of us with leg paralysis. Due to my competitive spirit, I was determined to be the kid who went the farthest before I fell … and the hospital attendants let you fall! A tape marked with my name was placed on the floor showing my progress compared to the other children. I was the kid that nine times out of 10 went the farthest before I fell. There was no prize, but there was polite applause from the staff for the kid who covered the greatest distance. And there was always plenty of encouragement and praise for all the participants.

The contests were both stressful and painful, but, in retrospect, I view this event as perhaps the initial challenge that tested my resolve and determination to overcome paralysis. I pushed through pain to go farther each time. I wanted to beat the other kids. My tenacity never wavered. The thought of wearing that heavy brace, or using a crutch, or being in a wheelchair the rest of my life spurred me on even more to fight and get better. It appeared that the Kenny treatments were working, and my leg was improving.

Even after my discharge from the hospital and return home, I wore a brace. The hospital clinicians adjusted the brace each time I returned for outpatient treatment. I was also fitted with a special Edwards orthopedic shoe that included a device inside called a Jones Bar. This was to lift my weak "drop foot" and support the weakened arch muscles. I couldn't lift the front part of my foot. I've worn a variation of the bar all my life.

While at home on temporary out-patient care, my mother registered me for enrollment in the first grade in the East Lansdowne Public School in January 1945. That school year began in September 1944 when I was paralyzed and in the hospital. She was hoping I could catch up during the second half of the school year to complete first grade with my friends. However, my hospital visits and long therapy sessions resulted in too many absences, and I was "retained" in first grade.

Rather than certifying that "Billy Matz is 'promoted' to Second Grade," my June 20, 1945, report card signed by the supervising principal, William D. Mower, read "… is retained in First Grade." I don't recall my parents being surprised, but I lived with the stigma of being the polio kid who was left down. Failing and having to repeat a year in school was embarrassing. But the academic setback at such a young age never resulted in low self-esteem or serious emotional or social difficulties for me. My every thought was focused and all my energy spent on physically recovering from polio!

In 1947, with the help of President Frank D. Roosevelt's GI Bill, my parents were able to buy their first house. My dad and I were setting up the bunk bed in my bedroom when he said to me, "Bill, you don't need that brace. I don't want to see it again. Put it under the bed. You don't want to be a cripple all your life." The brace and crutches went under the bed.

When my mother found out, she went ballistic. "Dr. Chance said he needs that brace, and he should be wearing it," she said. "Don't treat him like a cripple … don't spoil him," retorted my father. After a day or two of my mother and dad arguing, the brace remained under the bed. My dad prevailed. Except for wearing it for scheduled visits to the polio clinic for treatment and on long walks during early recovery, I never put it on again. My dad was right! I didn't need it anymore.

That same year, in 1947, my parents agreed to take in a foster child through the church. One-year-old Diane became a sister to Becky and me. I remember our excitement the day she arrived. We became a family of five. My parents raised Diane for 10 years until her mother could properly provide for her, at which time she returned to her mother's care. That was a sad, heartbreaking day for our family, especially for Diane, who would find ways to come back to the only family she knew and loved. Raising a child all those years and then having to give her up was difficult, although we knew that day would come. Our affection for Diane was tinged with grief. Despite losing her, we remained in close touch through the years.

The Kenny treatment continued working. As I was informed many years later, the motor neuron re-innervation process sparked muscle strength to the point where I could walk and even begin to run short distances unassisted. I often tripped and fell, but I got right back up! My treatment for the next several years consisted of reducing and finally eliminating the Kenny treatments while increasing massage therapy and swimming every Saturday. My parents joined the Polio Parents Club of Delaware County, and my father organized the "Polio Dads" in the surrounding boroughs and towns in the county, where they took recuperating polio patients to outdoor pools in the summer and the YMCA pools in the winter for swim therapy, which suited my father as he was a swim instructor in the Navy.

Long after my successful swim therapy, that method was the recommended rehabilitation treatment for polio patients and facilitated the same success rate for decades.

My mother joined the local March of Dimes (MOD) chapter to support and provide resources for the cure of polio. For seven years, she canvassed the neighborhood collecting dimes for the chapter. MOD, formerly called the National Foundation for Infantile Paralysis, was founded by President Roosevelt in 1938 to combat polio. Roosevelt, a survivor of polio himself, was permanently crippled in both legs.

The experience of having polio in the 1940s was isolating, depressing and most embarrassing for me. To gain strength and help to overcome that possibility, "many polio patients and their families turned to the example of Franklin Roosevelt."[6]

Thousands wrote letters to the president seeking encouragement. Even though Roosevelt failed to regain use of his legs, and his actual condition was hidden from the public, children and their mothers drew strength from his warm smile and strong public presence, as they watched his sterling example of a man who had polio and later became president. They took inspiration from his great achievements to spur their own efforts to recover from polio. I was no exception.

Roosevelt's responses were kind and encouraging. "By patience and perseverance … I am sure that you will in time win through to full recovery. Best wishes to you,"[7] he wrote in answering one of the many letters he received. I don't remember my parents ever telling me that either they, or I, ever wrote a letter to the president. Although it was something my mother would've readily done, as she devoted years of her life to my recovery.

The first polio vaccine was developed in 1955 by Dr. Jonas Salk. That vaccine, coupled with Dr. Albert Sabin's oral polio vaccine in 1962, resulted in the eventual eradication of polio in the United States. Thanks to massive vaccination campaigns around the world, polio cases dropped by 99 percent since the late 1980s, and, for a much longer time, remained endemic in only two countries. But sadly, the polio virus returned, showing up in wastewater in some U.S. cities in 2022, indicating the first case of paralytic polio in decades. What began for me in 1944 as a "crippling for life" disease, thankfully became less threatening.

In 1953, I had my last official medical checkup as part of my rehabilitation regimen. My HMS hospital-discharge document read, "Billy Matz—Normal recovery. Case closed." I was cleared to participate in all normal activities. I survived the potential "crippling for life" diagnosis, although my right leg remained two inches shorter and significantly less muscular and weaker at the thigh and calf than the left, causing me to walk with a limp, which I would try to disguise all my life.

I still had permanent weakness, disfigurement and stunted growth to my right foot resulting in a size 9½ shoe compared to my normal size 12 for the left foot. Unable to raise the front part of my foot, I compensated by raising my knees higher than normal to take a step to avoid tripping. Buying two pairs of shoes when needed was expensive over the years, but it was a small price to pay given the outcome of what doctors said was a miraculous recovery. It wasn't until 1990 that I discovered that

Nordstrom shoe department sold split-shoe sizes at no additional cost, providing a substantial saving.

Adversity had struck, but thanks to Sister Kenny, my mother and dad, and the good Lord, I didn't become that severely disabled person who would "never walk again without a brace or crutch." I thank the good Lord every night for the treatment I received! The journey wasn't easy. The mothers of my closest friends, for some time, would not allow them to visit or play with me out of fear they would contract the virus. That was upsetting to both me and my mother.

I also experienced bullying and harassment from some of the tougher neighborhood kids who attended Saint Cyril's, the local parochial school, as I walked alone to grammar school. "Let's get the polio kid," was the cry on the block. I dreaded that walk to school. On a couple occasions, they hit me and knocked me to the ground. My father told me to "suck it up" and to change my route to school and start walking with a couple of my close friends. It worked. The bullying stopped, and I learned early there is safety in numbers. I also remember humiliation, even anger, as some of my so-called friends, riding their Schwinn bicycles, mocked me, leaving me behind, while I was still riding an old-fashioned three-wheel tricycle, because my leg wasn't strong enough to pedal and control a bicycle; I needed the stability of a three-wheeler.

However, those early incidents of shunning, mocking, and bullying only toughened my resolve and determination to overcome my deficit, allowing me to compete with my boyhood peers. It wasn't a walk in the park, but I never remember a single time when I wanted to throw in the towel. I wanted to do what all the other boys were doing. I wanted to be like them. I fought hard to achieve some type of normalcy. I got tired of hearing my friends yell "C'mon, catch up, you're slowing us down." I refused to be left behind, stay at home and not participate.

During the early years of recuperating, I looked forward to visits with my older cousin, Tom, who lived in Connecticut. He was my "big" cousin, and I looked up to him. Our mothers were sisters, and the stigma and contagious threat of polio didn't prevent us from visiting. He was active and had a penchant for art, such as sketching and molding miniature toy soldiers. He had dozens of the tiny figures that we arrayed in battle formation as we laid on the floor of his home playing the game of war.

I was intrigued with the small soldiers, resplendent in their bright uniforms, rifles, and sabers at the ready, as I maneuvered my formation of infantry and dragoons across the floor. I played the game for hours on end, never wanting to stop. My early exposure to the soldiers and the game are indelibly etched in my memory.

My summers were spent in Wildwood and Ocean City, New Jersey, where our family vacationed. My mother was told the salty air and water would be healing therapy for my leg. I spent many days on the beach with family and friends. My mother and father walked me into the surf, one on each arm supporting me and my wobbly leg by holding my hands so the waves would not knock me over.

There were also special therapy sessions in the Flanders Hotel saltwater pools in Ocean City for kids recuperating from polio. My parents ensured I attended every session. I remember other less fortunate kids who were crippled being pushed in wheelchairs on the boardwalk by nuns and hospital attendants in white. I looked at them and thought "Oh, my God, how lucky I am."

When I turned 12, I joined the Boy Scouts. The year before, while a Cub Scout, our Cub Pack visited the second Boy Scout Jamboree at Valley Forge, Pennsylvania, in 1950. The first Jamboree scheduled for 1935 in Washington, DC, had to be cancelled because of a polio outbreak. It was eventually held in 1937. After such a long hiatus, the Valley Forge Jamboree was a big deal. My cub pack Cub Master, Mr. Norman Stone, was hesitant about my going because of my polio. "He won't be able to keep up," he said. My father insisted I be allowed to go. I did, and I kept up with the pack! Forty-five thousand Boy Scouts from around the world attended. I sat cross-legged, Indian style, on the ground when General Dwight D. Eisenhower visited and spoke. Years later, I would meet President Eisenhower while a student at Gettysburg College.

Our troop met Monday nights in the church social hall. My closest friends were also members, and every Easter we sold candy and flowers to earn money to pay our way to summer camp at Camp Delmont in Green Lane, Pennsylvania. The scout with the most sales would win an additional sum of money. Each year, it came down to my good friend, Art Hahn, and myself vying for the prize. Art had the highest sales for two years before I finally beat him our last year in scouting. I left the Scouts when I matriculated to high school. With high school activities, there was too much on my plate. To this day, I regret leaving Boy Scouts and not earning the rank of Eagle Scout.

When I was able to pedal and control a bicycle safely, my parents bought me a used Schwinn. The three-wheeler was mothballed, and I made up for lost time and rode my bike all over the county. I was adventurous and wanted to prove I had the endurance to travel great distances. Many a day and night, my mother and aunt, sometimes with the assistance of our local police, would scour the adjacent boroughs looking for me. A couple of my friends were always with me and what a fun time we had. No longer would I be left behind. I was limping along keeping up with them and was part of the gang. It felt good!

At 13, entering the eighth grade in our first-through-eighth-grade grammar school, I was looking forward to the challenges without the specter of polio. It was a good year. I was elected class president and played on the eighth-grade football team. Those early grammar school years of bullying toughened me in the eyes of my classmates and football helped the reinnervation of my leg muscle, providing yet another way to show I was a normal kid and could do what the other boys did. I matriculated to Lansdowne-Aldan High School, about a mile walk from my house, where I was on the football and track teams. My "polio leg" seemed to be getting stronger by the month. And my confidence in my ability to keep up and do more was gaining.

Physical recovery was only part of my rehabilitation. I also had to cope with the psychological challenge of coming to terms with my residual disability and the social stigmatization of having had polio. I worked hard during my formative and teenage years to not let it affect me. I was never timid, nor easily offended, so the occasional sometimes hostile stares and taunts rolled off my back. I would not let those bother me.

Living with a deformity brought with it a range of emotions. The embarrassment and shame I felt were constant. I did my best to hide and disguise my deformity and permanent limp but experienced the accompanying anxiety of being "found out." It was a learning process. I got pretty skillful at the art of disguise and at deflecting questions. I remember my parents, especially my dad, when I first came home from the hospital, not wanting anyone to photograph me if I wore shorts or a bathing suit. He did everything he could to divert attention from my thin, scrawny, and very weak leg, obsessed with the thought that his son could be a cripple. There are no photos of me wearing my leg brace.

My high school coaches were most helpful in re-building my self-esteem, and, with my father, who was always in overwatch, helped me overcome much of my self-consciousness and embarrassment. I was fitted with a special football shoe which made a difference and allowed me to play more aggressively. I learned to favor and protect my leg while competing. I wasn't a first team starter, but playing on the high school sports teams was invaluable and helped me, more than anything else, gain back the confidence and self-esteem that I had lost during my earlier years with polio. I gained a new set of friends for life and graduated from high school in 1957. Like the kid that earlier wanted to go the farthest without falling, I wanted to compete on the high school sports teams with my classmates.

I was soon off to college and applied to both Pennsylvania State University and Gettysburg College. I was accepted at Penn State in engineering, but I wanted Gettysburg. I visited both campuses and felt more comfortable at Gettysburg. Subconsciously, I think the smaller, compact, nature of the Gettysburg campus, coupled with a student body numbering in the hundreds versus the thousands, appealed to my senses and better judgment. Penn State's size overwhelmed me. For a kid whose lower limb muscles were still weak, a smaller, easier to navigate campus appealed to me. My parents let me make the decision.

After two attempts at taking the required College Entrance Examination required for admission, I finally received my letter of acceptance albeit late from Gettysburg in early August. My score the second time just barely made the minimum required for acceptance. And my grades in high school were nothing to talk about. My acceptance had been hanging by a thread and, to this day, I think my persistent father made a couple phone calls that sealed the deal.

My dad was pleased. He had attended Gettysburg in 1928–30 but had to leave shortly after the stock market crashed in October 1929 to help his siblings support the family through that terrible economic crisis and he never returned as a student. My going to Gettysburg gave him a sense of fulfillment, if not pure joy. His son would enroll in the college that he was never able to complete which was a constant source of mutual satisfaction for us. My life was ahead and fate, God's will, and my fortitude would determine my future.

CHAPTER TWO

Gettysburg College Years

Friday, September 13, 1957, was a warm, sunny day in early fall. My father drove me to Gettysburg College for my first day as a student. On that day, I joined 483 other incoming freshmen to form the Class of 1961. The ratio was three men to one woman.

The college, contiguous to, and just north of the center of the town of Gettysburg, Pennsylvania, was surrounded by the Gettysburg Battlefield. When I attended, it was known as the "oldest Lutheran College in America," founded in 1832. My freshman year marked the 125th anniversary of its founding. It is rich in history due to the Battle of Gettysburg in July 1863 and President Abraham Lincoln's visit that November to deliver his famous dedication address at the National Cemetery. The historic setting provided a unique and special backdrop for four years of college.

As my father's old Packard sedan approached the bucolic campus on North Washington Street that morning, we were met with many orange and blue welcome signs, the Gettysburg College colors. New freshmen arrived from all directions, converging onto the small campus. My father knew exactly where to go, and he found a parking place right in front of Old Dorm, 211 being my assigned room.

It was a spacious three-man room complete with a wooden mantel fireplace that I would share with fellow freshmen, Bill Floto and Bill Stevens. I couldn't believe they assigned three Bills to the same room! It was the same dorm room my father had his freshman year at Gettysburg in 1928–29. This wasn't by coincidence as I learned sometime later. He had arranged the assignment with his former Gettysburg fraternity brother, Stanley Hoffman, who was the college business manager.

Freshmen men had to report first to Glatfelter Hall to pick up their room key. My two roommates had not yet arrived. I remember my dad's excitement as we climbed the long iron staircase of Old Dorm to the second-floor portico entry. Richard Davidyock, the dormitory student proctor, greeted us as we entered. With key in hand, my father couldn't wait to unlock the door and enter his old room. He was in ecstasy!

Old Dorm in the center of campus, formerly called Pennsylvania Hall, was built in 1837 and at one time housed the chapel, classrooms, and dining hall. It was the most iconic building on campus. During the Battle of Gettysburg in 1863, the building's cupola was used by Union forces on the first day as a signal station and, later in the battle, by the Confederates for observation. At battle's end, the building became a hospital for both Union and Confederate soldiers.

After we moved my belongings into the room, I staked out my corner near the window. The room was dark and musty with a creaky wooden floor. We then descended that long staircase for my dad's return home. The entire time we were on campus on move-in day my dad reminisced about his time at Gettysburg and his memories of Old Dorm. When he said goodbye to me, he had a lump in his throat, and his eyes began to well up. He was experiencing strong emotion as he was reliving his days on campus, while bursting with pride and happiness for his son.

I remember vividly his telling me how proud he and my mother were. He mentioned my polio only briefly but turned to the challenges ahead. His stern advice at that moment was "concentrate 100 percent on your studies your first year, establish a good academic base and reputation with your professors, and all else will fall into place. Write us a note when you can and say your prayers." With that, we hugged, said goodbye, and off he drove. I watched his car disappear as he exited the campus. I turned and climbed the steps back to my new home, that moment burned into my memory.

Orientation week was well organized. It included registration on September 18 and formal opening exercises three days later. Classes began on Monday, September 23. During that time, I had a meeting with Mr. Butterfield, my freshman advisor. He was a mathematics teacher and a graduate of the Naval Academy. He helped me navigate the path of uncertainties in my freshman year. I still have the "Greetings to the Class of 1961" note dean of students, John Shainline, sent to each freshman. It talked about the strong tradition on campus for all to greet one another with a friendly "hello."

I was surprised, however, when I received a written invitation from the college choir director, Professor Parker B. Wagnild, to audition for the Gettysburg College Choir. It directed me to report to the Brua Lounge at 1:00 p.m. on September 16 or 17, for the audition. This was a by-name invitation not sent to all freshmen. I was puzzled.

Why would they want me in the choir? Two years earlier I had been cut from our high school choir because I couldn't hold a note. My soprano-voiced mother, who sang for years in the church choir at home, was very upset about that, and I believe to this day, she had something to do with the college audition invitation, hoping I might again sing in a choir. I never went to the audition ... but I kept Wagnild's invitation among my many treasures!

Freshmen customs were a tradition at Gettysburg. They included wearing your orange and blue hat, called a "dink," and sporting your name and hometown on a round, oversized, badge pinned to your shirt. Upper classmen could stop you anytime on campus and ask you to sing the College Alma Mater. None of this fazed me, although I hated wearing the dink and often got reprimanded by an upperclassman for refusing to do so. It did help me learn the Alma Mater.

During orientation week, the college sponsored a "get acquainted" dance in Plank Gymnasium for all freshmen. I met a freshman coed that night and after a few dances asked her for a date. She was attractive, friendly, and I thought the two of us hit it off right away. We went on a date to the movies a day or so later. When I asked her out again, she told me I wasn't sophisticated enough for her and that it would be best that we not date anymore. She was very polite, but very direct. I was stunned! I wasn't even sure what the word "sophisticated" meant. That was my introduction to Gettysburg coeds!

I am told the term "coed" was largely shelved from usage in the late 1990s with other politically incorrect castoffs. It is unfortunate that today's politically correct culture tends to forget about the past. When I was in college, the term coed referred to a woman student. There was nothing demeaning or derogatory associated with its wide usage. I always held in high esteem the women students at Gettysburg. They were bright and possessed more common sense and had better coping skills than any of the male students—including me!

Phi Gamma Delta

Greek societies were an important and integral part of college life. There were 13 national fraternities and six sororities. Gettysburg had a strong national reputation in the fraternity world. The school embraced the idea that although the academic classroom was central to an education, so was one's social learning, and the creation of mature, emotional and mental attitudes toward life. They stressed the fellowship and teamwork engendered by fraternities and sororities as they played an important part in the education of the total person. The college at that time promoted Greek life. The president's message in the 1957 welcome handbook encouraged each new student to "find the one (fraternity) which will enrich his life and bring full benefits to his college education."

Fraternity rushing began Monday, September 16, and ended the following Sunday. The week was devoted entirely to that. Classes began the following Monday. Students didn't rush the fraternities. The fraternities rushed the students. Each student was provided a "Freshmen Date Book" to record the date/time of the invitations for visits to the various fraternities. All week, brothers invited prospective members for meals, outings, etc., to "sell" their fraternity.

On Sunday night, we received formal invitations or "bids" to pledge. The following day, we went to the fraternity of our choice to officially pledge. I received bids from four fraternities and pledged Phi Gamma Delta, otherwise known as the Fiji House.

Nationally, we were XI Chapter. By the second day of rush week, I knew I wanted the Fijis as they impressed me the most and were where I felt the most comfortable. Brothers Jack Hathaway, John Smoot, and Del Warfel delivered my invitation to pledge that Sunday night. Hathaway would become my fraternity Big Brother and he and his wife Trudi became life-long friends.

On Monday, I went to the Phi Gam house and joined 15 other freshmen who also pledged Fiji. We received our pledge pins that day. That meeting of "the 16" was the beginning of what became a life-long friendship with my fraternity brothers. In January, another freshman, Tom McGready, joined, making our number 17.

Almost immediately, I experienced the spirit and feeling of brotherhood intrinsic to a fraternity. Fred Fielding was our pledge class president, and years later, was White House Counsel during the administrations of Presidents Ronald Reagan and George W. Bush. We met in my large three-man room in Old Dorm for our weekly pledge meetings. When we did, I asked my two roommates to leave. They were not happy that they had to leave their own room, albeit temporarily, for a bunch of Fijis. Neither one returned to Gettysburg after Christmas break, our freshman year, leaving me without a roommate. A fellow Fiji pledge, Ron Frederick, moved in with me for the second semester. Our pledge class bonded well, supporting each other while going through the next six months of pledge training, culminating in "hell week" in March.

During that period, we were assigned a Big Brother, an established member of the fraternity who provided advice and counsel when needed. We also had a live-in housemother, Mrs. Hilda L. Heldrich, a widow who was a lovely, gracious, lady. She assisted our fraternity pledge trainer in teaching etiquette, which included the graces and manners of everyday living. Every Friday, we were tested on fraternity history and rules of protocol. We were made to memorize the full names and hometowns of all brothers and ate all meals in the fraternity house.

Pledges set the tables and served the evening meal. My mother loved the fact there was a housemother and that her son was also getting some civility and etiquette training while at college. I learned later that she wrote thank you notes to Mrs. Heldrich after the annual Mother's Day weekends asking her to "keep a close eye on her son, Bill." In retrospect, we were fortunate to have a housemother. As a young man full of hormones, I didn't bridle against that. But I didn't fully appreciate the value of fraternity life and its lessons until years later. We had "quiet hours" in the evenings so brothers could study. Good behavior was emphasized, and we always showed respect, because Mrs. Heldrich lived there, also.

Every fraternity had a housemother who lived in a small, attached apartment. When I returned for college reunions, I always visited the Fiji Lodge. It is not the house I remember. It lacks the warmth, civility, cleanliness, and character that once filled its rooms. The brothers, in later years, hadn't taken the care we did, and the housemother was long gone.

To be initiated into the fraternity, a pledge had to attain a 1.000 (C) grade his first semester. After attaining a 1.589, I was initiated along with 10 of my fellow pledge brothers. The other six pledges either dropped out of school or didn't obtain the required C average. Phi Gamma Delta molded me, and my fraternity brothers were a supportive, inspirational group whose friendship in college and through the years has meant so much. Our pledge class bonded well from day one, and we had each other's backs.

Our fraternity was well-rounded, academically, athletically, and socially. We were not an animal house, and we certainly were not the nerdy, socially inept, studious type. Nor did we win academic honors. Most of the brothers participated in one or more intercollegiate sports. I was on the college lacrosse team. Teamwork was emphasized.

Christmas House Party weekend was highlighted by the annual competition of the fraternities for the best-decorated house. We took top honors in house decorations my freshman year with the theme "The Twelve Days of Christmas." For Homecoming weekend our football game was usually with Muhlenberg. Each fraternity built a float and pulled it around the track at halftime. We spent all our time with our fraternity brothers. Some would say, "Oh the Fijis, they're snobs. They don't associate with anybody else."

There was some truth to that. I regret not getting to know some of my other classmates better, because we were so closely bound to our own fraternity. Our pledge trainer, Howard Body, admonished me once for not being with my pledge brothers after I had gone to the town movie theater with a couple friends from another fraternity. He said, "Matz, when you go to the movies, you go with your Fiji pledge brothers, not the Phi Sigs."

My senior year, XI Chapter was awarded second place in the prestigious Baker Cup competition. The Baker Cup was awarded annually to the Phi Gamma Delta undergraduate chapter which excelled in "religious, ethical, and social service activities" in its daily conduct. The award is named after Brother Newton D. Baker, former U.S. secretary of war during World War I. It was quite a national recognition for our chapter. I was the chapter's social service chairman that year and was proud and appreciative of how we came together and worked so diligently on our many social service projects and programs that assisted the community. It was nice to bring home a national trophy. Indeed, it was a team effort!

Phi Gamma Delta is the venue that most molded and shaped my character and best prepared me for life immediately after school. Like being a member of a sports team or the ROTC program, it was a platform for socialization and friendly competition. And with our brothers, we competed to *win*! It instilled teamwork and pride in attaining common goals. We didn't let a fraternity brother down! I am indebted to my fraternity for this but would not have had the fraternity experience if not for the college. The two played a huge role in preparing me for the challenges ahead.

Gettysburg had a strong academic reputation. Its history department and its pre-medical and pre-law programs are among the best of the liberal arts colleges. It prided itself on "sending its graduates out prepared for a rewarding and useful Christian life" ready to "contribute greatly to their country." There was a spiritual dimension to the college that required all students to attend two periods of Chapel every week. These were 20-minute respites during the hectic academic day, led by the college chaplain, Pastor Korte. This provided a break in our academic day where we could sit quietly in the college's Christ Chapel and ruminate in a spiritual setting. It was good for the mind and soul.

The Gettysburg president during my four years was Willard S. Paul, a retired Army lieutenant general. He had commanded the famous "Yankee" Division, the 26th Infantry Division during the Battle of the Bulge in World War II. General George Patton personally praised him for his "splendid leadership in the counterattack that reduced the German salient." Prior to that, he rode with General John J. Pershing in the Pancho Villa Expedition into Mexico in 1916 and served in the trenches of World War I. A tough soldier, he was also known for his management and organizational skills.

He had been professor of military science at Johns Hopkins University. He was a firm believer and ardent supporter of the Army and Air Force Reserve Officer Training Corps (ROTC) programs at Gettysburg. Gettysburg was one of the first four institutions in the nation to be granted an ROTC program in 1916. General Paul emphasized Christian values and the importance of serving your country in every speech.

He wasn't always visible to the students, but I had a chance to hear him tell of his wartime service when the professor of military science (PMS), Lieutenant Colonel John H. Eddy, hosted a "chat with the president" session for a few of the ROTC cadets. It was a personal, intimate setting. General Paul was relaxed and open. We asked him questions. I was impressed with his combat record in two world wars and post-war record of public service including his time as a member of the Hoover Commission in the late 1940s.[1] Indeed, his experience, war record, and emphasis on values, character building, and service to the nation, resonated with me.

Army ROTC

When I attended Gettysburg, all male students had to enroll in either the Army or Air Force Basic ROTC program their first two years of college. The ROTC program officially opened to women on most college campuses in 1970. The only exceptions for male students not enrolling were those not physically qualified or those majoring in physical education. Academic credit was given for the military science courses. Some freshmen were disqualified because they had a history of asthma, emphysema, or other lung diseases. Polio was also a disqualifier.

When completing the medical questionnaire, I truthfully listed polio among my diseases and illnesses, knowing that might have disqualified me, but I wanted to be in the program. I requested and received a physical waiver for my polio and was able to enroll. Roughly 75 percent of the freshman and sophomore men were in ROTC. If you were not in the program, others would ask, "Why aren't you taking ROTC?" I'm not suggesting it was among the more popular programs or courses. It was a requirement that you take the Military Science I and II courses, so you made the best of it.

Every Tuesday was ROTC drill day. Students in the program became cadets that day. Cadet Puerta, Cadet Jacobs, etc. We wore our uniforms from daybreak to the evening meal. For two to three hours in the afternoon, we attended what the Army euphemistically called "leadership laboratory" or "drill period." The Army ROTC cadet brigade formed on the football field, and the cadet chain-of-command was in charge. Attendance was taken, announcements were made, and an in-ranks inspection conducted. We then marched off to Broadway in company formation at right shoulder arms to the drumbeat of a small cadet drum and bugle corps. Broadway is a wide street contiguous to the campus where we marched in ranks, conducted close-order drill, and practiced the manual-of-arms. Broadway and the football field were cordoned off every Tuesday afternoon by campus and town police for the exclusive use of ROTC drill and ceremony.

AFROTC drilled in the field directly behind Glatfelter Hall, later to become an academic building. Some sororities gathered along Broadway to watch the marching and drill. I never understood why. I guess they liked to see the men in uniform. Or maybe they were vying for nomination to be the Military Ball Queen! Whatever the reason, it gave us an added incentive to stand tall and look smart. Our women were watching!

Some of the townspeople, particularly the old vets, also came to watch us drill. President Paul sometimes walked the sidewalk waving to and saluting his gallant corps of cadets! He usually had one of the deans with him, Shainline, Jones, Wolfe, etc. It was one of the few times we saw the president. It was all very orderly and respectful. The campus was a sea of Army olive-drab green and Air Force blue the entire "drill day." When completed, we were glad to return our M1 .30 caliber Garand rifle to the arms room and discard our military uniforms for our white ducks and khaki chino trousers. ROTC "drill day" was met with mixed emotion. Some hated it, while others tolerated it.

There were a few gung-ho guys who craved it and who looked forward to it. I wasn't one of them, but I didn't mind it. After the inspection, and after we stepped off to the command of "forward march," I threw my shoulders back, picked up the pace, and enjoyed the cadence and sound of the drumbeat as I limped along.

One day, during my freshman year, a fraternity brother, Mike Pacilio, and I were cleaning our M1 rifles in the arms room when Captain Beirne approached us.

Beirne was a Korean War veteran and member of the ROTC staff. He was recruiting members for the Army Drill Team. "I'd like the two of you to try out for the Drill Team," he said. A brief discussion followed. The next thing I knew, Mike and I were at a try-out, and, after a short session of close-order drill, we made the team.

The Army ROTC Drill Team had a good reputation and had won best marching unit in several local parades and competitions in previous years. We practiced in the evening at the Gettysburg Armory on Seminary Ridge. We competed in parades and in the early spring, in the regional college drill team competition in Washington, DC. I enjoyed the camaraderie of the drill team, meeting new friends, and traveling outside Gettysburg, but I didn't continue with it after my freshman year.

After completion of the Basic ROTC program my first two years, I elected and was approved by the PMS to enroll in the advanced program for my last two years leading to an Army commission upon graduation. From 1940 through the early 1960s, men were drafted to fill vacancies in the U.S. armed forces that couldn't be filled with volunteers. At age 18, all men were required to register with the Selective Service System, same as today. It wasn't unusual for college graduates to be drafted after graduation.

Knowing this, it made sense to seek a commission through the ROTC program where I could serve as an officer upon graduation, avoiding the draft. There were limited vacancies annually for the advanced program, so it was quite competitive. Not all were selected. Once in the advanced program, I was paid $27.50/month, helping to defray the cost of my monthly fraternity bill and giving me beer money.

Among the requirements for entry into the advanced program, was passing a physical examination. The standards were higher than those for the basic program, and medical waivers were rare. I had to have a neurological specialist from the Department of Neurology at Walter Reed Army Medical Center examine me. He put me through a series of strength and flexibility tests to ensure the functionality of my polio leg. He reluctantly determined me to be commissionable, and I was enrolled in the program. I held my breath that I would pass my physical examination! Had I not, my life would have taken a totally different turn.

The summer between our junior and senior years we attended the ROTC Advanced Camp for six weeks. Gettysburg cadets went to Fort Meade, Maryland, for four weeks, followed by a two-week bivouac at Fort A.P. Hill, Virginia. One thousand cadets from 24 colleges attended. The camp assessed a cadet's ability to demonstrate proficiency in basic officer leadership tasks. It was the most significant training event in the ROTC program.

Training is complex and conducted in a stressful environment. I enjoyed summer camp. My tent mate for the two-week bivouac at Fort A.P. Hill was Barry Lindenbaum, a 6-foot 6-inch basketball player from Lehigh University. He was so tall that his feet extended beyond the end of the pup tent. He was a great guy, but he hated camp. It poured rain one night, and we hadn't properly "ditched" the periphery of our tent causing both of us to get soaked. At times, I thought he might quit so I spent considerable time keeping his morale up.

Teaming with him taught me that during tough times, you must dial up your inspiration quotient. We learned to operate in small units, e.g., fire teams and squads, and to look out for our buddy. We were critiqued and formally evaluated by platoon tactical officers and noncommissioned officers (NCOs) from the regular Army. The evaluation and final standing at the end of camp determined a cadet's rank and position in their senior year in the ROTC program.

The next to last day of camp, each cadet was called into the senior tactical officer's room and given his final camp rating. "Cadet Matz, you did well. You ranked fifth [of 42] in the platoon." The military services use order of merit lists (OMLs) for almost everything. "Good luck at Gettysburg your senior year," said Captain Lamp, the company tactical officer. I felt good. I had a sense of achievement! What I learned most from the experience was the importance of cohesion, teamwork and adaptability in small units.

The challenge of getting 42 men to put mission first and set their personal differences aside is no small task. We learned quickly who the "slackers" and "doers" were. More importantly, we learned who we could trust and count on when things got tough. It was a good experience in competitive interpersonal relationships with peers. I came away with a greater appreciation for the importance of mission, selflessness, and trust!

The second semester of my senior year I was selected to be the cadet brigade commander, the senior ROTC leadership position, with the rank of cadet colonel. My friend and fellow member of our Class of 1961 Reunion Committee, Bill Hockenberry, was the brigade commander for our first semester. The PMS made the selection with the approval of the Dean of Men. My duties required me to take command of the brigade's 450-man formation every Tuesday during drill.

I was responsible for planning the drill period, making announcements to the brigade formation, and leading the formation to Broadway for drill. The command experience I gained every Tuesday barking out commands and addressing cadets proved invaluable in my early years in the Army.

Some time ago, I was invited to participate in Gettysburg College's Vietnam War Oral History Program. During the interview, the archivist asked, "What did you pull out of yourself to become a leader of military men? Did you fall into it naturally?" I answered, "No! I'm not a natural leader. The Army teaches if you are not a born leader, and most are not—they'll make you one. They'll mold you, train you and teach you how to be a leader." The Army, of all the institutions, is the best at teaching and practicing leadership. I don't know of any special attribute that I had.

I never feared taking leadership positions, being in charge, out front. Often, I sought responsibility. Some told me I had a good "command" voice, and I was never bashful. Little things never bothered me, and I left college self-confident.

Some of these traits are considered leadership attributes, but I can't help thinking it was the ROTC program that provided the opportunity to hone these traits; to learn and develop other leadership skills necessary for a successful military career. My battle with polio instilled an insatiable determination to overcome any obstacle and push ahead.

Gettysburg athletic teams (nicknamed "Bullets") also had a good reputation among their competitor schools. I mentioned earlier that many of my fraternity brothers played one or more sports. My leg was becoming much stronger, and I was looking for a sport to play.

I was enjoying college, but restless for a physical challenge. The muscles in my leg strengthened, and I felt confident I could compete. The Gettysburg lacrosse team completed its first season as a NCAA college-approved varsity sport in 1958 and was looking for players. Many of the players were Fijis, including Dick Simpson, the captain and leader of the team. He was from Baltimore and instrumental in bringing lacrosse to Gettysburg. He approached me and other Fijis our sophomore year about trying out for the team. I had never held a lacrosse stick, but the game looked challenging and interesting. I signed up with three other members of our pledge class: Skip Yohe, Joe Baily, and Mike Pacilio. Skip had played lacrosse in high school and was a dominant attack man with good stick skills. He knew the game.

The college issued us sticks, and we began tossing and catching the ball. It didn't come naturally to me, but the more I practiced, the better I got. I played for three years, first as a midfielder, then as a defenseman. I thoroughly enjoyed it. It is a team contact sport but not a violent one. I enjoyed the physicality, running, and speed of the game. It was perfect for my leg, whose muscles were still re-innervating, as it toughened and strengthened the leg without fear of someone tackling or breaking it.

My fraternity brother and close lifelong friend, Ken Tholan, a transfer student from the University of Virginia, joined the team our junior year and was the "spark" we needed. He led the team in scoring during our junior year with 22 goals. He was a rugged attack man, with good stick finesse and always found a way to score. He played the crease, setting picks and scoring. When he hit you, you felt it!

Skip Yohe and Don "Dumps" Pearce, my fraternity Little Brother, rounded out the team's all-Fiji attack element. Dumps was the leading scorer our senior year. In my senior year, I tore a hamstring muscle and tendon in my polio leg in the Franklin and Marshall game. It was a grade two tear, and I was out the remainder of the season. I was devastated! I was concerned, but not panicked.

The injury was frustrating, and healing took a long time since it was difficult to give the leg the rest required. The college trainer, Rome Capozzi, had never treated an injury to a leg that had been weakened and atrophied from polio. "This is something I haven't seen before," he said. He treated it with ice packs, and because of my atrophied thigh, he was creative and extra careful in applying a compression wrap with an elastic thigh sleeve until it fully healed in mid-May. Graduation was three weeks away, and I wasn't going to do anything to further aggravate it. I would injure the same hamstring two more times while in the Army.

Gettysburg was a dry campus then. No alcohol was allowed anywhere. Fraternity parties were held off campus. We rented fire halls, hunting lodges, and similar-type venues in the surrounding towns. We usually always had some semblance of a band and always two or three kegs of tapped beer. Other than alcohol, drugs were unheard of.

Most of the brothers dated Gettysburg coeds. Some imported dates from nearby women's colleges like Penn Hall, Wilson, and Goucher, to name a few. Our coeds, as you might imagine, were not fond of this practice of "importing" dates. The parties were great, and we danced the Twist and drank beer! Everyone had a good time. There were always a few that had one drink too many and had to be helped back to the fraternity or dorm room. I don't remember anyone being an excessive alcoholic. Rarely did the parties include hard alcohol or spirits.

After the Saturday night parties, we all made it to the formal noon meal the next day at the fraternity house. Many brothers invited dates. A coat and tie were required. No one sat down until our housemother was escorted into the dining room and seated. Some of the brothers looked and felt better than others, depending on their beer consumption the night before.

On weekends in the spring and fall, when the weather was best, we invited a sorority to join us in "tapping a keg" at a selected hidden location along Rock Creek east of Culp's Hill on the battlefield. To keep the beer cool, we placed the keg in the middle of the creek so the cold waters could run over it. Many beers were enjoyed at this historic setting! We were kids having a good time. Looking back, I think how lucky we were that there were no serious automobile accidents driving back to campus in the wee hours of the morning after a long night of dancing and drinking. Providence was with us!

Some of my fraternity brothers and I had one incident with the law in the spring of my freshman year. I was underage and drinking beer at the Lee-Meade Inn on Emmitsburg Road outside Gettysburg when the Pennsylvania State Police and Alcohol Control agents raided the place on April 25, 1958. They had it well staked out, knowing underage college kids could go there and were being served. The agents descended upon the inn covering all exits. I tried to escape through a rear door, but a plainclothes liquor control board agent grabbed me and pushed me back inside. Joe Baily tried to get out a bathroom window and a state policeman grabbed him.

They took our names, age, and other identification and information. On May 19, 1958, I and 12 other Gettysburg students, including six of my fraternity brothers, were served a Commonwealth of Pennsylvania subpoena to appear in court on a liquor and beer violations charge. I still have the subpoena notice. It read, "We command you to appear before the Pennsylvania Liquor Control Board at a hearing on May 29, 1958, at 12:30 PM at Northwest Office Building, Harrisburg,

Pennsylvania, to testify on behalf of the Commonwealth in the citation hearing in the case of the Lee Meade Inn." It was signed by the Control Board's president.

We all appeared for the hearing. Sitting across the courtroom from us were the inn's two owners and bartender who were charged with selling beer and liquor to "minors under the age of 21." I don't recall the exact conduct of the hearing, but the result was the closing of the Lee-Meade Inn, the proprietors fined and losing their liquor license.

On May 29, we finished semester exams and went home the next day for summer vacation. My parents were upset. A copy of the subpoena had been mailed to them. My father tried to keep it from my mother, but she found out. It took the entire summer for her to get over it and finally forgive me. My mother was adamantly against alcohol, and it hurt her when she learned that her underage son was drinking at an off-campus bar. My dad handled it in a calmer, more understanding way. He was upset, but I sensed he began to mellow somewhat in how he dealt with my indiscretions. This was the end of my first year. His patience would be tested again. His prime concern was the health of my polio leg and the continued strengthening and re-innervation of the muscles. He knew what I needed most for a successful life beyond college!

January 1961 was my final semester of college. As a political science major, I had followed the presidential campaign and election the previous November. John F. Kennedy had been elected president, in a bitter contest against the incumbent Republican vice president, Richard Nixon. Kennedy was the first Catholic and youngest man ever to be elected president. Nixon's people tried to prove election fraud but were unable to do so.

Political science majors were glued to the TV on election night. We sat with Dr. Chester Jarvis, chairman of the political science department, in the student union TV room watching the returns and listening to NBC's John Chancellor and others report the outcome. It was a live, firsthand learning experience and complemented what we had been learning in our political science courses.

Saturday, January 21, 1961, was a cold snowy day in Gettysburg, but also a day of great anticipation! Ike and Mamie Eisenhower were "coming home," and I was among those selected to greet them at the town's Welcome Ceremony that day. I was accorded this honor because I was the cadet colonel of the Army ROTC program. My Air Force cadet counterpart was also invited. President Kennedy had been inaugurated as the 35th president of the United States the day before. The Eisenhowers drove to Gettysburg arriving at their farm late afternoon of inauguration day.

I was in my ROTC uniform standing on the steps of the Gettysburg Hotel as part of the welcoming party, which included the town and county "fathers," and President Paul and other college dignitaries who awaited their arrival. Ike and Mamie drove into the town square escorted by Pennsylvania State Police, got out of the car, and ascended the steps. They had big smiles on their faces as they topped the stairs and entered the hotel lobby.

I recall hearing Ike say to the crowd, "It is nice to finally come home to Gettysburg." The thought flashed through my mind that the day before, he had ended eight years of being the most powerful man in the world and was widely celebrated as a top general during World War II. Ike had been president the entire time I was in high school and college, 1953–61. He was now a resident of Gettysburg and would live there until his death in 1969.

Commissioning and Graduation

My final semester was rapidly ending. During industry interviews for seniors in April, I received a good offer from The Goodyear Tire and Rubber Company to take a position in their Sales Management Program at $5,160 per year. I thanked them but had to decline because of my upcoming Army obligation. Their regional personnel manager wrote and said they would honor the offer once my military service was complete. I was most appreciative of the sincerity and professionalism extended by Goodyear.

After our comprehensive exams in May, we prepared for commencement weekend. Baccalaureate service, commissioning and commencement exercises were held Sunday, June 4. It was tradition for the graduating seniors to go to a party the evening before and celebrate. The party was at a restaurant/bar east of Gettysburg on Route 30, called Lincoln Logs. I went with my fraternity brothers. Some families also attended. The place was packed that night with everyone celebrating graduation.

After too much to drink, Joe Baily, Ken Tholan and I left. It was late. We squeezed into Joe's new red MG roadster two-seater sports car his father had just bought him as a graduation gift and drove back to Gettysburg. How we all managed to fit into that small space, I don't know. Those details are blurred, but we ended up driving at excessive speed through town and then north on Harrisburg Road, almost to Dillsburg, before we turned around. We had entirely too much to drink. We changed drivers, and I drove most of the way. We were feeling no pain! Speeding back on Harrisburg Road, we returned to the campus and proceeded to climb the outside wall of Hanson Hall dormitory where many of our coed classmates were staying and we ripped out a downspout.

We entered the hallway of the dormitory when we heard the siren and saw flashing lights. It was the campus security chief, better known as "Nicky Nightstick," accompanied by a police cruiser from town. We were apprehended and hauled into the campus security office on the ground floor of Old Dorm. We were instructed to call our parents who were staying in local motels. My father and Ken's father arrived half-dressed and after some serious negotiation with Ramsey Jones, Dean of Men, also called to the security office, we were released to return to the Fiji House. It was almost 7:00 a.m.

The commissioning ceremony began at 8:30 a.m. in Christ Chapel. Baily was to be commissioned in the Air Force and I in the Army. We cleaned up, donned our uniforms, and were sitting in the chapel pew in time for the commissioning. The college authorities allowed us to be commissioned and to graduate. Our punishment was to pay for the damage we caused breaking into the dormitory and tearing up the adjacent manicured lawn and shrubs. My father was somehow able to conceal the early telephone call he received that morning and never told my mother anything.

I was commissioned along with 57 of my classmates: 34 Army, 21 Air Force, and 2 Marine Corps. My parents pinned the gold bars of a second lieutenant on my uniform. After my disastrous escapade the night before, I was recognized as one of the seven Distinguished Military Graduates honored at the commissioning exercise. General Lyman Lemnitzer,[2] chairman of the Joint Chiefs of Staff (JCS), was the presiding officer and commissioning speaker. President Paul, the PMS, and the college deans were present, and all wished us "good luck and God speed."

I don't know if President Paul knew about the incident earlier that morning. I was told later that Baily and I came very close to not being commissioned. I am sure the authorities weighed heavily on the decision to commission us or not. With families and the chairman of the JCS present, and only two hours until the commissioning exercise began, the college and the Army and the Air Force had to make a tough call.

It is unfathomable what could have been going through our minds that night that caused us to pull such a stupid, reckless, and dangerous stunt on the eve of commissioning and graduation. Much remains blank about that night. It was, indeed, a lesson on the profound effect alcohol has on behavior, memory, and judgment. It was the dumbest, most insane thing I ever did in my life! Thank God, no one was injured or killed.

Baccalaureate service was held at 10:00 a.m. followed by commencement at 2:30 p.m. Baily, Tholan and I skipped the Baccalaureate Service, being tired and not feeling well. We assembled at 1:30 p.m. for the academic procession from Plank Gymnasium to the Student Union building for the commencement exercise. By that time, I had been awake for 32 hours straight with a hangover. I managed to ascend and descend the stage for my diploma and was glad to return to my seat. Many of my relatives attended both the commissioning and commencement exercises with my parents and sister. My mother was so proud, smiling and hugging me throughout the day. My father gave me an occasional disgusted look that said it all. By the grace of God, I got through the day and my college years! Those four years went by quickly. I set my sights on the Army.

After graduation and before I reported for active duty with the Army, I had a few months to unwind and prepare. My fraternity roommate Joe Baily also had time before he had to report for duty with the Air Force. Joe convinced me to go into business with him during this period, selling sandwiches and sodas to fisherman and recreation boaters in the back bays and inlets of the Jersey shore. I was skeptical but joined him. We needed a boat, and Joe's father gave us an old 23-foot Bay Garvey with a small cabin and outboard motor.

We made fresh sandwiches each morning and put them with sodas into ice chests in the bottom of the boat. We launched at 10:00 a.m. so that we were in the bay in time for the lunch hour. Great Egg Harbor Bay and its tributaries were mostly between the towns of Somers Point and Ocean City, N.J., covering 9 square miles. Fishing was good, particularly for flounder and striped bass. The summer months were busy.

Joe piloted the boat while I was in the bow ringing a bell to get the attention of the boaters. An Atlantic City sign painter painted "Baily's Bay Buggies—Sandwiches and Sodas" in big letters on the side of the boat. When a boater signaled us, we pulled alongside for the sale, but it wasn't easy. Bay Garveys are flat-bottom boats and are not very maneuverable. This, coupled with the choppy waters and running tides, made it even more difficult when trying to come alongside another boat.

One hot afternoon, while coming alongside a beautiful 36-foot Chris Craft cruiser with a party of fisherman aboard that signaled us, we slammed hard into the boat causing damage to the hull. It wasn't serious structural damage, but the bow of the Garvey hit with such force that the hull was badly scraped and required repair. "What the hell are you doing? Don't hit us again. Back off!" yelled the skipper of the damaged boat. We exchanged contact information by shouting our identification and telephone numbers across the water to each other. They never got their sandwiches and sodas, and we ended up settling with the boat's owner for the damage. It was a failed business endeavor in more ways than one! We had other similar "close calls" prior to that, and I decided it was time for me to leave that enterprise.

My fraternity brother of four years and I departed on amicable terms. We often laughed about it years later, but I don't think Joe ever got over me leaving when I did. His brother joined him, and after just a few days, they also called it quits. Joe was stubborn and convincing him that it wasn't a good business model took some time. The business never made a nickel, and some fishermen complained to the Great Egg Harbor Coast Guard Auxiliary Police that our loud bell ringing was scaring away the fish. We were becoming the scourge of the bay!

I had to make up for lost time and lack of money. I took a job with the Carl Mitnick Construction Company building homes in Somers Point. In September, my friend, Lou Brown, and I took a month-long cross-country trip to California and back. We stayed in Phi Gam fraternity houses as we crossed the country and returned. I was biding my time until I went into the Army by traveling, working, and trying to keep in good physical condition. I was very comfortable living at home with my parents but bored with the mundane lifestyle and anxious to get on with the next phase of my life—the Army!

PART II

Military Service

CHAPTER THREE

Making of an Infantryman

My first assignment was Fort Benning, Georgia. Named after a Confederate general, the post straddled the Alabama–Georgia border contiguous to Columbus, Georgia, and was "Home of the Infantry." Much later, it would become the Army's Maneuver Center of Excellence, co-locating the Army's Armor Center and School, formerly at Fort Knox, Kentucky, with the Infantry Center and School.

I reported on January 6, 1962 for infantry officer basic training. After a 900-mile drive, I was happy to arrive. I was in the "deep" South and already enjoying the good weather and the warm welcoming hospitality. I joined 150 other newly commissioned second lieutenants. Among them was Bill Smith, from Fuquay-Varina, North Carolina. His father was a school principal and tobacco farmer, and Bill was commissioned from Wake Forest College, later Wake Forest University. We became friends for life and were both commissioned in the Army Intelligence and Security Branch (AIS).

I selected AIS much to the disappointment of my professor of military science, who wanted me to go infantry. I chose AIS because it was most compatible with my political science degree. All newly commissioned AIS officers were required to complete the Infantry Officer Basic Course (IOBC). While at Fort Benning, I received orders for the 82nd Military Intelligence (MI) Detachment of the 82nd Airborne Division. I was elated!

I volunteered for airborne upon arrival at Benning. Paratroopers were the toughest and proudest soldiers in the Army, and I wanted to serve with an airborne unit. Their élan, esprit de corps, and true grit resonated with me. I wanted the challenge, and I knew my polio leg was strong enough to handle the arduous duty. I received an extra $110 a month for hazardous (parachute) duty which was also attractive.

Within AIS, I was designated an Order of Battle (OB) officer and attended the 10-week OB course at Fort Holabird, Maryland, immediately following IOBC. Upon completion, I reported to the 82nd at Fort Bragg, North Carolina. My next school was the three-week Basic Airborne Course (known as Jump School), and I was in one of the last basic airborne courses taught at Fort Bragg.

Airborne and Ranger Training

I had to pass the airborne physical training (PT) test to be enrolled. It consisted of a two-mile run, push-ups, and deep knee bends. I had to do 80 deep knee bends in two minutes. I had difficulty squeezing out my last few bends and only completed 79 in the time allotted. My polio leg was giving out, and I couldn't complete the last one. The NCOIC, Staff Sergeant "Rocky" Coltrane, a tough old airborne trooper, saw me laboring and gave me a pass for the knee bends, mumbling, "Lieutenant, you owe me one." He could have failed me. Had I failed, my life would've taken a totally different turn. The course went by quickly. Our last of five required parachute jumps was from a C-123 Provider twin-engine aircraft on Sicily Drop Zone (DZ) at Fort Bragg. All five jumps were in good weather and with easy landings. I soon learned, all landings would not be this easy. Our parachute wings were pinned on us at the DZ by Major General John L. Throckmorton, the division commander. I was airborne qualified.

The following year, I enrolled in Ranger School, the Army's premier combat leadership course. The nine-week course had three phases: the patrolling phase in Fort Benning, the mountain phase in Dahlonega, Georgia, and the swamp phase in Eglin Air Force Base, Florida. It was a grueling test of one's mental and physical stamina and endurance under the most stressful conditions. Success or failure depended on grit, perseverance, fortitude, and resilience. The ultimate determinant was, "How badly do you want the Ranger tab?" For example, from 2004–16, only 49 percent who attempted the course graduated.

I was in the best physical condition of my life, and ready for the challenge. A slot became available, and I grabbed it. I had a tough time in Ranger School. It taxed me to the fullest, straining my every sinew of mental and physical strength! I knew it would be the ultimate test for my polio leg. Deep down I knew I could do it. There were some lieutenants at Bragg who had washed out of Ranger School and returned without the tab. If you were a combat arms officer, failing Ranger School often meant the end of your career.[1]

There were no menial tasks. All harassment and training tested your physical and mental stamina. What was your breaking point? Could you stay the course? The program of instruction (POI) was tailored to make you miserable. Depriving students of food and sleep was a feature, not a flaw, of Ranger School. We had one C-ration a day. A can of soup, a can of fruit, a pack of cookies, and maybe some peanut butter—my favorite.

The training was arduous and dangerous and was rarely canceled due to weather. Other than a brief "moment of silence" formation in the company street, training wasn't even canceled for President John F. Kennedy's assassination on November 22, 1963, or his funeral three days later. Over the years, the few student deaths that occurred during training were due to hypothermia, lightning strikes, drownings, and exposure. Only a few hours of sleep were allowed a night—some nights not at all.

On a night patrol raid on Santa Rosa Island in the Gulf of Mexico during the Florida phase, while running through the objective, I tripped and fell hard. A big strong Ranger classmate, Paul Stanley, picked me up and helped me to the assembly point. If I had stumbled and not been able to finish myself, I would not have successfully completed that final phase and would've either been recycled or dropped from the course.

My Ranger buddy was Jack Silvey, West Point Class of 1963. We always helped each other, especially during patrols and raids. That was why each student was a member of a two-man "Ranger Buddy" team, with someone who was always there to assist and help motivate. Often, we were soaked to the skin and slept on the ground. No raincoats, sleeping bags, or tents allowed. The weather in December in the mountains in Dahlonega was bitter cold with snow and freezing rain. I went days without dry clothes and frozen to the bone. It was all part of making you miserable and testing your psychological endurance.

While in such a weakened and sorrowful state of mind and body, and with little to eat, we were being observed and graded, judged on our action or inaction, whether we were in a leadership position, e.g., patrol leader, or more likely just a member of the patrol. Could a man still function, despite extreme deprivation? Instructors looked for such attributes and each student had to prove that he could. A grade of either "satisfactory" or "unsatisfactory" was given each time we were in a leadership position.

In the final week of Ranger School, I was so damned tired, I began to hallucinate. While resting against a tree, during a patrol break, I saw a battleship passing in front of me; I reached out for it. I have no idea why I saw a battleship and not a tank or something else! My brain cells were such that I was having hallucinations, seeing things that appeared real due to absolute exhaustion. I was probably in a state of sleep paralysis which suddenly disappeared when the patrol leader yelled, "On your feet!"

Dead tired and standing in formation in grubby field uniforms, we graduated on a field at the Florida Ranger Camp on January 29, 1964, and were awarded the Ranger tab. I was never happier to finish something than the day I completed Ranger School!

When I look back across my career, I can say, unequivocally, Ranger training was the toughest thing I ever did in the Army … perhaps in life! I knew it would be a miserable experience, and it was even worse than I expected. Given the choice between combat and recycling through Ranger School, nobody who has ever been through it chooses Ranger School.

The Department of Military Instruction director at the United States Military Academy at West Point, New York, Colonel Robert "Tex" Turner, famously said, "I woke up in a cold sweat. I had a nightmare that I was still in Ranger School. Thank God that I was in Vietnam. Compared to Ranger School, combat was easy."

A Paratrooper with the 82nd Airborne Division

The next two and a half years were an interesting and exciting time for the 82nd Airborne. It was the Cold War, and President Kennedy had wanted the U.S. to be able to "respond swiftly to threats against peace in any part of the free world." United States Strike Command (STRICOM) was established in late 1961 as a unified combatant command with the mission of responding to global crises, commanded by General Paul D. Adams, a World War II Ranger.

The 82nd and 101st Airborne Divisions were the principal subordinate Army commands. For six weeks during the summer of 1962, I deployed with the 82nd as it joined other Army and Air Force Strike units in Joint Exercise *Swift Strike II*. Its purpose was to test the Joint Task Force (JTF) concept of rapid deployment and the ability of the Army and Air Force to operate together. Over 100,000 soldiers and airmen converged across the Carolinas from multiple bases to participate in this war game.

It consisted of opposing red and blue forces complete with referees to determine who won or lost the battles. The 82nd was the friendly (Blue) force; the 101st was the aggressor (Red) force. It was the largest airborne maneuver ever in peacetime. For six weeks, I was in the field spending most nights on the ground under a small pup tent called a shelter-half. We maneuvered across tobacco fields and watermelon patches, through small towns and across rivers and streams. Air land operations by Army C-7 Caribou aircraft occurred on improvised airstrips. Parachute assaults were conducted on drop zones across the fertile Carolina farmlands.

For the most part, we were welcomed by the local populace. This exercise was repeated in the summer of 1963. By then, I was quite familiar with the topography and landscape. I remember the small town of Joanna, South Carolina, where I parachuted in with an airborne battle group during *Swift Strike III*. Watermelon patches dotted the landscape, and we would help ourselves to these delicious fruits when we could.

The *Swift Strike* exercises were an eye-opener for me. I learned to quickly appreciate the rapid deployment capabilities of airborne units, and the toughness of the individual troopers. And, as a young officer, I was witness to Army tactical aviation coming of age and the coming together of Army-Air Force coordination.

We no sooner returned from Exercise *Swift Strike II*, when the fall of 1962 brought some unexpected events. In September, elements of the division were deployed to Oxford, Mississippi, to assist civil authorities in maintaining law and order during the integration of the University of Mississippi. As the division OB officer, I found myself supporting the five battle groups and briefing them on conditions in the area of operations (AO).

A federal appeals court had ordered the University of Mississippi to admit James Meredith, a black student. When he tried several times to enter in mid- and late

September, the white populace rioted. The president ordered the 82nd and other units to deploy to assist local civil authorities in restoring law and order. The governor, Ross Barnett, declared he would "go to jail before he would allow the university to be integrated." He was willfully obstructing the enforcement of federal laws, and it wasn't clear exactly what he would do; it was clear what the 82nd might have to do.[2]

Our units flew from Pope Air Force Base to final staging sites at Millington Naval Air Station in Memphis, Tennessee, and Columbus Air Force Base in Mississippi. General Hamilton Howze, commanding general of XVIII Airborne Corps, was the overall commander in Oxford. Major General Creighton Abrams, DCSOPS (Deputy Chief of Staff for Operations) of the Army, was the principal liaison between Howze and the Army chief of staff. Abrams later commanded all U.S. forces in Vietnam and then became Army chief of staff. At that time Robert Strange McNamara was secretary of defense (SECDEF).[3]

President Kennedy addressed the nation on TV on September 30, saying, "There is no reason this case can't now be quickly and quietly closed in the manner directed by the court." At that very moment, Oxford was experiencing gunfire, tear gas, and Molotov cocktails. Two people were killed, and many were injured. Police cars and military vehicles were destroyed. The university's all-white student body was adamant that a black man would not be enrolled. As the decision to use federal troops was discussed, many issues surfaced. Should the troops deploy with weapons? Should they have ammunition?

The question arose as to how to use our black soldiers. This was debated at the highest level. Some didn't want black troops to have any contact with the civilians in Oxford to avoid unnecessary incidents. We were informed that the issue was being discussed. I was shocked by this. Our black troops were every bit "airborne" and did everything the white soldiers did. Every trooper was ready to go. The final official decision from the White House was, "There would be no changes in racial composition of units to be used for the mission ... except where a black soldier might face serious danger." Black soldiers would deploy with their units.[4] I hate to think of the morale problems we would've faced had the decision been to not deploy our black soldiers.

Federal troops were deployed, and Oxford was an armed camp. Mississippi Senator John Stennis declared he "couldn't believe" reports of black soldiers being assigned to patrol around the university. Meredith was finally admitted on October 2, and he required an armed guard for the next several months.

I learned later a serious morale problem resulted when one of our units, the 1st Airborne Battle Group, 503rd Infantry, left all its black soldiers back at the airport while the rest of the unit moved to Oxford. It was an isolated incident, and I never learned the details of what happened. The Oxford incident lasted from October 1–10 for 82nd units.[5] For my part, I had never been so busy moving back and forth between division headquarters and the five Battle Group headquarters and to Pope

Air Force Base (AFB), distributing maps of the AO and university campus and providing departing commanders planeside briefings on the most recent intelligence and activity in the Oxford area.

For someone who had spent his entire life in the North, and had attended racially integrated public schools, I had a difficult time comprehending what was happening on that Southern university campus. I learned the depth and absolute viciousness of Southern racism.

As if we were not busy enough, a few days later in October, the Kennedy administration was confronted with the Cuban Missile Crisis. For 13 days in October 1962, the U.S. and Soviet Union engaged in a tense political and military standoff over the installation of nuclear-capable Soviet ballistic missiles in Cuba. Spy satellites and aircraft took photos daily. The entire 82nd was alerted for deployment, and I was manifested with an airborne battle group for a parachute assault on a drop zone south of Havana, Cuba.

Kennedy and his national security team met to consider options.[6] They debated three courses of action: a quarantine (blockade), air strikes, and an invasion. The president chose the quarantine. Preparations were also made for the alternative invasion. The 82nd and other Strike forces were alerted on October 16. General Howze ordered his XVIII Airborne Corps' major unit commanders to Fort Bragg for a briefing on October 19.[7]

The 82nd MI detachment provided aerial photos of the missile sites, airfields, and detailed order of battle of the Cuban military for the briefing. Some 82nd units began immediate movement to intermediate staging areas in Florida. My battle group remained in the Bragg-Pope staging area. The Air Force began moving transport aircraft to support the parachute assaults by the 82nd and 101st. The airborne assaults would be from an altitude of 700–900 feet. Some DZs were in sugar cane fields with cane stalks reaching as high as 7–10 feet, presenting a landing hazard for the parachutists.

I remember special instructions were given for the treatment of Soviet personnel should they be captured. They were to be "carefully handled and taken into protective custody." We were told to minimize any casualties regarding Soviet advisors. On October 22, Kennedy spoke to the nation. The U.S. military was put on a DEFCON 3 status, a defense condition level marked by an increase in force readiness above that required for normal readiness. DEFCON 5 was normal peacetime readiness. On October 24, we went to DEFCON 2—maximum alert. DEFCON 1 meant we were at war.[8] The 82nd's objective was to seize the San Antonio de los Banos military airfield southwest of Havana and the Jose Marti International Airfield just outside Havana.

The 101st would take the airfield and port of Mariel. The Marines were to come across the beach and link-up with the 82nd. It was a tense time! During the alert period, much of Fort Bragg was enclosed in concertina wire with the troops sealed in their barracks or marshaling sites and sentries posted.

As the division combat intelligence/order of battle officer, I worked under the supervision of the division G2 (Intelligence). Each day during the alert, the division staff would brief the division and battle group commanders on the latest intelligence and operations using the most recent aerial photos of the missile sites, DZs, and objective areas. My portion of the daily briefs consisted of giving an updated order of battle of the Cuban forces and Soviet advisors to include unit designations, strengths, locations, weapons, etc.

I also provided information on the weather, condition of DZs, the sugar cane fields, the Almendares River, and other obstacles and terrain features. The Almendares flowed through the city of Havana, had a considerable current, and we determined it to be a formidable obstacle. Some of our units would have to cross it. One battle group commander demanded to know the speed and depth of every foot of the Almendares, where the best crossing sites were, etc. I stayed busy poring over photos, maps, and terrain studies gathering this data, never leaving the division area for a week of being on heightened alert and ready to deploy.

Briefings were conducted at division headquarters in a small windowless and secure briefing room. We assembled and awaited General Throckmorton, and his assistant division commanders, who sat a few feet from the briefing officer. Captain Jim Lindsay, who later commanded the division, gave the G3 (Operations) update. He was crisp, articulate, and highly confident. I followed with the intelligence update. Throckmorton, a World War II and Korean War veteran, was an astute listener who valued good intelligence and wanted the facts. He rarely smiled and had a way of making briefing officers very nervous.

At one update, I stumbled on a word when briefing about the many Cuban "peasants" in the villages and towns in our AO. I said the word "pheasants" when describing their activity, threat, etc. I used the word several times. He finally interrupted: "Lieutenant, we are not going pheasant hunting. The commanders are not interested in how many 'pheasants' there are. Would you tell us what threat the 'peasants' pose?"

The incorrect word "pheasants" was even displayed on the large order of battle chart at the front of the room which I prepared. The room exploded in laughter! My face turned red. I wanted to leave but couldn't. Those few seconds standing before those seasoned and experienced generals and colonels with a long briefing pointer in hand were torturous. Embarrassing moments are hard to forget, and I have never forgotten that one.

I caught an occasional glimpse of the news on TV. We were not allowed to send or receive notes or postcards. We were on edge and wanted to go. It was like standing in the doorway and waiting for the green light to come on before you jumped. It never did, and we eventually stood down.

On October 28, the Soviets ordered the dismantling of missile sites. Nikita Khrushchev had backed down. The quarantine and threat of invasion had worked and marked the closest point the world had ever come to nuclear war. It was, indeed, the hottest point of the Cold War, bringing us to the brink! The Cuban Missile Crisis and the Oxford, Mississippi, integration incident brought national attention to the division and tested our ability to task organize, rapidly assemble, and deploy as America's STRICOM force. It was a very exciting time for me, with only nine months in service. I was getting my feet wet fast!

These events had a profound effect on me in another way. While I enjoyed the fast-moving tempo and felt I had contributed to the mission, something was missing. I wasn't at the tip of the spear, the cutting edge. I wasn't doing what my fellow lieutenants were doing: leading troops as rifle platoon leaders. I was tired of distributing maps, briefing enemy orders of battle, and poring over classified documents. Lieutenant Colonel George Bland and Lieutenant Colonel William J. Patch, both infantry officers who had served as the 82nd Airborne G2, encouraged me to branch transfer to infantry.

I got to know them well. They took me under their wings and became mentors for life. With their encouragement and support, I applied for a regular Army commission in the infantry. I went before an XVIII Airborne Corps selection board and was approved and appointed a first lieutenant in the regular Army with a branch transfer to infantry. I was confident that I would be approved. With the brief time I had remaining in the division, I would not serve in my first infantry assignment until my next duty station.

While at Bragg, it wasn't all duty and work. There were three to four other lieutenants with whom I bonded well. We were all bachelors and enjoyed the "happy hours" at the 82nd Officers Club on Friday nights. We also drove to one of the many beaches or lakes on weekends or traveled north to Raleigh or Greensboro for dates at one of the women's colleges. Most of the Southern girls I dated were outgoing and fun to be with. They were polite, charming, and flirty but could be scrappy. I always got along fine with them until I met their parents. For the first few dates, I met them at their college dormitory.

When they invited me to their home or when I picked them up at their home, the parents would size me up. They were all "Daddy's little girl," and it was usually the father that was most inquisitive. On more than one occasion the father or grandfather of my date would say, "You're a Yankee, where up north are you from, son?" When I answered "Philadelphia," that was like driving a nail in the coffin. The only worse answer would've been "New York City."

The grandfather of one of my dates, while I was in the parlor waiting for his granddaughter, Martha, told me, "All the cannons in the South are still pointed north, young man." I nodded respectfully. I'll never forget that. Meeting Southern girls' parents was an education for me. I never knew what the opening salvo would be. The women I dated didn't care where I was from. They enjoyed dating a paratrooper and going to the 82nd Officer's Club … and found a Northern boy to be different and interesting! At least that's what they told me. Most of the mothers didn't seem to be bothered, either.

They were warmhearted and more interested in whether I was going to stay in the Army and who my parents were. The fathers and grandfathers of those "Southern Belles" though, were still fighting the Civil War from a century earlier! They were never nasty or in any way combative. They simply couldn't forget the "War between the States" as they referred to it and felt it was their duty to remind me about it every chance they could. All my upbringing and education didn't prepare me for such moments. I was learning and beginning to understand the American South's culture of honor!

I met Betty Ruth Maness in 1963. She was a junior at Methodist College in Fayetteville, and we dated off and on. Her father was Reverend M. W. Maness, Minister of the historic Camp Ground United Methodist Church in Fayetteville. He liked me, probably because I was a Methodist and often attended his church service on Sunday.

She was on her way to Wilmington one weekend with two of her girlfriends to meet me for a weekend date at the Azalea Festival when their car overturned, and she was killed. Amazingly, her girlfriends lived. It was tragic. She was from a well-known and highly respected Fayetteville family and had won some local and state beauty contests. I attended her funeral service and met with her parents at their home to pay my deepest respects and offer condolences. They were heartbroken at the death of their daughter. I felt terrible and blamed myself for her death. It was a very tough time for me. I had a remorseful, unhappy feeling, grieving, thinking it was my fault. It took me a long while to get over it.

Each summer, when I was able to get a three- or four-day pass or a week's leave, I drove to Ocean City, New Jersey, to visit my family. On Labor Day weekend 1963, I was in Ocean City when a longtime friend arranged a date for me with one of his girlfriend's sorority sisters. Drexel University's Delta Zeta sorority was staying in an Ocean City motel enjoying the beach and supporting a sorority sister who was involved in the Miss America pageant production in neighboring Atlantic City. I went to the motel on 9th Street and Atlantic Avenue to meet my blind date, Judy Einstein, the daughter of an Army colonel stationed at Carlisle Barracks, Pennsylvania.

Although I was supposed to go out with Judy, instead, while waiting in the motel courtyard, I met Linda Heal. I was immediately attracted to her when she appeared outside the motel, having just taken those huge rollers out of her hair. I couldn't take my eyes off her. There was a strange, vibrant clicking feeling inside me that I had never experienced before. I looked at her and said, "I bet you were a cheerleader." Although she had been a cheerleader in high school, she knew this was a guy's way of starting a conversation. I asked her out that evening, even though I had a blind date with someone else, but she had other plans. I didn't give up. I asked her for a date for the following Tuesday evening. I had to drive two hours to her home in Hamilton Square, New Jersey. She told me later she was sure I would not show up but stayed home from her bi-monthly Grange meeting just in case I did.[9]

We went to a movie and had a snack at the White Horse Diner. It was late, and I didn't want to drive the two hours back to Ocean City, so I asked about a local motel. Without hesitation, she said, "I'm sure you can stay here, let me ask my mom." Mr. and Mrs. Heal had returned from the Grange meeting and were asleep, but her mother got up and prepared the pull-out couch in their sunroom.

It was a small room used to prepare and store her mother's Avon orders, and it smelled of every fragrance imaginable. The many pungent aromas permeated my clothing and stayed with me for three days all the way back to Fort Bragg. Her mother made me a great breakfast. I was most appreciative of her kindness. I sent her a "thank you" bouquet of flowers, and Linda's mother never forgot that gesture. This was one of the smartest things I ever did!

However, her mother was skeptical about her daughter dating an "Army guy." She had told her to be cautious of dating soldiers. Indeed, I worked hard to dispel this notion. I didn't see Linda again until I came home at Christmas. At that time, I introduced her to my fraternity brothers and their wives, and we dated through the holiday. While apart, we continued to date others, but whenever I came home on leave to visit with my parents, I always saw Linda. And I was always on my very best behavior when in the company of Mrs. Heal.

After a year and a half of on again-off again dating, I knew that I wanted to spend my life with her, and we made plans to get married upon my return from Korea in 1966. I was leaving for Korea in January 1965 for a year. She arranged with the Dean of Women at Drexel to take 21 credits per quarter to graduate early.

While in Korea, I kept in touch with Linda and her roommate, Peggy, who looked out for her and kept me informed of how she was doing. They became great bridge players as she refused dates with admiring suitors. Linda was popular and had been nominated for various campus queen contests. She was also a chorus line dancer in the annual fraternity musical show where she designed and made the costumes. She was more faithful and consistent with her letter writing than I was. I still have all her letters bundled and stored in a container.

While at Fort Bragg, I joined the Masons, the oldest and largest fraternal organization in the world. My father, a 32nd Degree Mason and Shriner, encouraged me to look into masonry. Its principal teachings of brotherly love, relief, and truth appealed to me.

In 1970 I became a member of Cairo Temple Shrine in Rutland, Vermont. Shriners International was a fraternity based on fun and fellowship, emphasizing philanthropy, patriotism, and charity, which are among the affirmed values of the Shrine. Its official philanthropy, Shriners Hospitals for Children, was one of the largest pediatric health care systems in the world. I still contribute to these children's hospitals and to the March of Dimes, as both support providing excellent care for children with health disabilities. All my life I felt an obligation to give to organizations that help disabled children. I would not have beaten polio without the March of Dimes and charities like Shriners Children's. I will forever be indebted. Whenever I can help a young child, I do.

On June 20, 1964, I wrote infantry branch requesting an assignment to Vietnam as an advisor to a Vietnamese infantry unit. Our involvement in Vietnam was steadily increasing and many 82nd officers were receiving orders to Military Assistance Command, Vietnam (MACV). Two of my friends at Bragg received orders for Vietnam. Five days later, a certain Major Abt, infantry branch, wrote, "You are commended for your desire to serve in Vietnam, but you must first become branch qualified by serving eight to twelve months in an infantry position." That fall I received orders for South Korea.

Soldiering in the Demilitarized Zone

On January 5, 1965, I arrived in South Korea and was assigned to the 1st Cavalry Division. The division's G1 staff officer said, "Lieutenant Matz, you're a bachelor and Ranger qualified, we're sending you to the DMZ." I was assigned to the 2nd Battalion, 8th Cavalry Regiment, one of three U.S. battalions north of the Imjin River.

The Demilitarized Zone (DMZ) was a strip of no-man's land close to the 38th parallel running 151 miles across the Korean Peninsula, from the Sea of Japan in the east to the Yellow Sea in the west. In the center was the Middle Demarcation Line (MDL) or border that separated North from South. The U.S. sector was 18½ miles in length. The Republic of Korea troops were responsible for the rest of the border. Within the U.S. portion, my battalion was responsible for maintaining watch over a 6-mile sector.

Our mission was anti-infiltration and surveillance—killing or capturing North Korean line-crossers and agents as they infiltrated south. We set up ambush stakeouts and patrolled the DMZ night and day. We also manned four observation guard posts 24 hours a day. It was the best peacetime assignment in the Army for a young company grade officer. We handled live ammunition every day and kept our weapons and ammunition next to our bunks. It was classified a "hardship" tour and the chance of getting wounded or killed on DMZ duty was much greater than serving in a rear area unit further south.

During the 1960s, there were many attempts to infiltrate the south by North Korean agents. The Korean War ended in 1953 with an armistice, but not a peace treaty. We were in a truce and technically still at war and remain so today. There was a lot of tunneling under the DMZ and daily propaganda broadcasts by North Korea. During that year, the 8th Cavalry intercepted and captured a couple enemy agents. I spent New Year's Eve 1965 in North Korea, with a small five-man patrol.

We were only a few meters inside North Korea when we set up a hunter-killer ambush. There was snow and it was bitter cold.

We didn't get anyone that night, but vigilance was never relaxed. All communist activity north of the DMZ was observed and analyzed. The atmosphere of combat readiness was heightened since there were no civilian communities in our sector north of the Imjin. It was a live mission close to the enemy. We could see and hear them. Their broadcasts would taunt American GIs and encourage them to defect to "a better life" in North Korea. Four American soldiers defected to North Korea in the early/mid-1960s. There was never a dull moment.[10]

It was the challenge and excitement of the duty that appealed to my senses. I was drawn into it and welcomed it. Officers and enlisted soldiers alike shared the hardships of living north of the Imjin. My first sergeant and mortar platoon sergeant, Davis and Mehaffey, were experienced NCOs and had fought with the 1st Cavalry Division during the Korean War 12 years earlier.

Each U.S. infantry company had a KATUSA platoon of 30–40 South Korean soldiers. KATUSA stood for Korean Augmentation to U.S. Army. KATUSA soldiers were tough and enjoyed serving with an American unit. I was a mechanized infantry company commander and enjoyed working with Allied soldiers early in my career. The South Koreans *hated* the North Koreans. My KATUSA soldiers always volunteered for additional stake-out or patrol duty. They wanted to catch and kill a North Korean, and they were indeed a force multiplier for U.S. infantry companies operating in the DMZ. Commanding and working daily with South Korean soldiers was an incredible, unique aspect of this assignment.

I was fortunate during this tour to have been both a company commander and battalion S3 (operations officer). When our S3 departed in the summer of 1965, the battalion commander selected me to become the new S3, which was a major's billet, and I was the most junior captain in the battalion. The battalion commander, Lieutenant Colonel Francis A. Santangelo, promoted me to captain one month before while in the field on a training exercise. The 13-month assignment on the DMZ solidified for me my decision to make the Army a career as an infantryman.

I was infantry-branch qualified, and my tour ended in February 1966. I received orders to the Ranger department at Fort Benning, replacing my good friend, Captain Antonio Mavroudis. I next received orders for the Infantry Officer Advanced Course with a report date of January 4, 1967.

Upon return from Korea, I linked up with Linda and we got engaged. She had completed all her academic credits toward graduation and was finishing her student teaching requirement in home economics when I returned. Her graduation ceremony was June 18 (her birthday), and we were married on July 2 in the Hamilton Square Baptist Church in New Jersey. I had my partner for life! Her parents gave us a beautiful wedding with the reception at the Forsgate Country Club.

It was a 104-degree, record-setting, sweltering-hot day. The church wasn't air conditioned. Linda's dad arranged for the wedding party to travel in air-conditioned cars. Everyone wanted to get to the air-conditioned reception. An issue for my friends was the fact that Baptists didn't drink. Linda's family never had a drop of alcohol, so the reception was completely dry—not even champagne for a toast. Her mother saw to that! She also refused to print my military rank of "Captain" on the wedding invitations, instead, referring to me as "Mr." She couldn't accept that her daughter was marrying a soldier!

My groomsmen were high school, college, and Army friends. My best man was Bill Hemsing, my fraternity brother and roommate for two years in college. I was a groomsman at Bill and Nancy's wedding, three years earlier. One of my thirsty groomsmen, Ken Tholan, located the club's open bar and had a cold beer. Ken was most comfortable when he had either a lacrosse stick or a beer in his hand. Others soon followed him, much to the dismay of Linda's mother. Before I knew it, many of my friends joined Ken at the bar in another section of the club and had to be rounded up and returned to the ballroom for the toasts, served meal, and cake cutting. I tried, but never made it to the bar. It all ended well, but my mother-in-law never forgave Ken for adding alcohol to her daughter's wedding reception.

We honeymooned for a week in Jamaica. On the flight, my young bride, donning a beautiful dress, hat, and corsage, was sobbing for most of the flight. The flight attendants knew it was a special day for us and were perplexed as to why she was crying, as was I. She was sad about leaving her very close-knit family with whom she had spent all her life. It struck me also that leaving home for the first time can be an emotional experience. I was sympathetic and did my best to help her adjust not only to a completely new and different culture, the military, but also to living in the Deep South far from home and family. Once we got to the Half-Moon resort in Montego Bay, she was fine.

After the honeymoon, we drove south in my 1965 Corvette Stingray, pulling a trailer packed full of wedding gifts, and arrived at Fort Benning. We moved into a small single-brick house in the Custer Terrace housing area. Linda was excited about making it a home. I was happy to have the comforts of a home again after 13 months of spartan living in Korea.

Linda's introduction to the Army was the Ranger department. The department leadership couldn't have been more welcoming, and the officers and their wives made her feel at home. She was now an Army wife and adjusting to the culture and unique requirements of living on a military post.

One setback early in our marriage occurred while shopping at the PX (post exchange); we purchased an ironing board and some long drapery rods, and they would not fit in the Corvette. Linda, by then pregnant, said, "Bill, this car is not practical. Where are we going to put the baby? It's got to go." She was right, and I sold it within a week. We soon became proud parents of a wonderful little boy, Bill III, who was born at Martin Army Hospital.

That summer, I received orders for Vietnam assigning me to the 101st Airborne Division which was operating in the Central Highlands. Seventy percent of my Infantry Advanced Course classmates went to Vietnam, some for their second tour, and some of my friends had been killed. I was ready to go. That military aphorism about a soldier "marching to the sound of the guns" was in play. In mid-1967, we were continuing to build our force levels, and more individual replacements and units were being assigned to Vietnam.

The fighting was becoming more intense, and Linda knew that I would be leaving. We had talked about Vietnam while we were engaged and that I would be "going over" soon. I departed for Vietnam in October. It is never easy saying goodbye to your wife and child. Linda and Billy, then six months old, would live with her parents in New Jersey. She couldn't have had better support. It was a great comfort for me knowing they would be there. We said our goodbyes, and the airplane left Philadelphia International Airport. It was a long flight to South Vietnam with a lot of time to think about family, and what to expect when I got there. Suddenly, the war took on a different, closer, and more personal meaning for me.

CHAPTER FOUR

Vietnam War—Infantry Combat and the Tet Offensive

On November 1, 1967, I arrived in South Vietnam at Ton Son Nhut Airbase, about 6 miles northeast of Saigon. I flew in on a MAC (Military Airlift Command) charter flight from Travis Air Force Base, California. The Pan American Boeing 707 was packed with 200 teenage soldiers, field grade officers, and NCOs. We made one refuel stop in Guam and when we approached the coast of South Vietnam, I looked down for the first time and saw the war zone.

When we landed, and taxied to the gate, I saw some silver metal transfer cases neatly lined up on the tarmac containing the remains of American soldiers which would be put on airplanes returning to the United States. I learned a few days later that my good friend and roommate from Fort Benning, Captain Tony Mavroudis, was KIA (killed in action) and posthumously promoted to major a few days before I arrived. He had been a company commander in the 101st Airborne, and I often wondered if his remains were in one of those transfer cases that I saw that day.[1]

I learned as soon as I reported to the small reception building at the air base that my orders were changed from the 101st to the 9th Infantry Division (9th ID) in IV Corps Area. I was not happy! My former boss and mentor in the 82nd, Lieutenant Colonel George Bland, was a battalion commander in the 9th ID, and he wanted me to be one of his company commanders. A bus collected those of us going to the 9th ID and took us to Camp Bearcat.

Camp Bearcat was the 9th ID headquarters base located about 20 miles east of Saigon. It was a large camp built around a small Vietnamese village called Cat, so they called the camp "Bearcat." At the Replacement depot, they told me I was assigned to the Second Brigade, the Army component to the Joint Army-Navy Mobile Riverine Force (MRF). I spent a few days there going through the Old Reliable Academy, an in-country orientation program for new troops. I received equipment, got the required shots, and finished in-processing. I then flew by Huey (UH-1 Iroquois helicopter) to the MRF anchorage in the Mekong River Delta.

The Mobile Riverine Force and the Mekong Delta

The MRF was a joint U.S. Army–U.S. Navy force organized specifically to fight in the Mekong Delta. The Army met an unusually complex challenge in Southeast Asia, as we were a modern conventional army called on to fight in an undeveloped environment against an indigenous enemy. We had to adapt our conventional doctrine, tactics, etc., in order to fight a largely local, unconventional enemy. The Viet Cong (VC) were a formidable guerrilla force who were masters of irregular warfare. They were equipped and advised by the North Vietnamese and Chinese communists.[2]

The VC command-and-control structure was less rigid than ours but well suited for irregular warfare. They were armed with Chinese- and Soviet-supplied weapons, and operated in small teams, sometimes in platoon strength. They specialized in ambushes, booby traps, and sabotage, and were always intimidating the local populace. They moved at night by foot and small boats, called sampans. They travelled lightly and were fast and elusive. Their method of warfare was what I call "harassment through surprise." On the other hand, the American Army was a large, mobile, and conventional military force burdened initially with equipment that slowed us.

The VC were South Vietnamese supporters of the communist National Liberation Front (NLF) of South Vietnam. They were allied with North Vietnam and Ho Chi Minh's troops who were fighting to create a unified communist state. The VC had near absolute control of the Mekong Delta, and other than a few U.S. Army Special Forces detachments and some army and navy advisors to Vietnamese units, there were no other American troops in the Delta, so the situation was ripe for exploitation by a U.S. riverine force.

VC fighters were tenacious and resourceful. They were willing to pay a heavy price and sacrifice lives in a war America wasn't prepared, militarily or politically, to fight. They used small units and armed political cells to maintain their hold in the countryside and villages. They killed their own people to make a political and psychological statement to intimidate the locals.

Some VC defected to South Vietnamese and U.S. units. They were called *chieu hois* (loosely translated as meaning "open arms"). The Vietnamese government established a policy of an "open arms" program to entice communist guerrilla fighters to leave the VC. The 9th ID supported the program and called their *chieu hois* "Tiger Scouts." The U.S. Marines up north called them "Kit Carson Scouts" and also used them.[3] The defectors underwent a stringent reeducation at *chieu hoi* centers. Their training included the English language. Once certified by the Vietnamese government, they joined the ARVN (Army of the Republic of Vietnam) and U.S. units and were paid by the U.S. government.

We learned firsthand just how ruthless the VC were towards the very people they claimed they wanted to liberate. My *chieu hois* told us that the VC would threaten, kidnap, and even kill local villagers if they didn't cooperate and allow the VC to stash arms and medical caches in their villages, or volunteer to set up booby traps to kill Americans.

Once, after we swept through a Vietnamese village and cleared it, we took a rest break. It was a sweltering hot day, and the troops were thirsty and tired. Some young Vietnamese boys, probably age 9–12, offered them bottles of cold Coke. As it turned out, the bottles were booby trapped, containing slivers of crystal-clear glass. Some of the men had their mouths and throats cut gulping down the cold Coke, thinking it was ice. On-the-spot interrogation by my *chieu hois* determined that the VC made them do it and after a search we found more bottles of glassed Coke.

My *chieu hois* and a couple soldiers beat the living hell out of them before we evacuated them as POWs. We called a dust off helicopter to medevac the troops with mouth and throat wounds. "Dust off" was the term commonly used to describe a helicopter that evacuates the wounded and dead from the battlefield. Incidents such as those were not that unusual in the Delta. In fact, such surreptitious sabotage against our troops became a regular event throughout all South Vietnam, not just in the Delta. I learned more about just how cruel and devious our communist enemy was as time passed.

The Delta extended from Saigon south and west to the Gulf of Thailand and the Cambodian border for about 40,000 square kilometers, which was one-quarter of the total land area of South Vietnam. In the Delta, there were 8,000,000 inhabitants, which was about half the country's population. It was created by the huge Mekong River and its distributaries and was bordered on the east by the South China Sea. The other large river was the Bassac River, a distributary of the Tonlé Sap and Mekong rivers; it started in Phnom Penh, Cambodia, and flowed southerly, crossing the border into Vietnam near Chau Doc.[4]

The Delta was mostly a flat area where much of the surface was covered by rice paddies, one of the world's most productive rice-growing areas. Laced with many navigable rivers, streams, and canals, it was well suited to riverine operations. Observing conditions in the area, I was thankful for the techniques I learned in Ranger School regarding rubber boat rafting and how to construct rope crossings over deep, treacherous swamp streams. Due to its anomalies, the Delta was the most important region in Vietnam in that it was heavily populated and contained the heaviest concentration of VC units. There was one major hard-surface road, called a "hardball," which was Route 4, and it ran from Saigon south to Ca Mau. The road network was, therefore, limited for military operations.[5]

The Delta was very good for air operations, with abundant landing zones (LZs) and good flying weather.[6] It was, however, very inhospitable! High humidity and temperatures made it unbearably hot for soldiers. We constantly operated in knee-deep or higher mud-sucking water. The canals and rice paddies contained microorganisms and fungi that caused serious skin problems.[7] As a result of the prolonged exposure, there was a high incidence of foot and leg problems with soldiers contracting skin diseases and experiencing other dermatological issues.

One condition that we experienced was immersion foot, or trench foot, which was the most debilitating. It occurs when feet are wet for a long period of time. No medicine could treat Delta's immersion foot. Doctors' sick slips read, "Soldier can't operate in the water." In fact, the putrid water and mud was as debilitating as the VC, just not fatal. Sunlight and wearing sandals were the remedies. I contracted a Delta fungal strain sometime in February 1968, and it remained for several years. It was itchy, unsightly and occasionally became infected. At times it oozed and bled. For years afterward, I often wrapped the affected area with gauze, so it didn't soil my pant leg or bedding. Linda worried that it might be contagious.

My own medical diagnosis on December 23, 1968 read, "Matz: fungus left leg 8 × 10 contracted in the Delta." It persisted, and no medication could cure it. A medical exam diagnosis on December 26, 1970 read, "chronic dermatitis, not responding to antifungal topical steroids or sulfur treatments, biopsy tuberculoid leprosy." Tuberculoid leprosy manifested itself in widespread sores and lesions affecting nerves. It went away, then months later, flared up. It was finally and fully cured by treatment at Walter Reed Army Hospital in the 1990s.

What amazed me was that the skin ailments didn't appear to affect the enemy. The VC and farmers went barefoot or wore open sandals and dried their feet out each night. Vietnamese who lived in the Delta had a toughened leathery skin which protected and immunized them from the fungal and bacterial infections. As a result, monitoring U.S. infantry company "paddy" strength was a priority. We had to keep at least 120 physically fit soldiers or 73 percent of 164 men as authorized strength in the field. I had to report my paddy strength daily through channels to division headquarters. U.S. units operating in other parts of Vietnam didn't experience or suffer this dermatological, casualty-producing enemy unique to the Delta!

In fact, it was the second enemy of the infantrymen, and many reported to sick call with permanent physical profiles (medical conditions preventing them from going on operations) as a result of severe leg and foot problems. The bacterial and fungal infections were due to prolonged periods of operating in the filthy brackish waters. Once men hit the river/canal bank, if the landing craft ramp wasn't far enough over the bank, they sank up to their waists in the filthy, dark glue-like mud. Men who became wounded in that quagmire remained stuck, encumbered with combat gear, and evacuating them was very difficult.

Then there was the putrid smell of rotting fish and vegetation and the stench of the Delta swamp gas, something I will never forget. Finally, there were the malaria-bearing mosquitoes and leeches. Oh, the leeches! They bred in the standing brackish water: huge, dark, slimy, blood-sucking globs that attached to and burrowed into exposed skin and under the uniform. The only way to remove them was by burning them out with cigarettes or spraying them with a special insect repellant.

We were issued 2 oz plastic bottles of smelly, sticky insect repellent that we called "bug juice." Next to 5.56 mm and 7.62 mm ammunition and water, it was the most important item carried. We used it to remove leeches, light fires, clean weapons, and cook C-rations. A cigarette burn or a squirt of "juice" made the leech curl up and fall off. If they were deep into the skin, they had to be cut out or pulled out. They rarely caused permanent damage, but they could cause infection, nausea, and be painful! During breaks in movement, we checked each other for leeches. Sometimes, they were found on your anatomy's most surprising places. I enjoyed watching the troops from the rural areas mock the "city boys" who were deathly afraid of the greasy leeches. Without a sense of humor, the Delta would've been unbearable!

We operated under strict rules of engagement. U.S. commanders had to maneuver their forces so as not to interfere with the normal flow of civilian life. I agreed with these precautionary measures to a certain extent. Civilians should not suffer as a result of being caught between two belligerents. Population centers were primarily concentrated along waterways, and the people lived in areas where there were trees, undergrowth, and nipa palm. Nipa palm was a thick species of mangrove palm common along Vietnam's coasts and rivers, with leaves 30 feet in height offering excellent concealment and protection to the enemy.

Due to the nature of the terrain and the enemy's movements, as well as our operations against them, engagements often took place in populated areas, so we were faced with the task of finding and killing the enemy while simultaneously not hurting the locals, complicating planning and execution of operations.[8] We labored to prevent collateral damage and harming civilians, whereas the VC used collateral damage as a propaganda tool and intimidation factor. We were held strictly accountable by our chain-of-command for any civilian casualties.

The terrain, demographics, and location of population centers favored the enemy causing us to adapt and change our tactics. We soon became very knowledgeable in the lay of the land. We also had to deal with the influence of sea tides along the waterways. The enemy was excellent at mining and obstructing the waterways, using below-surface stationary and drifting mines to prevent movement and entry. On November 1, 1968, the MRF's USS *Westchester County*, an LST (Landing Ship, Tank), was mined in the Mekong River with a loss of 25 killed, 27 wounded, and 4 missing, the work of VC swimmers and sappers.[9] It was also very difficult to base large concentrations of troops as there wasn't enough hardstand, so a floating base offered an alternative against the enemy forces in the Delta.

The VC knew the environment; they had both a political and military organization in the U.S. IV Corps area. They operated out of large, well-supplied base areas. It was their backyard, and they knew the terrain and the people, putting us at a huge disadvantage. There were about 85,000 VC troops in the Delta where we operated. All were combat and support troops and local part-time guerillas, with a political cadre of Chinese and North Vietnamese. They were well organized into regiments with three battalions each and 19 additional separate battalions. They had good equipment, such as 7.62 mm weapons from China and ammunition from Russia. They had the AK-47, SKS carbine, Russian belt-fed RPD light machine gun and the RPG (rocket-propelled antitank grenade launcher), which was a potent weapon against Navy riverine craft.[10]

They were highly capable of infiltrating our radio frequencies voice circuits, often entering our net jabbering in Vietnamese. For instance, during the December 24–25 Christmas truce, while occupying a defensive position several kilometers from Dong Tam,[11] the large U.S. base on the Mekong River, the VC clogged my company and battalion command nets with continuous jabbering. My battalion commander spent that night with us. Again, on 30–31 January, right before the Tet Offensive, our radio nets began receiving rapid and unintelligible chatter. I had my radio telephone operator (RTO), Specialist Benjamin Underhill, from Tennessee, switch to alternate frequencies to avoid it.

Their tactics were primarily small-unit operations, but they were also very capable of mounting battalion-size attacks, if needed. Before I arrived, the VC ambushed and destroyed the South Vietnamese 33rd Ranger Battalion and the South Vietnamese 4th Marine Battalion in the Delta's Ap Bac area—a major VC base.

South Vietnamese forces in the Delta consisted of three divisions: the 7th, 9th, and 21st Infantry Divisions. There were also five Ranger battalions and three ACR (Armored Cavalry Regiments) for a total of about 40,000 ARVN troops in the Delta.[12] In April 1964, U.S. Ambassador Henry Cabot Lodge had optimistically said, "I would not be surprised to see the Mekong Delta totally cleared of communist forces by the end of 1965."[13] (I was in Vietnam in 1967–68.) How naïve could this man be!

This comment was a prelude to further incorrect statements and wishful thinking by civilians and politicians who had no skin in the game, and who were ignorant of the situation on the ground, making ridiculous predictions that were reported in the news in hopes of guaranteeing success. When the military fell short of these unrealistic expectations, which the Army had no part in making, we took the ridicule and public scolding.[14]

Lodge was dead wrong. In late 1966, one third of all VC action against our troops in Vietnam occurred in the IV Corps area. The VC controlled over 25 percent of the population in the Delta, and the night belonged to them. They were also choking off the flow of rice to markets in the farther North. Far from being totally cleared of communist forces in 1966, the Delta was increasingly coming under VC control. Something had to be done.

This situation had forced the creation of the American MRF in late 1965. In a message to CINCPAC (commander in chief Pacific) Admiral Ulysses S. Grant Sharp, on May 11, 1966, General William C. Westmoreland stated, "enemy access to Delta resources must be terminated without delay."[15] Westmoreland, who was COMUSMACV (commander United States Military Assistance Command, Vietnam) and Major General William E. DePuy, his operations officer, developed

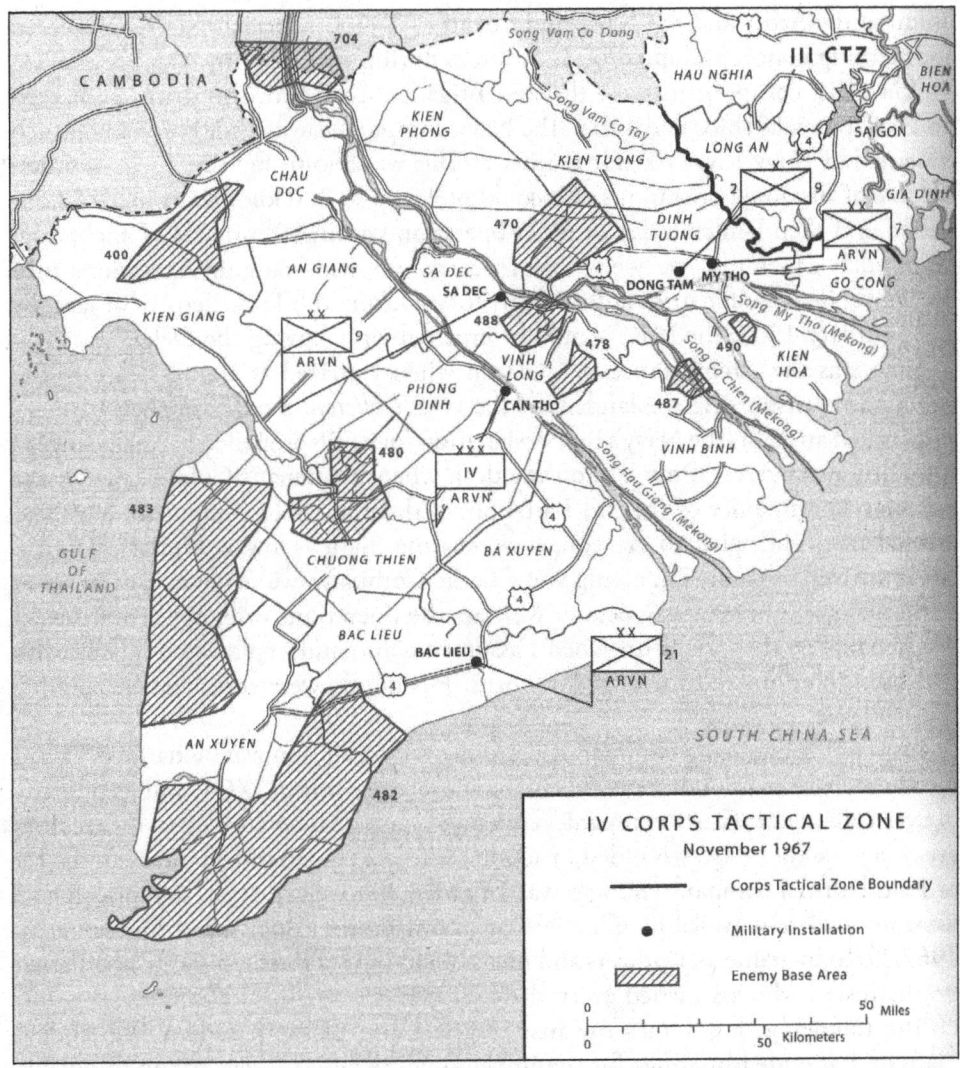

Map of enemy base areas and provinces/cities in the Mekong Delta where my battalion operated in 1967–8. (Map from U.S. Army Center of Military History, "Staying the Course")

the original concept of an Army/Navy force consisting of an Army infantry brigade and a comparable Navy organization.[16] Sharp and the JCS approved the concept and implementation.[17]

The MRF mission was to conduct offensive operations against the VC, isolate key food-producing areas from VC control and interdict VC supply routes. The MRF was a unique force, and the most significant organizational aspect of the MRF concept was the permanent integration of Army and Navy units to provide a fighting force uniquely tailored to the nature of this area of operations. Indeed, it was a fortuitous union of the Army and the Navy, and I believe, the most specialized organizational and fighting concept adopted by U.S. forces during the Vietnam War.

The MRF components were the 2nd Brigade, 9th ID for the Army, embarked on Navy barracks ships and LSTs. The Navy component was Task Force 117 which consisted of Navy River Assault Flotilla 1. This was a joint fighting force of 5,000 capable of sustained operations. We could move up to 200 kilometers in a 24-hour period and then launch a day or night operation within 30 minutes of anchoring. With such capability, we were able to carry out wide-ranging operations into previously inaccessible or remote VC territory. During the Tet Offensive of January and February 1968, the MRF would be credited with "saving the Delta."

Such was the circumstance in Vietnam when I arrived at Camp Bearcat. The helicopter transporting me landed on the USS *Colleton*, a self-propelled barracks ship, called an APB (the Navy's hull designation for a self-propelled barracks ship).[18] I got out and stretched my legs on the deck when Lieutenant Colonel Bland, the battalion commander of the 3rd Battalion, 47th Infantry Regiment came up and greeted me. "Bill, glad to see you, how was the flight? How is Linda?" Before I could answer, he said, "I'm giving you Charlie Company, we've got an operation in 48 hours, get squared away below. We start the operations briefings tomorrow." I was aboard less than 24 hours when I assumed command, replacing a captain who had been in command for several months. His nerves were shot, and he couldn't wait to leave.

I was a rifle company commander, and my radio call sign was "Charlie 6." I had four lieutenant platoon leaders, a lieutenant executive officer (XO), and a company of over 160 men. Infantry companies contained extra men because we took casualties every day—either KIA, wounded in action (WIA) or due to fungal injury to the leg or foot. Normal company strength was 145 men, but we kept 160, so replacements were immediately available. Charlie Company's battle roster, dated December 1, 1967, listed by name 161 officers and men. Blood type and religion were also shown on the roster, which I carried everywhere.

The following day, I met my first sergeant and platoon leaders. We studied maps and attended briefings for the impending operation. I was trying to quickly orient myself, and learn about the area we were operating in. Lieutenant Colonel Bland introduced me to the other company commanders, Captains Ray Sanders, A Company; Craig Boice, B Company; and Al Sabitch, D Company. They had been in command for several weeks. I was the new, untested guy. They welcomed me but viewed me askance. I was the new guy and had to prove myself.

Boice stood out. He was older, more mature, and gave me the lay of the land. He was very savvy and always asked probing questions at our operation order briefings. Craig and I worked well together in the succeeding months of combat. I had never felt so overwhelmed in my life. A sudden onset of anxiety was triggered by a plethora of new things: riverine warfare, a brand-new environment, new faces and names, and my first combat operation only hours away. I wasn't even exactly sure where I was.

I had just met and learned the names of my key personnel, but I had no time to get to know them. At 0200 the next morning, November 16, I was descending the ship's gangway to the loading barge below to board the armored troop carriers (ATCs). It was dark, and we were in blackout mode. I felt a push from behind: "Move out, you're slowing us down!" yelled a voice in the dark. It was a soldier carrying an M-60 machine gun. He wasn't the least bit bashful, not knowing I was the new company commander, and I had no idea who he was while breathing down my neck. The ATCs were 50-foot LCM-6s, and we called them Mike boats. Each company was allocated three boats which moved us to the assault beaches. With 3,000 kilometers of navigable waterways and no adequate roads, the principal means to our objectives and targets were the boats. It took the company 20 minutes to complete loading.

Once loaded, the ATCs proceeded to a rendezvous point where they waited for the remainder of the battalion to load. Upon completion of the loading, the ATCs crossed the start point and our column of boats with the minesweepers out front headed to the objectives. It was a pitch-black night, and we observed strict radio silence. The River Assault Squadron commander was aboard the lead boat. During these operations, my battalion commander was either in his C&C (command and control) boat or in a helicopter above. Armored gun boats called monitors with 20mm and 40mm guns and 81mm mortars supported the companies.

While underway, the Navy Assault Squadron commander exercised command and control, and company commanders monitored both the Navy and battalion command nets. When we neared the assault landing sites, the field artillery preparation fires were lifted or shifted. The monitors provided direct fire support onto the beaches as the infantry disembarked from the boat's bow ramp.

The operation was codenamed *Coronado IX*.[19] The *Coronado* series of operations began in June 1967 to deny the enemy areas where they had previously operated with total freedom. They were named after the Navy's California training base in Coronado on San Diego Bay. Our objectives were the 502nd Local Force Battalion and the 267th Main Force Battalion in Base Area 470 along the Kien Phong and Dinh Tuong border. We had the South Vietnamese 5th Marine Battalion and elements of the ARVN 7th and 9th IDs working with us on the operation. On the second day, November 18, we made contact with the enemy. My battalion was on the flank and didn't engage the enemy directly. We did uncover a large VC cache of medical supplies and detained some suspected VC. Most VC casualties were from helicopter gunships.

Total losses for all U.S. and ARVN units were 26, while the VC suffered 178 killed. My first operation as commander of Charlie Company, 3rd Battalion, 47th Infantry Regiment (C/3/47) was over. I got my feet wet quickly and literally … and was introduced to the vast, unforgiving terrain, stench and climate of the Mekong Delta and the tough, wiry enemy soldiers we would be fighting the next year. I saw firsthand the courage, grit and stamina of the American infantryman as he operated in this treacherous and unforgiving environment. All they wanted to do was kill the VC! I asked myself, as their new company commander, "I wonder what they think of me, if I met their expectations?" Time would tell.

Two days earlier, while in the well deck of the ATC steaming towards the objective, I couldn't find my first sergeant. When I asked where he was a platoon leader said, "Sir, he doesn't go on operations. He stays on the ship and monitors the net." I was pissed off and befuddled with this answer. A company consisting of scores of 18-year-old riflemen and young sergeants on a combat operation with the company's most experienced NCO, the first sergeant, remaining on the ship safely monitoring the radio net didn't sit well with me. I made an immediate change.

Within a few days, I had a new "top" soldier, First Sergeant Richard Vasquez. Following a brief respite at the Dong Tam Anchorage, we entered the Cam Son Base Area in Dinh Tuong Province on November 23. This was the home base for the 263rd and 540th VC Battalions. We entered by riverine assault immediately following a B-52 strike which we heard from a distance. These were Arc Light missions flown at 30,000 feet. Upon reaching the bombed area, we found the terrain pockmarked with large craters. Nothing was moving. The highly mobile VC forces made accurate bombing difficult. We found pieces of destroyed sampans and a few dead enemies, their bodies shredded, along with caches of equipment.

From November 27–30, 1967, operations were conducted to clear a series of canals near Dong Tam.[20] Opening the upper reaches of these canals would allow the use of the MRF's assault boats to attack the VC's Ap Bac base. It was also the home base of the ARVN 7th ID under the command of General Huynh Van Cao and American advisor Lieutenant Colonel John Paul Vann. They engaged the 261st Main Force Battalion of the NLF five years earlier and we were returning to that area.[21]

We encountered only a few enemies but were able to destroy upwards of 60 bunkers along the canals. After two days of slogging through the paddies and mud with no meaningful contact, I was frustrated. I reported to battalion: "This is Charlie 6. We have swept this area twice and no VC."

Major Dick Healy, the battalion S3, told me to move to the next canal where there was a suspected bunker complex. We discovered a huge, unmanned network of recently completed fighting and storage bunkers. I called my two engineer demolition teams forward. As a precaution, I placed infantry on both banks of the canal for security. The engineers, using C4 plastic explosive, placed stringers of block charges at critical points throughout the bunker system.

After a couple hours of continuous ear-busting explosions, we destroyed the bunker complex. At my debrief the following day, the brigade debriefing officer asked, "Captain Matz, are you listening, can you hear me?" My ears were still ringing from the loud blast noises the day before, and I was struggling to hear. The next day my hearing was normal, but I prayed for no more loud explosions!

My first exposure to Agent Orange occurred during my first few operations in the Ap Bac area. Agent Orange was a strong chemical herbicide and defoliant used by the U.S. military to control vegetation and eliminate thick undergrowth used by the VC to hide and stash weapons. It was usually dispensed by C-123 aircraft across a large swath of heavy vegetation or nipa palm. We sometimes operated close to recently sprayed areas. If we moved through or even near a recently sprayed area, the sticky, gooey spray on the plant and palm leaves would get on our boots and pants and it had a nauseating smell.

The troops hated operating in areas struck by Agent Orange, and I maneuvered my company to avoid those areas around the defoliated patches whenever I could. The chemical contaminant was dioxin, a highly toxic carcinogen and known to cause cancer. In 2009, I was diagnosed with prostate cancer and had a radical prostatectomy at Walter Reed Army Medical Center in January 2010. My surgeon, Colonel David G. McLeod, MD, attributed my condition to my months of exposure to the toxic dioxin chemical while operating in the Delta.

My first two and a half weeks went by quickly. I was settling into the fast-paced routine of MRF operations. I was also getting acclimated to living aboard the *Colleton* and familiarizing myself with Navy customs and terminology. Despite inter-service rivalries, the brown-shoe Navy took excellent care of us. They fed us well and always had some 3.2 beer waiting on the AMMI pontoon barges (finger piers, pontoons, and barges were referred to as "AMMIs" in Vietnam) when we returned from operations. The 3.2 percent beer had less alcohol content than standard beer with 4 or 6 percent alcohol. It wouldn't make you unsteady or dizzy. It was ice-cold, and the troops loved it. Some disdained the beer, being too weary and emotionally drained to care about anything but getting some sleep.

Our standard operating procedure (SOP) upon returning from an operation included the Navy hosing us and our equipment down with their fire hoses. We reeked of a terrible odor and were filthy from the brackish water and mud. Lightly wounded troops went to the ship's sick bay where our battalion surgeon treated them. The KIAs and seriously wounded were evacuated by dust offs during the fight. I reported immediately to the battalion and brigade debriefing teams. Casualty reports received during the operations were verified and finalized.

Each 11B infantryman carried an M-16 rifle and 15–20 magazines of 5.56 mm ammunition. Attached to his LBE (load-bearing equipment) harness was his bayonet, first aid pouch(es), fragmentation and smoke grenades, several canteens of water, and poncho. Around his neck were his dog tags and a P-38 can opener (his best friend). Infantrymen also carried a riverine waterproof light assault pack containing C-rations, extra socks, batteries, and extra medical dressings. Some carried a light entrenching tool, concussion grenades, and M-18 claymore mines. Those carrying machine guns, 90 mm recoilless rifles, and the M-79 grenade launcher (aka "thumper") carried variations of this equipment. Body armor vests (flak jackets) and steel pots (helmets) were worn.

Night resupply by helicopter brought water, rations, small arms ammunition, and 81mm mortars. Every man carried insect repellant and water purification tablets. The tablets were not very effective in purifying the germ-filled Delta water. The total weight carried was around 40 lbs as we traveled light and without heavy rucksacks.

Weapons and equipment were cleaned and inspected in platoon areas. Wet and damaged ammunition and dry cell batteries were disposed of. For the platoon leaders and NCOs, their work never ended. They were responsible for ensuring all tasks were accomplished in their respective squads and platoons. Hot showers and chow followed. I personally inspected my platoon leaders' M-16s, and they inspected my AR-15. Crew served weapons (those taking two or more men to operate them), e.g., machine guns, 90 mm recoilless rifles, and mortars received special attention. Those were the tools of our trade, and they had better function properly. When all was done, if there was time, I visited my wounded troops with the other company commanders. The battalion commander was good at getting a chopper to fly us to the three or four field, surgical and evacuation hospitals where our wounded were taken.

One of the toughest duties as a company commander was writing the next-of-kin of my dead soldiers, almost always a mother or wife. I couldn't send the letter until I was told in writing that the next of kin had been personally notified by the stateside casualty assistance notification team. The Army teams always consisted of two people, and the notification experiences affected families as well as the notification team members, long after the war was over. A dying soldier's last words, as he lay gasping for his last breath, were often a cry out for either his mother or God. With Charlie Company, I heard these cries all too often.

One clear example regarding the effect on families comes to mind. On July 2, 1994, I received a letter from Polo Merguzhis, the brother of Private First Class Herman Doelger-Landwar, one of my soldiers, who was killed in Dinh Tuong Province on March 17, 1968. Polo wrote to me after he briefly talked to me on the phone. I still have the letter. It read, "I deeply appreciate talking with you today about my brother. The letter which you sent to us on the occasion of his death will be forever appreciated, as it was of great comfort to my mother, whose grief was beyond description." This is why I always took great care in writing these letters; they had to be personal, precise, and above all else, penned with compassion and with a comforting touch.

Troop morale had its low and high points. After five or six days in the Delta mud and paddies, no shower, and eating nothing but K-rations (surplus stocks of combat food rations from World War II) and C-rations, and being shot at, returning to the comfort and relative safety of your shipboard home was always a high point! A great part of troop morale was soldiers knowing that their leaders cared for them.

They didn't want to be coddled or loved. Soldiers who know that their lives are considered valuable by their commander will do extraordinary things in combat. It is comforting in their eyes and soul, when the leader is physically beside them when it matters most, sharing the hardship and danger, and leading by example. This is especially true of the infantryman. German Field Marshal Erwin Rommel perhaps said it best: "When leading men into battle, if they can't see the back of your head, then you are in the wrong profession."

There is always some amount of fear and/or anxiety in all facets of infantry battle. What I worried about most, every waking hour, was losing a soldier! It ripped me apart when I put the still warm bloody body of one of my young soldiers on a helicopter and sent him back. That stays with me today. I never feared for my own safety—not that I was invincible or immune to fear. I wasn't. I was simply too damned busy tending to the myriad of things required of me as a company commander to think about myself. My priority was doing what was necessary to accomplish our mission with the least number of casualties. There was no time to fret about myself. The responsibility of command carries with it great burdens, and I put aside everything else for the mission and my men.

Ambush on the Rach Ruong Canal

With the opening of canals in the Mekong Delta, we moved west to the provincial city of Sa Dek and began operations to find and destroy the VC 267th and 502nd Battalions. We had good intelligence that they had refitted and were in the Ap Bac base area. The ARVN 5th Marine Battalion remained attached to the MRF. The battle that followed was the most significant the MRF had yet experienced. We embarked on our boats early that morning, December 4, and steamed in formation down the Cuu Long river and turned north into the Rach Ruong canal.[22]

I had a strange, eerie, almost frightening feeling that morning. I was more nervous than usual. We were on radio silence. There was uncomfortable stillness among the troops in the well deck of our ATC as we steamed slowly up the canal. There was no activity in the small villages as we passed through. It was a nightmarish quiet. From the well deck below, I popped my head up through the coxswain's cockpit. "Do you see anything, Chief?" I asked. Daylight was beginning to peek through the morning darkness when suddenly, all hell rained down on us. A well-concealed VC force attacked the boats with rockets and machine guns.

The pinging and bursting noise against the boat's armor plate was deafening. We were still moving slowly to our assault beaches. The ARVN Marine battalion was in the lead and passed through the first killing zone. I screamed at the Navy coxswain in the cockpit of the pilot house to beach us: "Beach us Navy! Goddammit, beach us now!" as we were sitting ducks in the middle of a narrow canal. He was hesitant to turn the boat into the beach. The Navy Assault Squadron commander was the only one who could order the beaching. He was in his command boat at the rear of the column. That morning, the Navy coxswains beached my company's three boats on my order. To hell with the normal landing procedure. We were in the middle of an ambush and had to get out of the stream and onto the beach!

The Marine battalion was able to land north of the kill position with light casualties thanks to a flame thrower aboard one of their ATCs, which had both a killing and a psychological effect on the entrenched VC. The fighting was intense, and my RTO, Sergeant Mike Wardell, was wounded. The brigade commander, Colonel Bert David, inserted the on-call 4/47th Infantry Battalion by air assault west of the VC position to form a blocking force. The firing was continuous.

They were hitting us broadside with direct fire from small arms, automatic weapons, and B-40 propelled rockets. Ricocheting enemy rounds had a spalling effect. Enemy bullets were breaking up into small metal splinters (spalls) as they ricocheted around the ATC's metal well deck. Metal splinters and fragments were everywhere, hitting the troops in the well deck. The noise was deafening. Loud and lethal. Just as he turned the boat into the canal bank ambush site, the coxswain was hit in the neck and shoulders. His body slumped forward and he fell dead into the well deck. We rammed the beach hard, and a sailor lowered the bow ramp.

We scrambled quickly out of the boat, climbing over each other and splashing through two feet of mud and water, making it to dry land. After all three of my boats beached, we attacked across the canal bank through heavy VC fire. The deep dikes gave us some cover. While my infantry was moving against the VC fortified positions, I was bringing in artillery and gunship suppressive fires with my forward observer (FO), Lieutenant Jim Hasselmann. The battle raged all day.

The small-arms fire was intense, rounds were sparking and zipping all around us. The Navy monitors were putting direct fire into the canal banks, chopping off heavy nipa palm branches overhead. A couple of my men froze in the well deck and wouldn't move. I relied heavily on Staff Sergeant Julio Diaz, a seasoned platoon sergeant, who was an acting platoon leader that day. He was one of the most savvy and courageous leaders in the company. He went on to become CSM (command sergeant major) of the 1st Infantry Division, the Big Red One. He kicked ass and screamed at them in Spanish and English. They were out of the boat in no time and followed him up the canal bank. I pressed inland with my RTO and FO calling in artillery fire as we moved. It was slow, moving only a few feet at a time. My other platoons were spread out to my right and left. Second Lieutenant Dick Evans of 2nd Platoon had just joined the company.

It was Evans's first operation. I stayed closest to his platoon throughout the day. Evans was a calm, cool, Alabamian who led his platoon that day with courage and skill. Within six months he would command a rifle company. He was that good! We remained close until his death in 2014. He had serious post-traumatic stress disorder (PTSD), and years later, I was able to assist him in getting good VA care. I also helped his family with his burial at Arlington National Cemetery.

Late that morning my battalion commander, call sign Langley 6, called and told me to carve out an LZ for evacuation of battalion casualties. This meant we had to shift fires to the east bank to allow the dust off helicopters to approach and land. I wanted to keep my infantry attacking and keep up the effective fires on the targets to the west but was told, "No, no, we need the dust off." We were also able to move some of our casualties to the Navy aid boat beached on the canal, and fortunately our battalion surgeon was aboard.

The after-action report (AAR) stated, "Without the immediate accessibility of the waterborne aid station, the battalion would've sustained several more deaths before the casualties could have been evacuated by air."[23] I wrestled often with the decision determining when to slow or stop the fight to evacuate casualties. Stop too soon, your attack momentum is lost, along with fire support and any progress made. Wait, not acting soon enough, more wounded might die! Soldiers bled to death when help couldn't get to them quickly enough. No one wants to lose a soldier. It was always a tough call to make! I constantly weighed that decision: when to stop the fight to evacuate the wounded? I'm not sure I always made the right call, but I never wavered.

Captain Ray Sanders with A Company was on my right. We were separated by a tree line that had numerous enemy bunkers. Both companies were receiving fire from the tree line. We had to coordinate closely the calls for artillery, airstrikes, and gunship fire to ensure we didn't hit each other's troops, which was a major problem all day. The battalion AAR, when addressing indirect fires into this tree line, stated, "The airstrikes were delivered before C Company got too close to the objective."[24] We continued to push the VC west and north, encountering enemy booby traps which slowed our movement. I had three men badly wounded that day from booby traps alone.

Diaz and Evans wanted Stingray gunships every minute of the fight. The Stingrays were a new Cobra helicopter outfit based out of Bien Hoa. Their miniguns were very lethal, and their pilots were the best! They provided the most effective suppression fires, and they did a good job that day refueling and rearming while providing maximum time on station. They were with us all day. "Bill, you took over those gunships. Every time I checked, you had them shooting for Charlie Company," said Bland a few days later. He called me aside and during the ass-chewing, reminded me there were three other companies in the battalion!

At about 2020 hours, we set up a battalion night defensive position and received a FRAGO (fragmentary order) for the following morning. At 0500 the next morning all companies executed a 100 percent "stand-to," where all soldiers were up, armed, and ready for action. A small element had become separated from A Company the day before and was missing. The following morning, during our sweep of a connecting canal, we found the body of one of A Company's missing men. He had been underwater since the day before and was bloated and decomposing. The body was removed by dust off. All companies continued to sweep the AO taking a few wounded prisoners and finding significant amounts of ammunition and a few weapons discarded by the fleeing VC.

At 1730, all companies loaded aboard the ATCs and began movement back to the MRB (mobile riverine base). The operation developed as two separate battles: one fought by the Vietnamese Marines to the north, and one fought by the 3/47th. The Marines and our battalion were never able to link up due to the enemy's major bunker complex separating us. The Marines suffered the most casualties and accounted for the most enemy losses. I learned later that they charged directly into the bunkers. They were tenacious fighters and fought well that day. With over 200 killed, the 502nd VC Battalion was rendered ineffective for a while.

We learned valuable lessons from the December 4–5 battle. The fight showed the riverine forces' vulnerability to ambush when moving slowly up a narrow canal. It also demonstrated the importance of quick decisions by an afloat force commander when under fire to land immediately. UPI dispatches the following day stated it was the "biggest battle ever fought by the U.S. riverine force special Army/Navy strike group created to fight in the canal-laced Delta."

One of the enigmas of the Vietnam War was our military's fixation on enemy body counts as the measure used to determine a win or loss. That was attributed to Defense Secretary Robert S. McNamara and his closeted Washington whiz kids, who prophesized that the war could be mathematically measured and won through attrition. Were they ever wrong!

After every operation, I gave my best assessment of enemy casualties. I can't remember ever feeling pressure from battalion or brigade to change or inflate the number of enemy casualties. I can't speak to any pressure that might have come from division and above. I abhorred the use of body counts as the measure of battle success. Body counts were sometimes used to show that the U.S. was winning the war. We believed that eventually the VC and the NVA enemy would lose enough soldiers through attrition warfare that they would quit! History shows how very wrong we were and is a subject for another day.

Reports of VC casualties were rarely accurate. VC losses in the Delta environs were impossible to confirm accurately, as we were often unable to confirm artillery and gunship kills, the biggest killers on the battlefield. The canals and deep muddy paddies prohibited quick closure on the objectives, thus giving the VC time to flee with their casualties. The better measure of success was, "Did we drive the enemy from the fight? Did we push him off the objective or out of the town/village? Did we destroy his bunker systems and cache sites? Did we disrupt his plans?" If we did, we were successful! Albeit sometimes, only temporarily. The frustrating part was, it was rarely permanent, and in 2–3 months' time, we would be fighting a refurbished VC unit in the same area. That was the nature of the guerrilla war fought in the

Delta. And it was terribly frustrating to the infantryman! We never stayed and held the ground we won.

The fight on the Rach Ruong canal dominated war communiqués during December 4 and 5 1967 and shifted the main theater of combat from the central highlands to the Delta. A week later, I received a clipping from *The Trentonian* newspaper that Linda had sent with a front-page headline that read, "235 VC Killed by U.S. Riverine Force Biggest Fight in the Delta." As stated previously, the media rarely got the facts correct, hence the disparity in actual figures and the reported numbers.[25]

Regardless, the enemy 502nd Battalion suffered heavy casualties and were ineffective. From December 6–19, we continued to make both riverine assaults and air assaults where enemy sightings were reported. We had little meaningful contact during that period, and our last operation ended on December 22 when we returned to Dong Tam to refit and prepare for Christmas truce operations.

We were told only defensive operations should be conducted during Christmas, December 24–25. I spent Christmas Eve and Day in a patrol base in Long Dinh District near Dong Tam. We experienced only occasional harassing fire during the truce. In late December, the *Colleton* left the MRF to refit at Subic Bay, Philippines. The USS *Benewah* (APB 35) replaced her, and Charlie Company embarked on January 21, 1968, ending Operation *Coronado IX*. We began to prepare for the upcoming Tet truce, a three-day cease-fire between January 30 and February 1, coinciding with the celebration of the Vietnamese Lunar New Year.

With my platoon leaders and platoon sergeants, I reviewed the past 10 weeks of continuous riverine and air assault operations, and recounted the obvious lessons learned about our enemy and our own tactics. We learned the VC were being taught that anyone captured by U.S. units would be killed, so they fought hard. But when captured and roughed up by my infantrymen, they spilled the beans.

We faced ever-present booby traps and concealed VC in very dense nipa palm groves. The VC hid and flourished in the most rotten and inaccessible areas. We had to go in and dig them out, and booby traps on the trails and rice paddy dikes were a major cause of casualties.

One of the casualties was Sergeant Jim McIlvoy, of Michigan, one of my squad leaders, killed by booby trap on December 30, 1967, while walking point on a dike. McIlvoy was well-liked, and the company took his death hard. The troops were told constantly to stay off the dikes, but after hours of exhausting plodding through thick mud and two feet of water, some would invariably give in to temptation and return to the dry ground of a 12-inch-wide dike where walking was easier. When my polio leg got so weak I could no longer lift it out of the sucking mud, I disobeyed my own order and returned to the dike; it wasn't the example I wanted to set. The result all too often was a loud boom followed by the chilling yell, "Medic! Medic!" The VC anticipated American soldiers' behavior and used it to their advantage. On December 4, I had three soldiers seriously wounded by booby traps and evacuated. They lost limbs and never returned to the company.

I also learned quickly that some of our Rules of Engagement (ROE) were not practical and favored the enemy. We were not to destroy villages or hooches (an informal name for Vietnamese village dwellings). An example of impractical and ambiguous ROEs can be found in a monograph titled *Vietnam Studies-Riverine Operations 1966–1969*, written by Major General William Fulton, which states, "The MRF was instructed that operations should be defensive during the Christmas truce period; troops could fire on groups of enemy soldiers who seemed to be trying to breed contact or who were more than platoon size in number."[26] I asked my battalion commander what in the hell does "seem to breed" mean in this instance? I received no reply. This is an example of some of the dumb, nonsensical rules infantry soldiers were told to fight by. As the days wore on, I paid less and less attention to the rules and did it my way!

The rules we lived with instructed we couldn't destroy or set hooches or villages on fire. The VC knew it and stashed supplies, and ammunition and weapons caches in and under the villages. Some villages, where the ground allowed for deep digging, had networks of tunnels. I had several soldiers, small in size and wiry, who volunteered to be "tunnel rats," going down into the spider holes and connecting tunnels with pistols after first dropping concussion grenades (MK3 series). We had to root many VC out of spider holes during *Tiger Coronado XI* in Can Tho on February 14–20, 1968.[27]

When the town/village structures offered cover, the VC fought from them. They fought from building doorways, windows, and rooftops during the Tet battle in My Tho on February 1–2. The only way to flush them out completely and ensure the sites couldn't be used again by the enemy was to destroy them. And we did … no questions asked! We knocked a few buildings down in My Tho with our 90mm recoilless rifles. And when our *chieu hois* told us a particular hooch/building was manned by VC, we rendered it ineffective.

On this issue, I defer to the Geneva Convention of 1949, in that once an enemy converts a civilian or other protected dwelling into a military base of operations, let alone offensive operations, such a structure loses its protected status. The onus was upon the VC to protect civilians and not bring them into harm's way. But the VC didn't give a damn about civilians. Their list of violations of the laws of warfare would be another book.[28]

During a village clearing operation, one of my lieutenants threw a concussion grenade into a hooch that was full of frightened women and children, inflicting terrible, indescribable casualties. He was clearing the hooches and thought VC were inside. He was so shaken by his action that I had to evacuate him. He never came back. It impacted me, also. We treated the injured Vietnamese and evacuated them to field hospitals. This was a tragic mistake by a U.S. soldier fighting this insidious guerrilla war.

However, given the circumstances, it wasn't an illegal action, nor did it violate the conventions of war. He was searching for the enemy while protecting the lives of his men and that is superior combat leadership. But even when morally and legally correct in an action, experiencing such a tragedy tears at the heart of the most seasoned, battle-hardened warrior. Tragically, civilians always pay the highest price for the folly of war. Vietnam was no different.

At times, we had to take such actions, and I received verbal reprimands and an occasional ass-chewing, but everyone in my chain of command knew what had to be done. Sometimes, I had to wait to get permission from higher authority to bring artillery fire on certain targets. While awaiting permission, we continued to take fire and the enemy was *didi mauing* which, in Vietnamese, means "bugging out."

After a while, I stopped asking and waiting and called in the fires, rules and consequences be damned! The AAR for the December 4–5 battle noted, "the battalion experienced some delays in receipt of fire; in particular, suppressive fires on the east bank of the Rach Ruong."[29] My heart ached for the innocent Vietnamese who, all too often, bore the brunt of the battle.

Scrutiny of the eyewitness photos of the city of My Tho during the Tet Offensive illustrate the absolute destruction and homelessness rendered at the cruel and murderous hands of the VC. They devastated the city, killing and maiming many women and children. Lieutenant Colonel Travis Blackwell, 9th ID division surgeon, flew in with a team of doctors and immediately set up a 24-hour surgical schedule in My Tho's small hospital. Thirty-five lifesaving operations were performed the first day. Amputations were performed on an outside ramp of the hospital while savage fighting flared throughout the city.[30]

I did what I had to do, what I thought was right, and I lived with the consequences of my decisions. I had a mission to accomplish while ensuring the safety of my men. In my opinion, if any infantry leader did anything less, he would be in dereliction of his duties and unworthy of his commission. We couldn't fight and win these kinds of engagements with one hand tied behind our backs because of overly restrictive ROEs emanating from Washington.

Tet Offensive

Shortly before the beginning of the three-day Tet cease-fire, the MRF moved to Dinh Tuong Province to take up positions astride major VC communication routes. We were getting intelligence that the VC were building up and repositioning forces. We weren't sure why, so we established company-sized bases from which we could provide wide surveillance.

IV Corps intelligence was warning that during the Tet cease-fire the VC were going to "resupply and move into positions for a post-Tet Offensive." Some reports said the VC might deliberately violate the truce; for example, the IV Corps senior advisor reported that in a message dispatched on January 29, 1968.[31] We were on heightened alert and were told to look for any signs of enemy deception.

Charlie Company was operating from a patrol base in in Dinh Tuong Province just west of My Tho. I remember how still and quiet it was. On the morning of January 30, 1968, I received a call from Major Healy, our S3, saying that the truce

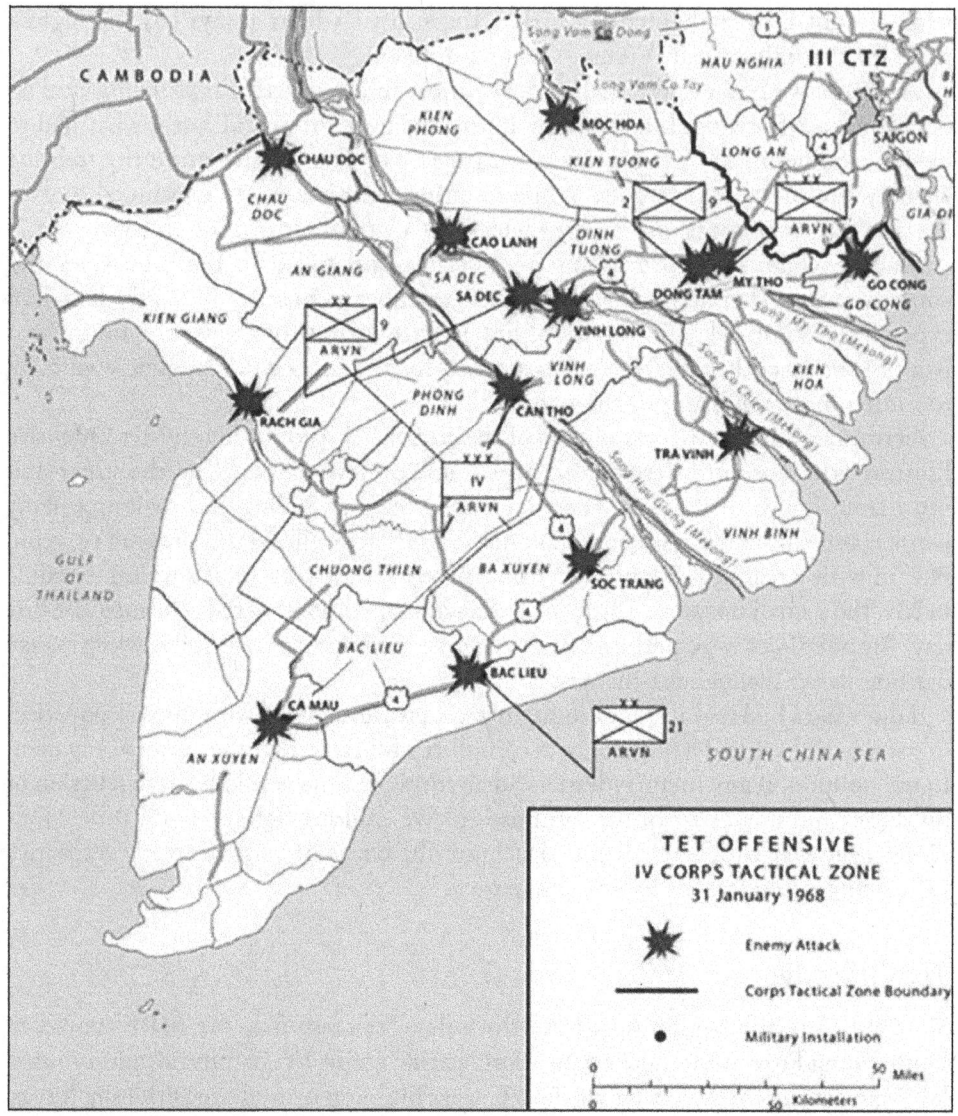

Viet Cong (VC) attack locations in the Mekong Delta on the first day of the 1968 Tet Offensive. (Map from U.S. Army Center of Military History, "Staying the Course")

had ended because of VC attacks throughout Vietnam. We were told to stay close to waterways, remain near good pick-up zones, and conduct saturation patrolling in our respective AOs. At that point, we didn't need any more intelligence about what the VC were going to do; they were already attacking in other parts of the country.

I sensed something was wrong; my intuition told me something was going to happen. We were at a canal crossroad where we normally saw sampan traffic and other Vietnamese activity. There was nothing moving, and my men were restless. I asked my Tiger Scouts what they thought. As best I remember, they answered, "The people are scared and hiding, because VC are all around and going to fight, and there will be no holiday."

I trusted my Tiger Scouts implicitly. They called me *dai we* in Vietnamese, loosely translated as "boss" or "leader." They knew the terrain, equipment stash points, and ambush sites, making them even more valuable to us at the time. As an indispensable combat multiplier, I integrated them directly into my platoons. My scouts were strongly anti-VC. They were often older Vietnamese who, with their families, experienced the ruthless brutality of communism and wanted to fight with the other side. They had good knowledge of the Delta and were familiar with enemy tactics and communicated easily with the local Vietnamese. Once integrated into a unit, they were very loyal. I was always asking for two or three more to be assigned to Charlie Company.

I relayed this information to Healy at battalion headquarters. On January 31, the Delta erupted as all the major cities were attacked by the VC. I got a call to saddle up and move to a PZ (pick-up zone) for extraction by helicopter and return to Dong Tam to await further orders. Within minutes of closing Dong Tam, I received a FRAGO directing me to embark on Mike boats for water movement to My Tho. The provincial capital of Dinh Tuong was under attack and in ruins. One portion was enveloped in flames and families slaughtered by the VC lay dead in the rubble. Three VC battalions, the 261st, 263rd, and 514th, had hit the city with unrelenting force.

We landed mostly unopposed on My Tho's shore on February 1 and moved north into the city. Our battalion's two lead companies met heavy fire. D Company was pinned down and unable to advance. With my mission of "follow and support," I was ordered forward to assist the beleaguered units as fighting raged in the streets ahead. Snipers fired from rooftops and windows as I moved to the front of the company with my FO.

My machine gun teams established a base of fire as we moved quickly through Delta Company. I made contact with A and D companies as soon as I could. The commander of D Company was on R&R in Hawaii, and the company was led by a young former National Guard lieutenant who had just recently arrived from the States. Although very capable, he was inexperienced, and the company was unable to advance and fight through the small-arms fire. The situation was the same in A Company. I grabbed the handset from Underhill and told them, "I'm coming through your lines, on your flank and rear and moving to the front." I got a "roger" from A Company, and as I recall, no acknowledgement from D Company.

It was tight moving through the A and D Company positions. I spread my platoons as best I could and pushed forward. At one point, I almost tripped over a dead dog. There was debris and glass everywhere. Hostile fire was intensifying. Buildings were burning. As we passed through built-up areas, we began to take increased fire from buildings and rooftops. I was closest to Evans's platoon, and his gunners swept the top floors and rooftops with bursts of automatic weapons fire. I told Lieutenant Keith Mietz, of 3rd Platoon, to hang behind to mop up any VC that might still be in buildings behind us and to maintain contact with D Company. I wanted a safe rear area clear of enemy.

When we got to the My Tho reservoir, near the center of the city, I saw several wounded soldiers from A Company pinned down between the reservoir wall and a small canal dike. One of the men was screaming. A platoon leader, Lieutenant Lynn Smith, was killed moments earlier trying to rescue them, and his body laid there. Another soldier tried to get them, and he was hit by small-arms fire. The noise was deafening.

"Sir, they are hit and down there!" yelled an A Company soldier, pointing to the wounded men.

The pinned-down wounded soldiers caused the battalion to halt its forward momentum in its sweep north through the city. Heavy fire was coming from an entrenched enemy in the cemetery to our front and buildings to our left. Two of the wounded were ambulatory, and two would have to be dragged and carried to safety. I couldn't find the A Company commander and D Company was still to our rear.

The VC were taking advantage of the stalled situation by pouring fire into our lead elements. I sized up the situation instantly, and without giving any further thought, dropped my heavy flak jacket, load-bearing equipment, and steel pot so I could move faster. I ran across 30 meters or so of open area to the small canal dike that led to the wounded. VC fire intensified; the zinging bullets kicked up puffs of dirt and jagged stones as I ran. Once my machine gunners identified targets, they returned fire with a staccato of constant bursts.

I reached the wounded and was pulling one of the men alongside the upper part of the canal bank when I felt a sharp stinging burn in my arm. I was grazed by enemy fire, but I returned for the other three wounded. Two could move with me, the third was hit badly. He was in pain and let out an occasional piercing scream. I think shock was setting in. I dragged him to the edge of the dike and across the open area where a medic and another soldier met us and helped get him to safety. I returned and assisted the other two to safety. These were four young, wounded, and frightened soldiers from A Company. They were medically evacuated, and we never saw them again. One died of his wounds.

The cacophony of loud, continuous, automatic weapons fire, the sight of smoke filling the air, dirt and stones flying everywhere, and men screaming, best describes the hellish scene I shall never forget.

The flesh wound to my right arm from small-arms fire was stinging the hell out of me. I remained at the front of the company directing fire on the VC positions. I was exhausted, and the VC fire wasn't letting up. I moved my 90mm recoilless rifle crews to the front and had them fire on the VC bunkers and buildings. The 90mm was man-portable and had an effective range of about 250 meters, and we carried both HEAT (high explosive anti-tank) and AP (anti-personnel) rounds.

Three of the 90mm crews were very effective that day in busting the enemy defensive complex, including the cemetery's concrete wall, and in knocking out the walls of buildings concealing VC gunners. It was rare that we had cause to use our 90s. My weapons squad gunners saw a lot of action that day and talked about it for weeks. One of the 90mm crews never fired a shot, and I'm not sure why. In my excitement, I got caught in the powerful back blast area of one of the 90mm shots and was knocked down. A couple of my sergeants ribbed me about that later.

The fight was unlike any we had seen. We were used to moving slowly through water and mud, fanning out in squad/platoon formation or moving in open rice paddies and assaulting river beaches. The My Tho battle was close city fighting on hard surfaces moving through VC-infested urban streets by bounding overwatch, infantry fire teams leapfrogging one another. Infantrymen went house by house, building by building, rooting out the enemy. By evening, we had neutralized the VC force in the large cemetery used to hide weapons and ammunition prior to Tet.

They had also prepared hasty fighting positions among the gravestones. While we were firing on and driving the VC from the cemetery, we were also fighting VC who were firing machine guns and rockets from buildings on our left flank. As the evening wore on, and we thought they were eliminated, they would reappear and begin firing again from hidden positions. No one slept that night as we continued to use fragmentation and concussion grenades to eliminate the remaining pockets of VC in the buildings and cemetery and along the canal bank.

I hated that night, and the memory of it stays with me. Fearful and anxious, my soldiers described it as "creepy." We spoke in whispers knowing there were still enemies hiding in the nearby buildings. Soldiers would shoot or throw a grenade at the first sign of movement or noise. Evans's platoon cordoned off and staked out the cemetery that night while two other platoons cleared and occupied the buildings. 3rd Platoon leader Lieutenant Mietz said, "It was like watching Atlanta burn." Fires were everywhere. He did a good job that night controlling his men's fears and anxieties, at the same time snuffing out the last of the VC from the buildings. In the murky darkness, his men mistakenly shot and killed a friendly elderly Vietnamese man. That weighed heavily on Mietz who I remain in contact with today.

Our combat medics patched up the lightly wounded. The seriously wounded and dead were evacuated by dust off. A medic poured disinfectant on my wound and bandaged my arm. Other than a persistent stinging sensation, I was okay.

The next day, we were to continue our sweep to the north. I was nudged early that morning by Lieutenant Evans. "Sir, I think we need to do one more good sweep of the cemetery. We are hearing movement." Underhill and I accompanied his platoon at daybreak as they swept the cemetery finding several dead VC and a few weapons. The retreating VC took the other weapons. There were shell casings and spent rounds on the ground. I took a few as souvenirs and gave them to my Navy buddies on the MRF ships. We found some gravesites that had been opened and used to stash weapons and ammunition. In a subsequent brigade debrief, I learned that was a tactic used by the VC to pre-position and hide weapons and supplies in city and village cemeteries prior to the Tet attacks.

I looked east across the reservoir and saw some ARVN 7th ID units sweeping their side of the reservoir, on our right flank. I learned later that a good friend of mine, captain of the 1960 West Point football team, Captain Al Vanderbush was an adviser with the ARVN 7th ID, Mietz's platoon went through the remnants of the buildings on our left flank to ensure they were clear. The buildings were smoldering, and a putrid stench began to fill the air. Other than a few scraggly dogs rummaging through the debris, there was no movement. I radioed, "This is Charlie 6. We are ready to continue the sweep north, request permission to move out." Our S3, Major Healy, was a good troop; he knew I wanted to get the hell out of there. "Roger, Charlie 6, move out."

B, C and D Companies began their sweep north but were soon halted and ordered to return to the river for backload on the ATCs. Tactical airstrikes were effective against the retreating VC. We loaded onto our ATCs for the 25-kilometer move west to the Cai Lay district to cut off the enemy escape routes from Eastern Dinh Tuong Province. We embarked and left My Tho that afternoon never to return.

Can Tho Battles

By the middle of February, the Tet Offensive had been broken in the My Tho and Vinh Long areas. However, the VC continued their attacks on Can Tho, the capital of Phong Dinh Province, further south. As a result, the MRF was employed to counter the attacks. Thus, *Coronado XI* (the last of the Coronado series of operations) was initiated. The MRF moved south of the Bassac River for the first time.[32] Enemy units there had a good fighting reputation: the 306th and 303rd Main Force battalions, Tay Do and U Minh battalions and their Military Region III headquarters. Our mission was to conduct riverine and air assault operations to find and destroy them. The MRF traveled 109 miles from Dong Tam to the Can Tho anchorage, arriving on February 13, 1968.[33]

The next morning, we landed by boat and began reconnaissance-in-force operations along the canal south of Can Tho. For seven continuous days, we engaged platoon and company-sized VC units. On February 14, while moving along a feeder canal, my company came into heavy contact with a well-concealed enemy force armed with rockets and automatic weapons.

We had been moving for several hours through thick tree-lined areas. Boice's A Company was on my left with a feeder canal separating us, and the movement was slow and methodical. We were searching hooches and spider holes, and finding bandoliers of ammunition, an occasional weapon, documents, and weapon-cleaning kits. The area was uninhabited, empty of people, which seemed strange. We knew the enemy had been in the area and were close by. My point element spotted three Vietnamese males running in the opposite direction and fired at them. Around the same time, an A Company Tiger Scout tripped a booby trap and was evacuated by dust off.

"I don't like this," a worried Lieutenant Hasselmann said. "Something is up there." I agreed. Moments later, we began receiving sniper and light automatic weapons fire, first from the front, then from the right flank. "Get the machine guns up here!" I yelled. We were in a fight. Hasselmann was fixing final coordinates to call for fire. My platoons quickly maneuvered where they could return fire. The trees and lush green nipa palm were thick, and we couldn't see the VC. There was a ditch a few feet ahead, which my RTO, FO and I moved to for cover. We were receiving rocket and heavy machine gun fire. It sounded like a U.S. .50 caliber machine gun firing at us.

My RTO, Specialist 4 Benjamin Underhill, had his head ripped off by heavy machine gun fire. I was lying in the ditch two feet from him and had just turned away from him to say something to Hasselmann when he was hit. His helmet and remains splattered on my neck and shoulder. A medic got to him right away, but he had been killed instantly. The fire was loud and continuous as I left the ditch and moved forward with my alternate RTO and Hasselmann who was calling in artillery fire on the enemy. Like many close fire fights, our voices couldn't be heard on the tactical radio. My RTOs and I often had to scream into the push-to-talk handset to be heard above the din of battle.

The fire mission was "danger close" (a call for artillery fire where friendly troops are in very close proximity to the target). Our direct support (DS) artillery battalion, the 3/34th Artillery, was masterful that day putting rounds on target. The fires were close with the rounds impacting directly in front of us. I leaned my infantry squads into the fire, and we could hear and almost see the hot splinters of shrapnel flying about. A couple of my men were hit by friendly fire. We were also able to get a light fire team to make a couple passes over the tree line. The dust off couldn't land right away because of the heavy volume of fire. When the enemy was neutralized, we were able to finally evacuate our casualties. I lost three men, including Underhill, and had several seriously wounded. Those with minor wounds stayed. The fight ended with the VC withdrawing along the many paths and canals leaving behind some wounded and equipment.

The battalion AAR cites the effective use of field artillery support. It also highlights the effectiveness of the enemy "in locating and killing RTOs and leaders as principal targets."[34] Losing Underhill was like losing one's eyesight. I was shaken. I know it was the heavy machine gun fire that killed him. Others close by thought it was a VC mortar round. He had been with me since I had taken command of the company. He was my battle buddy, always one step behind me carrying the company commander's most important tool in combat: the PRC 25 radio. I loved the kid. I felt lost, naked, without him.

My battalion commander told me later that the only time he ever detected a quivering or tremor in my voice on the battalion command net was when I reported Underhill as KIA. He died one day after his 22nd birthday. I still think about him after all these years ... and my last few moments with him are seared into my memory.

Through all the intense activity, my polio leg was holding up well. But the leather orthotic Jones Bar rotted and fell apart. Fortunately, a Navy machinist mate fabricated arches for me that fit into my jungle boot, replacing the Jones Bar. The arches eventually rotted also, and I was constantly replacing them. The fabricated arches were a lifesaver as they gave my polio leg and drop foot the lift support they needed. We continued to operate in the Can Tho area, intercepting enemy sampans and capturing large caches of VC medical supplies, equipment, and food stocks.

On February 22, we beached along the Kinh Lai Hieu canal and took the VC by surprise. It was rare that we surprised them, but in this case the fighting ports of the bunkers faced the rice paddies in anticipation of air assault attacks. The MRF had never operated in the area, so the enemy had not experienced waterborne attacks. The VC ran into the open fields away from the canal where we brought effective artillery and gunship fire on them. It was a killing field.

They suffered heavy casualties, but many were able to escape using cross-canal trails and sampans before we could catch them. We had no blocking force that day and they knew the area well. The operation marked the deepest penetration into the Delta by U.S. forces. Combat operations ended on March 3 and the MRF departed Can Tho for return to Dong Tam.

While at anchorage in the Mekong, just south of Dong Tam we were attacked on March 22 by enemy mortars and 75mm recoilless rifles. It wasn't unusual for the MRF ships to take occasional fire from the shoreline. The USS *Benewah* received two hits causing minor damage and USS *Washtenaw County* registered some near misses. My company was divided aboard the two ships. In addition to the two APB barracks ships, *Colleton* and *Benewah*, and the towed barracks ship, the Navy always positioned an 1152 Class LST with the MRF. Of interest, later in the fall of 1968, *Benewah* would mark the 10,000th helicopter landing on her deck.

I got to know Lieutenant Commander Alfred "Al" Dillon, the skipper of USS *Washtenaw County* (LST-1166) well, a true Irishman and Villanova University graduate. When we embarked, he took great care of my troops. He was always wanting a souvenir or memento from our combat operations, and I often gave him items we took off a dead or captured VC. One time, I gave him a captured AK-47. In turn, he ensured the troops had an extra steak or dessert for their meal, or he made sure the returning troops received an extra iced can of beer. It was "interservice cooperation at its best!"

I was told later that Al even made a secret place available on the ship for those who wanted to share a quick joint, not that I approved. The LST stored most of the MRF ammunition and other supplies. It also received the dead soldiers as medevac choppers brought the dead directly from the field to the LST. A small mobile morgue was installed in the forward part of the tank deck below. The bodies remained on board until the mortuary choppers from Saigon picked them up. The wounded were flown to the MRF base where they were treated in the sick bays aboard the two APBs or the 3rd Surgical Hospital at Dong Tam. When embarked, we conducted memorial services for our dead soldiers on the decks of the Navy ships. There was always an Army or Navy chaplain to conduct the service.

There was excellent camaraderie among the Army troops and the Navy crews. After seeing the infantrymen return from four- to six-day operations in the leech-infested Delta mud, every sailor thanked his lucky stars he had visited the Navy recruiter and not the Army one. Whenever I had a spare moment, I went to my cramped state room aboard ship and dictated a brief message to Linda and my young son using a small Sony tape recorder. She, in turn, would send a message back. She also wrote nearly every day. Mail call was an important event. It wasn't unusual to go two to three days without one.

Letters from Linda, my parents and sister allowed me to forget about all things war related and for just a few moments immerse my thoughts in the happiness of home and family. Letters from my high-school friends and college fraternity brothers meant a lot. I was pleasantly surprised when I received a letter in January 1968 from Dr. Chet Jarvis, my political science professor at Gettysburg. It was a sincere and welcome message filled with praise for the American soldier, thanking me for my service and wishing me a safe return.

I left the company at the end of March and was assigned to the G1 (Personnel) Section of the 9th ID Headquarters at Dong Tam. Two weeks later, I took my R&R for Hawaii where I met Linda. It was an escape, however fleeting and brief, from the tension and rigor of war. My skin was bronzed from the sun, I had lost weight and had an ugly seeping fungal infection on my leg but was otherwise healthy. I processed through the Army's Fort DeRussy R&R Center on Waikiki Beach, and Linda was waiting for me. We both did our best to prepare for the emotionally charged moment. It was an instant return to civilization, and I couldn't wait to see my wife. We stayed at the Ilikai Hotel, and our daughter Heather was conceived there.

Linda noticed how I rattled and jumped when the Waikiki Beach fireworks went off in the evening. The loud bursts when incoming mortar rounds land have a way of staying with you. I was startled easily. It took a few days for me to relax and settle down. The seven days went by quickly, and I was back in Dong Tam before I knew it. R&R was a wonderful break, but my platoon leader Lieutenant Keith Mietz said it best: "It took the edge off. When you returned you had lost your warrior momentum, and it was difficult to get back into the groove of Delta operations and fighting." I agree.

On September 18, at a ceremony in Dong Tam, General Creighton Abrams,[35] the MACV commander, presented me with the Distinguished Service Cross (DSC) for actions on February 1, 1968 during the heavy Tet fighting in My Tho as commander of Company C, 3rd Battalion, 47th Infantry. Some of my Charlie Company soldiers who were with me that day were standing in formation, and Major General Julian J. Ewell, 9th Division commander, was also present. It was a humbling moment that left a mark on my emotions and feelings. General Abrams, who was legendary during World War II as a battalion commander, would later become chief of staff of the Army.

Leaving Vietnam

As my time in Vietnam was winding down, my thoughts constantly returned to my time as Charlie 6. Memories of my soldiers and their bravery dominated my thoughts. At any given time, I had 160 soldiers, a combination of draftees, volunteers, and lifers. All ethnic and racial groups were represented. They were mostly young 18- to 21-year-old Americans serving their country. We all had a job to do, and we did it. My soldiers fought for each other to survive and return home. To quote a controversial World War II Waffen SS officer, General Leon Degrelle: "They were men to the marrow of their bones."

There were a few whiners, but I had no problems with drugs. I experienced no fraggings (the deliberate or attempted killing of a soldier, usually a superior, by a fellow soldier) or insubordination. The consequential use of drugs and incidents of fragging did become an issue later in the war, but my soldiers concentrated solely on finding and destroying the enemy. One of the most disingenuous, hurtful, and persisting myths about Vietnam was that our troops were murderous malcontented drug addicts. I can tell you firsthand that was far from the truth. They were great Americans doing a tough job, in a very unpopular war, and I was honored to have served with them.[36]

My Puerto Rican soldiers stood out. They were tough as nails and always volunteered to walk point or go down into the spider holes—the two toughest and most dangerous jobs in infantry operations in the Delta. Staff Sergeant Julio Diaz ensured we had good tunnel rats. He recruited from among the Puerto Rican and ethnic Spanish-speaking soldiers. Private First Class Felix Rios-Ramos was one of the best. He reminded me of a thoroughbred jockey: short, skinny, wiry, and tough, although he was about 100 pounds, and jockeys weigh in at around 118 pounds. On a couple occasions, the tunnel rats found frightened women and children in hiding. They had signs in Vietnamese reading that said, "We are friendly. We will help you." They brought them to the surface where the *chieu hois* interrogated them on the spot. They were either released or detained and sent to Dong Tam for further questioning.

My troops had each other's backs, were conscientious, kept their weapons clean and never missed movement. Unfortunately, there were some whom I never got to know. They reported one day, and four or five days later they were killed, or wounded and in a hospital, and I never saw them again. The platoon sergeants and first sergeant collected and inventoried their personal items and had them shipped to the soldier's next-of-kin. I lined through their names on my company battle roster. To this day, I carry a paper in my wallet with their names and KIA dates. Such was the callous, fast pace of the war we were fighting in the Delta.

Damn few American infantrymen made it through a 12-month tour in the Delta of Vietnam without sustaining a battle wound, fungal infection, or death. We all came home with a scar or two—if not in the flesh, certainly in the mind and soul. No one serves months in infantry combat that isn't scarred in some way. We carry with us invasive memories of traumatic events, feelings of guilt, scenes of dead comrades and dead Vietnamese, and innocent young Vietnamese mothers holding babies while trembling with fear. Many live a life with post-traumatic stress. I'm proud to have known, suffered with and fought beside these men. They answered the call when their country needed them. Most didn't want to go to war, but they did. They didn't run off to Canada or fake some injury. They were America's best!

President Ronald Reagan said in 1980 of America's Vietnam War veterans, "It is time we recognize that ours was, in truth, a noble cause. They fought as well and as bravely as any Americans have ever fought in any war. They deserve our gratitude, our respect, and our continuing concern."[37]

They did their duty. God bless them all!

CHAPTER FIVE

From Middlebury College to Navy-Marine Amphibious Force Pacific

We were on final approach to Philadelphia International Airport. The airplane banked slightly to the left, and I was able to get a glimpse of the ground below. I recognized the Delaware and Schuylkill rivers and some landmark buildings. The airplane touched down, and I knew I was home.

Hours before, the pilot of the MAC charter flight that flew me and 200 other war-weary soldiers from Bien Hoa Airport in Vietnam to Travis AFB, California, had broadcast a "welcome to the world" announcement throughout the cabin. I don't recall my exact thoughts as we taxied to the gate, but I knew Linda, with our 18-month-old son, would be there to meet me. As I climbed down the airplane's stairway, I spotted Linda and Billy immediately!

United Airlines allowed families to wait on the ramp to meet their returning servicemen and women. A few steps behind Linda were my parents, my sister, her husband and their one-year-old baby, Jennifer, and my cousin, Nancy. What a wonderful homecoming! It was truly a moment of joy and relief!

Linda remembered, "I was carrying Billy and after a huge hug and kiss from my wonderful husband, Billy leaned forward to be picked up by his dad. I had Bill's picture prominent in my parents' home where we were living. I spoke to him about his Daddy often, and always, before putting him to bed, we said our prayers together for his safe return. He knew his Daddy."

The "spitting on returning Vietnam War veterans" and "baby killer" yelling anti-war protestors we heard so much about were nowhere to be seen on that warm fall day at the airport. As we moved through the terminal corridors to our ground transportation, the people we met and passed by couldn't have been more cordial and appreciative to those of us in uniform. Warm welcoming smiles donned their faces at the height of the war, and they knew we were Vietnam veterans returning home.

There were no banners waving or bugles playing, and I didn't expect that. Nor did I want that. My family was there, and that was all that mattered! I know some troops experienced an unpleasant welcome when they came home, being greeted with anger and hostility by anti-war protesters. I think those incidents were rare, and like so many other stories about the Vietnam War, were blown out of proportion by the corrupted and biased media to serve their agenda and are more myth than reality.

Unbeknownst to me, while in flight from Vietnam, on October 24, I learned that was the date that my promotion to major was effective. My assignment order gave a report date to Middlebury College of not later than December 5. The ROTC detachment was short officers; therefore, I was needed now, and the PMS, Lieutenant Colonel James C. Hefti, asked that I report immediately. That meant foregoing the 30 days' leave authorized and moving to Vermont as soon as possible.

I drove to Middlebury to find a place to live while Linda quickly arranged to have our few household goods packed and moved from New Jersey. After a day's search with a local realtor, I finally found an old farmhouse in the neighboring town of Shoreham (population: 940 and one general store), 13 miles from the college campus. The owner, Gladwyne Fuller, a nurse, moved from Shoreham each year to a small apartment in nearby Brandon to be close to the hospital where she worked during the winter months. Therefore, her property would be available until her return in June 1969. In-town apartment and house rentals were scarce that time of year, and we were delighted to get a house. I signed the rental lease and thus began our adventure living in rural Vermont. There was no heat on the second floor, only that which passed from the first floor through ceiling vents to the second. Our water supply was from a large stone cistern in the attached barn fed by rainspouts on the roof. With freezing temperatures, water had to be trucked in periodically. It wasn't potable, so I brought home large jugs of fresh water from the college for drinking and cooking.

The winter of 1968–69 was very cold with temperatures consistently in the single digits. With a toddler and Linda seven months pregnant, she couldn't wait to get settled and begin life in our new home. Despite the rather spartan and hardy living conditions, we were thankful and happy to be together again as a family. We made the most of it that cold snowy winter as our coping skills and adaptive powers were tested beyond belief!

When I arrived home, my wife noticed a change in me. She said, "You are much more intense. You are not the Bill I knew before you left. You have very little patience and are much more critical of things." She was right. I knew I was different, and as I grew older and gained more exposure to other Vietnam veterans, I understood that such a transformation is part of the human condition. Some cope with adversity and the horror of war better than others. The reason I managed to adjust reasonably well upon return was because of Linda.

Her understanding and patience, not passing judgment and her emphasizing to me the importance of a bonding father–son relationship with our young son was crucial to both my readjustment and that of our family. Indeed, she knew what was important after a year's absence in a war zone. She was the consummate Army wife, so calm, patient, and understanding in helping me to quickly settle and readjust.

Middlebury was a private liberal arts college and enjoyed an outstanding academic reputation nationally. It was one of the "little Ivies." The undergraduate population was about 1,800. The nine-person ROTC detachment was a separate academic department, and the officer cadre enjoyed full faculty status including voting rights at faculty meetings. We had a particularly close relationship with the college's History department and sometimes would "team-teach" military history courses. Military Science courses received full academic credit.

There was a small faction within the faculty that disliked anything military. They clearly kept their distance from the uniformed ROTC instructors and you could feel their ire when sitting in faculty meetings. At every opportunity, a few would find a way to slander or demean the military, particularly the Army. The small group began to grow in number and become more hostile as the anti-war protests gathered momentum nationally.

We wore our Class A green uniform every day and were encouraged to always display our military service ribbons and badges. We stood out and were easily recognizable. On the other hand, there were faculty members who had served in uniform during World War II and the Korean War, and they were not happy with their vocal anti-military colleagues. Among that greatest generation of war veterans were some of ROTC's strongest supporters.

The president of the college, Dr. James I. Armstrong, a Princeton graduate, had served with the U.S. Army in the Pacific in World War II and in Europe during the Korean conflict. Although he stated publicly, "Perhaps in no time in the history of the country has military service been so central to the concerns of the nation," and emphasized that "ROTC programs offer the student the opportunity to acquire in the context of liberal education the training and skills for military service," I was never wholly convinced he was an ardent supporter of our program. Nor were the college deans strongly in favor of ROTC.

The residents of Middlebury and Addison counties were very supportive of the ROTC detachment and appreciated all we did to aid with their parades and other town/county events. The Middlebury American Legion and VFW posts were most hospitable, offering us special membership. A major segment of our social life was centered around their activities. We attended dinners, picnics, dances, and usually stopped by their "happy hours" on Friday nights for a beer or two and to share war stories with the older veterans before going home. Occasionally, a professor from the History department would join us.

We also enjoyed the excellent skiing offered by the college's Snowbowl ski area 13 miles to the east in the Green Mountains. The ski shop on Main Street fitted me with a specially designed ski boot and skis to accommodate my polio leg and foot. One ski was shorter than the other to accommodate my weaker, less agile leg. After several adjustments to my new equipment, I was negotiating the trails and slopes in relative comfort and ease. I skied the beginner and intermediate slopes, always protecting and being very careful not to injure my weaker leg. One bad accident on the ski slope could have benched me for life!

One day, I took a wrong turn and found myself on the last leg of a long "expert" slope/trail with narrow chutes and varying pitch and width. I finally reached the bottom, but only after slow, careful snowplowing, sliding on the seat of my pants and a few falls. The expert skiers, including one of my ROTC students who was on the college ski team, went racing by as I cautiously and feebly made my way to the bottom. Word got around among the ROTC students about Major Matz's pitiful descent on the slope, and I heard about it in class the following days.

The PMS, Lieutenant Colonel Hefti, was a burly rugged Special Forces officer with a chiseled, weather-beaten face. A master parachutist and Ranger, he crossed the Normandy beach, fought in the Battle of the Bulge, and was a Korean and Vietnam War veteran. He played professional football three years with the Washington Redskins and was selected to be the commander of the Special Forces Detachment at President John Kennedy's funeral. He was a hulk of a man, modest and every bit a soldier, more comfortable in a field environment than walking the halls of academia. He kept his thoughts and criticisms to himself and guarded closely his prerogatives. He had absolutely nothing in common with the average Middlebury College professor and administrator, and, frankly, held some in low esteem. He and Dr. Armstrong were polar opposites if there ever was—and they both knew it.

Hefti and his wife, Sally, were the most down-to-earth, kindest, and most fun-loving couple ever. Their care and concern for the detachment cadre and our wives was unequaled. They had five children of their own yet took care of us like we were their sixth child. Hefti promoted me to major in a very nice ceremony in his office shortly after we arrived. He was uncomfortable with anything having to do with academics, including teaching in a formal classroom setting. However, he was very relaxed and most productive when in small intimate informal discussion groups with the ROTC cadets. They enjoyed his stories of his experiences in three wars.

The unique aspect of the Middlebury ROTC program was its organization. With the arrival of Hefti in the fall of 1968, the unit soon became the first ROTC unit to be redesignated a cadet Ranger Battalion. We had both served in airborne units and were graduates of the Army's Ranger School. We wanted to bring the esprit, élan, and warrior ethos of the Army's Airborne and Ranger units to Middlebury's cadets. The black beret became a standard part of the uniform. Utilizing the natural rugged environment of Vermont's Green Mountains, we moved the sterile classroom instruction outdoors whenever practical.

I got several local landowners to agree to allow us to use their land to set up a six-lane land navigation course. Day and night land navigation training with compass and map became part of the program of instruction. The emphasis on outdoor training was received enthusiastically by the cadets.

In December, while on Christmas leave visiting Linda's parents in New Jersey, our daughter, Heather, was born almost a month early on Christmas Eve at Walson Army Hospital, at Fort Dix, New Jersey. She didn't wait for our return to Middlebury. She was a beautiful healthy baby—the best Christmas gift ever!

The young men enrolled in the ROTC program were the most balanced on campus. Some were top athletes, and others held student leadership positions. During the first year, I taught both the juniors and seniors. Knowing I was fresh from the battlefield in Vietnam, the cadets were interested in my experience as a rifle company commander and eager to learn my perspective on the war. They were bright and asked probing questions, particularly in light of what they were hearing from the nightly news broadcasts and what they were beginning to experience firsthand on campus. I enjoyed the frank and honest discussions with them, sometimes extending into the evening hours.

On June 1, 1969, 47 Middlebury cadets received their commissions as second lieutenants in the U.S. Army at a ceremony in Wright Memorial Theatre. The 1968–69 academic year was over. I spent the summer of 1969 at ROTC summer camp at Fort Indiantown Gap, Pennsylvania, as a company tactical officer. Before I left for camp, however, we moved from our rental home in Shoreham because the owner was returning on June 1.

We found another farmhouse in East Middlebury that had been recently renovated inside. It belonged to a long-time Middlebury farm family named Fenn, who lived on the adjacent farm, and was only 6 miles from the college campus. We had survived the harsh conditions of living in our chilly rustic farmhouse in Shoreham and were glad to be moving into a house with heat on all floors and an abundant supply of potable well water.

I enjoyed my seven weeks at the "Gap" training and evaluating ROTC cadets. It brought back memories of when I was a Gettysburg College cadet going through ROTC summer camp at Fort Meade, Maryland. One weekend, Linda took the train from Trenton to Harrisburg, Pennsylvania, to meet me while I was on break.

Shortly after her arrival, she was hospitalized at the Harrisburg Hospital with a possible miscarriage. The doctor determined she could safely manage the train ride back to Trenton, but prescribed bed rest, no lifting, and no exercise. She was to take a drug known as DES (diethylstilbestrol—a lab-manufactured estrogen) throughout the remainder of the pregnancy. DES was banned for pregnancies in the U.S. by the FDA in 1971 after studies showed a high risk of cancer.

At the time, we had a toddler and a seven-month-old. Upon return to Vermont, Linda visited Dr. Bertrand J. Andrews, the town doctor who delivered the babies. He said to her as only a country doctor would, "Throw that medicine away [DES drug] and continue to do what you need to do. I don't like delivering babies of someone who has not exercised." He threw out the Harrisburg doctor's prescription. He was right. Our beautiful baby girl, Rebecca, was born by natural birth at Middlebury's Porter Hospital on January 22, 1970.

Dr. Andrews was on the ski slopes and didn't make it in time for the noon delivery, but I did! I had been dressed in a white hospital gown and was allowed to stay in the delivery room during the birth. Even so, I was no help to the attending nurse who, after the birth, immediately had to assist me to a waiting chair. I was woozy having just witnessed my first live birth. Our other two children were born in Army hospitals where the waiting father wasn't allowed anywhere near the delivery room and was sent home to await the birth.

Following Army tradition, the detachment presented us with an inscribed silver baby cup in honor of the birth. It was our second cup in 13 months at Middlebury, and as he handed it to Linda, Lieutenant Colonel Hefti said, "We're running low on funds, no more babies!"

I was looking forward to the students returning for academic year 1969–70. I got to know the former year's juniors well and would be principal instructor for their senior year and responsible for their commissioning in the spring. We also received two additional captains over the summer, giving our detachment its much-needed full complement of officers. Our XO, Major Herbert Koenigsbauer, departed for Vietnam, making me the detachment XO.

The Anti-War Movement on Campus

The war in Vietnam remained ever present in the hearts and minds of nearly every American. The inauguration of Richard Nixon in January 1969 led to a reevaluation of the U.S. role. We were at peak fighting strength in June and with U.S. casualties mounting, the United States and South Vietnam agreed on a policy of Vietnamization (the U.S. policy of withdrawing its troops and transferring the responsibility of the war effort to the government of South Vietnam).

ARVN forces began expanding and equipping to take over more of the ground fighting from the departing U.S. troops which began in July. The morale of American ground forces was beginning to fray. Incidences of fragging and combat refusal were increasing. The anti-war movement took on new meaning, and public opinion turned increasingly anti-military. The My Lai massacre in March 1968 was fully revealed to the American public in November 1969 resulting in an ever-wider anti-military movement.[1] Those in uniform were labeled by some as "baby killers" and "murderers" of innocent civilians.

As the academic year progressed, more professors and students became actively involved in anti-war activities. Even some of the MS III (junior) and MS IV (senior) cadets began asking, "Why is the U.S. fighting in Vietnam? Is this a 'just war'?" I became increasingly uncomfortable when wearing the uniform on campus. It wasn't uncommon for ROTC instructors to be shouted down or to have projected images of napalm-ravaged villages and dead Vietnamese children waved in our faces.

I had several objects thrown at me while passing a dormitory. The student culprit was a poor shot, but nevertheless, it put me on edge. Tensions were building. My wife and three young children also endured some of the anti-war sentiment. One afternoon, a few firecrackers were thrown onto the front porch of our farmhouse located close to the road. On another occasion, a group of student activists from campus drove down our lane shouting harassing statements. Linda, alone in the house with three babies, was fearful. The activists remained a short while, then moved across the road before finally leaving. Harassing stunts like this would continue.

One day in early spring, Hefti received an anonymous phone call that an anti-war group was planning to break into the ROTC arms room on campus to steal the weapons. We took the call seriously and coordinated with the Vermont National Guard to have them store the weapons in their armory in Vergennes. In a late-night move, with the National Guard's help and a Vermont State Police escort, the weapons were safely transported to the armory. The attempted break-in never happened, but harassing calls continued.

At that time, Middlebury wasn't a radical campus. The administration, although not in favor of the war or enamored with the presence of ROTC on campus, worked hard to maintain a tranquil, non-hostile campus. They reassured the parents and community that the college-wide activism was moderate in tone and that they had things under control. However, a few students with activist reputations were gaining notoriety, as was their following. Our cadets kept us informed of the potential student troublemakers and what their plans might be. Any negative incident could spark them into action. I sensed that, and we kept our scouts out and our guard up.

On May 4, 1970, protests against the war and bombing neutral Cambodia engulfed Kent State University in Ohio. The Ohio National Guard was called in, and in the ensuing confusion, shot and killed four students and injured nine others. Their deaths gripped the nation and brought home the horrors and the futility of the war that the battlefield deaths of American soldiers in Vietnam never had.[2] In the words of one student, "the spark of radical anti-war activism finally reached the sleepy town of Middlebury."[3] The killings at Kent State were the catalyst that aroused Middlebury's activists into action.

During the ensuing two days, student activist leaders began to take control and make their demands known. The dean of the college, Dennis O'Brien, was told by the activist leader that the students were striking, and all classroom doors would be sealed shut. They demanded all classes end and ROTC be removed from campus. The College Council and faculty, rather than challenge the young activists, succumbed to their demands and suspended all classes for the rest of the week to protest the war and grieve for those killed at Kent State. Hundreds of other colleges and universities across the nation were also protesting.[4] The students had taken over. I was beyond furious at how quickly Middlebury's leadership had capitulated to the demands of a few intolerant students!

Those students continued demanding that the college end its 18-year relationship with the military and remove the ROTC detachment from campus "immediately." Then, in the early morning hours of Thursday, May 7, a disgruntled Middlebury College student using gasoline set fire to Recitation Hall, a wooden building on campus that held some ROTC equipment. The fire destroyed the building and its contents.

It was the third attempt to vandalize ROTC offices that week. Public safety officers, assisted by some ROTC cadets, had stopped previous break-in attempts at the ROTC headquarters office. Nighttime guards were placed on buildings where ROTC offices and classrooms were housed. The strike and fire were enormously upsetting. O'Brien said, "Things were reasonably tranquil in 1964," but "the Vietnam War, combined with the presence of ROTC on campus changed all that."[5]

Some of the more vocal Middlebury students and faculty canvassed throughout the town, trying to educate the largely conservative, patriotic community about the horrors of war and the ROTC program. Town residents were fearful of violence, and tension emerged between the college and the community.[6]

Classes resumed on Monday, May 11. The dean made exceptions for students who wanted to take the rest of the semester off to protest the war. They did so without any academic penalty. I was stupefied by that decision. Why would anyone agree to such a policy? Each day, the administration yielded to another student demand. With a few exceptions, the ROTC department continued their classes through the strike period. Hefti and I made the decision to ignore the college's "strike" edict to cancel all classes, and we continued to teach. O'Brien wasn't happy.

The cadets supported our decision and attended their Military Science classes. One of the cadets said to me, "Sir, the students causing all the problems are wimps, and we don't follow them." First Army Headquarters, our higher command, authorized us to wear appropriate civilian attire when teaching if "we felt our safety, or the safety of the cadets was at risk." Indeed, the events in early May were the catalyst for widespread student and faculty anti-war action at Middlebury in the years that would follow. It was a major reason for the college's ultimate decision to remove academic credit for Military Science courses and "relegate ROTC to an off campus extracurricular activity in 1976."[7]

On May 4, 1970, I received orders assigning me to the Navy's Amphibious Ready Group Three in San Diego, California. It was a 24-month assignment which included two six-month deployments to WESTPAC (Western Pacific). My report date was June 18, Linda's 26th birthday. It was a very busy time for our detachment. We were closing out the academic year and planning the commissioning ceremony for the MS IV cadets who would be graduating.

While working to accomplish these tasks, we had to contend with not only the anti-war, anti-ROTC activism permeating the campus, knowing we were their principal target, but also with an increasingly unnerved and frightened administration who were ceding more control every hour to the activist students. It was humiliating, if not downright painful for me and Lieutenant Colonel Hefti as we watched these spineless administrators succumb to every student demand. There were days when I wondered how it would all play out.

The greater number of Middlebury students returned peacefully to the classroom, but there was a larger activated core within the student body who remained committed to the growing anti-war effort and the removal of ROTC from campus. They couldn't be reasoned with. We had to endure their wrath and unpleasantries directed at us personally. Despite the toxic atmosphere, on May 31, 31 Middlebury cadets received their commissions as second lieutenants in the U.S. Army at a ceremony at Wright Memorial Theatre on campus. Extra security was posted that day to ensure a peaceful uninterrupted ceremony.

The newly commissioned lieutenants were special to me. I had worked closely with their class for a year and a half, spending countless hours advising them and sharing with them my experiences in one-on-one and small-group settings. We did a good job preparing them to meet their obligations as commissioned officers. A couple served in Vietnam, and some made the Army a career. A few stayed in touch with me for several years. They were the cream of the crop at Middlebury. They took the road less traveled and never wavered in their support for the ROTC program when it was becoming the least popular entity on campus.

With 18 days to go before reporting for my joint assignment with the Navy in San Diego, there was much to do. Linda and I said goodbye to our Middlebury friends, packed, and with the kids, drove south to Linda's parents in New Jersey. As I look back over the years, my assignment with the ROTC detachment at Middlebury at the height of the Vietnam War clearly was more challenging and instructional than I realized at the time. Those 20 months showed me firsthand another dimension of the war and its effects, and I came to appreciate the ambivalent realities of the war. Indeed, Middlebury was a different kind of battlefield!

The May "strike" and burning of the building at Middlebury revealed to me the power of collective action by students and how pathetically weak academic leadership was during those war years. Those in charge wouldn't stand up to student criticism and demands; they showed no backbone! Thousands of students joined protests that shut down classes at over 700 colleges and universities around the country. Buildings housing ROTC detachments were set on fire or damaged. Middlebury was among the eastern vanguard of that national strike, which, in my view, caused colleges and universities to, in turn, strike out against national policy. The nightly news and other media were stoking the flames.

One academic said, "It was an unparalleled crisis in higher education." As a professional soldier with eight years' active duty and a year fighting in Vietnam for what I believed was a just and noble cause, I quickly realized increasing numbers of my fellow Americans didn't see it that way—particularly, the 18- to 20-year-old men the country was counting on to fight the war. It didn't dampen my enthusiasm for my upcoming tour which would take me back to Vietnam, nor did I lose my warrior spirit, but it gave me serious pause as I prepared for my assignment.

Navy Amphibious Force Pacific

"I am confident you will find duty with an Amphibious Group Staff a challenging and rewarding experience," wrote Captain J. W. O'Neill, chief of staff, Amphibious Group Three, in his welcome letter. He continued, "Although the staff is homeported in San Diego, our time is equally divided between the Eastern and Western Pacific." O'Neill was an old school Navy surface officer with a formal demeanor and rigid shipboard presence. His every action was formal and official. He exuded no warmth. But I respected and liked him!

Amphibious Group Three traced its origins to World War II, where it operated throughout the Pacific Theater. Its mission during Vietnam was to provide amphibious expertise and a deployable staff for combat and contingency operations by amphibious and coastal forces. It was commanded by a RADM (rear admiral) with a Navy staff of 35 officers, including five Marines and one Army officer, and approximately 90 enlisted sailors and Marines.

When deployed to the Western Pacific, we became U.S. Seventh Fleet's amphibious force or CTF-76. When not deployed to WESTPAC, we became COMPHIBGRUEASTPAC (Commander Amphibious Group Eastern Pacific)—and I thought the Army's acronyms were confusing! The group was leaving for a six- to eight-month wartime deployment to WESTPAC beginning June 22. I got Linda and the kids settled into a rental home near her parents in New Jersey. I then flew to San Diego to join the group's staff and from there flew to Buckner Bay at White Beach, Okinawa, where we embarked on the USS *Eldorado* (LCC 11).

When we arrived at the pier, a group of U.S. Navy Filipino stewards, dressed in their whites, met us and began carrying our bags up the gangplank and onto the ship. One of the Navy staff members yelled at me, "Hey Army, give them your bags, that's their job to carry them aboard for you. You're now in the Navy."

"No, I'll carry my own bags," I answered. They still took my bags and carried them to my stateroom. That was my first culture shock with the Navy as a member of the afloat flag staff. Within 24 hours, I was counseled by the staff's flag lieutenant on Navy protocol at sea and, in particular, Navy stewardsmen and what their duties are on board a flag-configured ship. I never did get used to the stewards carrying my bags, making my bed, folding my laundry, and doing other menial duties required of them. Whenever I balked, I was scolded by my Navy shipmates. My wife later blamed the Navy for my "spiritless" attitude toward routine housekeeping and daily personal tasks.

The *Eldorado* was the Amphibious Ready Group (ARG) command ship until we changed to the USS *Paul Revere* (LPA 248) in late August. *Eldorado* had been flagship for amphibious commanders in both World War II and in the Inchon invasion in 1950 in Korea, and later was flagship for the 5th Marine Expeditionary Brigade during the Cuban Missile Crisis in 1962. She was a combat veteran, indeed!

Seventh Fleet's amphibious force was assigned an expanded mission in command and support of river and coastal warfare in Vietnam. Included in those requirements were the *Market Time* and *Game Warden* operations along Vietnam's coast and in the Mekong Delta. In late 1966, the Navy requested an Army infantry major be assigned to the Pacific Fleet's amphibious force deployed staff to assist in the planning of those operations.

With my recent experience with the Mobile Riverine Force (MRF) in riverine warfare, I was selected to replace Army Major Dudley Budrich whose two-year tour was ending. "The only thing you have to know about shipboard life is to salute the stern of the ship, then the Officer of the Deck when coming aboard and the opposite when leaving," Dudley wrote in his welcome letter to me.

As assistant plans officer on the staff's N3 division (operations and plans), I was the principal advisor in all "matters related to the planning and conduct of restricted water and riverine operations and in providing Army input to seaborne amphibious and coastal operations." My duties included "the formulation of orders, tactics and doctrine for execution of joint and combined amphibious operations and contingency operations."

We were the planning and command staff for Seventh Fleet's Amphibious Ready Group/Special Landing Force (ARG/SLF). The ARG was the Navy component and typically consisted of four ships including an amphibious assault ship (LPH), a dock landing ship (LSD), an attack transport (APA) and an LST (tank landing ship).

The SLF was the Marine component and consisted of a Marine Battalion Landing Team (BLT) composed of a reinforced infantry battalion, plus a Marine helicopter squadron. Some operations called for an Army infantry battalion to replace the Marine battalion. In an effort to maintain their unique capabilities as an amphibious force, and not become a second land army, the Navy/Marine Corps Team developed a series of "multiple amphibious raids to be conducted along the Vietnam coast against Viet Cong and North Vietnamese Army units."[8]

Seventh Fleet's ARG/SLF had the mission of conducting the operations. From 1965 through 1967, various operations were conducted, primarily in II & III Corps Tactical Zones employing heliborne and amphibious assaults. These *Dagger Thrust* and *Deckhouse* series of operations met with varying degrees of success.

In 1970, the mission of the SLF reverted to what it had been before 1965: the amphibious strategic reserve force for the Pacific. The Marines continued to provide the BLTs for the SLFs.[9] From 1970, until U.S. forces withdrew from Vietnam, we continued to steam off the coast of Vietnam. We were constantly looking for targets and writing plans to support various contingencies. I was the principal planner and writer for all basic plans and annexes for these operations. In late 1970, the SLF was renamed the 31st Marine Amphibious Unit (MAU). The following year the MAU was given the mission of feigning an amphibious landing along the coast of North Vietnam in order to draw away NVA forces in the south.[10]

Back in 1968, General Westmoreland, backed by Admiral Sharp, was considering an amphibious landing north of the DMZ. Intelligence showed up to 30,000 enemy in the objective area. The plan was code named Operation *Durango City* and Sharp requested authority to conduct the operation on or about June 1.[11] It would have included both Army and Marine ground forces in a joint and combined amphibious and air assault campaign against the north.

SECDEF Clark Clifford and his newly appointed task force (known as the Clifford Committee) that was reviewing the various military options, ruled against the bold operation saying, "Sending more American soldiers would increase U.S. casualties and war costs and have divisive political effects at home."[12] He was largely influenced by several current and former Defense Department civilians, notably Paul Nitze[13] and Paul Warnke,[14] who believed "further military escalation," especially sending U.S. troops into North Vietnam "would be futile."[15]

I called them the "surrender" group. They wanted U.S. forces to wind down and stop fighting. Clifford, himself, thought the war was unwinnable and from that time on, decided the United States should "level off our involvement ... and work toward gradual disengagement through negotiations."[16] The U.S. would continue to support some air campaigns, but the conduct of further large land campaigns by U.S. ground forces gave way to the administration's policy of disengagement and turning the fight over to the South Vietnamese Army. As an aside, Secretary of State Dean Rusk[17] disagreed with Clifford and thought "that the war was being won by the allies" and that it "would be won if America had the will to win it."[18]

As history shows, we didn't have the "will to win it." Military leaders—Wheeler, Sharp and Westmoreland—wanted to take the ground campaign into North Vietnam at a crucial time in the war. We had just beaten the enemy badly on the battlefield in Tet (although U.S. news sources reported the Tet Offensive as a Viet Cong victory), and we had excellent intelligence on where the NVA was regrouping and licking their wounds just across the DMZ in North Vietnam.

Operation *Durango City*, with its formidable supporting naval and air campaign, could have been easily launched. CTF-76 was on station and would have executed the amphibious phase of the operation. When I joined CTF-76 in 1970, I reviewed the earlier plans hoping we still might execute a *Durango City* type operation. Intelligence still showed large concentrations of NVA units just north of the DMZ. Instead, our civilian leadership, influenced by the "unwinnable" and "surrender" group in the Pentagon, and driven by an increasingly anti-war American media, gave up. We knew where the post-Tet enemy units were, that they were ill-equipped and leaderless at the tactical and operational levels. Morale was low. We could have hit them hard with air, ground, and naval forces and killed many, but we didn't execute. Indeed, a lost opportunity! In my view, it was the operational turning point that decided the outcome of the Vietnam War.

We also had attached for operational control and planning, Underwater Demolition Team Eleven (UDT-11), the Navy's frogmen, a tough, proud group of warriors whose history traced back to World War II and Korea. UDT-11 embarked with us most of the time. Their commander was Lieutenant Commander Charles "Chuck" Le Moyne, who later rose to the rank of rear admiral and was deputy commander of United States Special Operations Command in Tampa, Florida. As shipmates, Chuck and I became instant friends during the deployment.

We had much in common and shared stories of our respective earlier combat tours in the Mekong Delta. Not only did we work together closely when planning amphibious and naval special warfare operations, we also became close off-duty "liberty" buddies. He and I later attended Harvard University's Senior Executives in Government/Management Course in 1988. He was the first SEAL officer to achieve two-star rank. I admired him and respected him as a man of honor and integrity. He died of cancer in 1997.

When we weren't steaming off the Vietnam coast, we were conducting amphibious training operations throughout the Western Pacific. Early during the deployment, I was given the nickname "The Force Troop" by the Marines. One day, while in port at Subic Bay, Philippines, I was summoned to Captain O'Neill's stateroom. "Major Matz, the admiral wants you to be the physical fitness officer for the staff." He explained they had seen me doing PT both aboard ship and while in port and thought I would do a good job. Admiral Rubel was somewhat of a PT guru himself and wanted to keep his staff fit while at sea.[19]

The combination of excellent chow and long stays at sea in cramped quarters wasn't conducive to staff officers maintaining a fit and healthy body. They needed physical exercise. My August 3, 1970 Amphibious Group Three appointment letter read, "Effective this date, you are designated as Physical Fitness Officer for the Staff of Commander Amphibious Group THREE. You will be guided in your duties by provisions of BUPERSINST (Naval Instructions) 6100.2 series ..." For the next eight months, I led the staff in PT. I approached my collateral duty cautiously and with no air of snobbery or superiority.

I sensed immediately that some resented the "Army guy" leading them in PT (particularly the Marines), and I didn't want to make it any worse by rubbing salt into the wound! I even tempered my enthusiasm. I discarded the BUPERSINST and used the Army PT manual. Before long, the staff were participating in routine on-deck and on-shore workouts including short runs. Pushups, crunches, and burpees became the norm. Some of the Navy officers had never done any of the exercises. The admiral loved it and rarely missed a PT session, the Navy officers tolerated it (they had no choice) and the five Marines never got over the fact that the Army "Force Troop" was their PT instructor.

Any resentment that might have resulted was short lived. The Marines took me under their wing and from day one schooled me in the ways of the Navy. I shared a stateroom with Major Harry Ling, a Marine intelligence officer on the staff. The senior Marine on the staff, a crusty, red-faced lieutenant colonel, said, "Stick with me during this deployment," implying that he and his Marines would take care of me. I was constantly reminded by my Marine friends that every Marine is first an infantryman. In their minds, that made me one of them. I learned quickly that they were not terribly fond of their naval counterparts, particularly the lieutenant colonel.

Members of other military services wanted to give them a nickname that was sufficiently pejorative and germane. The Marines called them "squids." Some say that was in retaliation for the Navy calling the Marines "jarheads" during World War II, a term that wasn't endearing to Marines. Navy officers on the staff hated the name squid because it is a slimy, spineless creature and otherwise devoid of personality. Most of the Marines used it in their everyday conversations when referring to ship's company and the sailors on the staff, especially the lieutenant colonel. He didn't hold back or curb his thoughts or words in any way. I didn't like it, and as the Force Troop, I did my best to maintain some semblance of neutrality and distance when it came to name calling and interservice rivalry between the Marines and sailors.

I was unique in that I was the only Army guy on an all-naval staff, and, in a way, both my Navy and Marine shipmates were vying for my attention and support. Commander J. B. Harwood, the N31 plans officer, and my immediate boss, who was a rather reserved and quiet deep thinker, told me, "If you want to piss off a Marine tell him to turn over his ID card. It clearly reads, 'Department of the United States Navy.'"

Indeed, the Navy officers were very effective at retaliating in their own polite and unobtrusive way. I marveled at how some Navy officers stuck it to the Marines without them even knowing they had been had! Despite an occasional flash of anger or loss of temper, it was mostly a good-natured interservice rivalry. And we all knew, when the chips were down, we held each other up; we had each other's backs.

I thought I knew my geography until that tour with the Navy. I never knew how many seas there were in just the Western Pacific alone. On multiple occasions, we sailed the eight seas in our area of operations. We also sailed the East and West Caroline Basins and crossed the Equator.

Military customs have long been an integral part of the Navy. The granddaddy of all seagoing ceremonies is the "Shellback" initiation when the ship crosses the Equator ("the line") in which slimy "Pollywogs" (sailors who have not previously crossed the line), become "Shellbacks" (fit subjects of King Neptune). It is a ritual that dates back over 400 years. On September 9, 1970, USS *Paul Revere* crossed the Equator and entered the domain of Neptune Rex, Ruler of the Raging Main.[20]

As a Pollywog, I was duly charged with repeated use of the landlubberly terms "walls," "floors," and "latrine" when aboard a ship of the realm and ordered to appear before King Neptune and his Court to be initiated in the mysteries of his royal domain.[21]

After a day of hazing and trial, including a couple saltwater showers, I survived the supervised, yet uncomfortable and embarrassing ceremonial ritual on the mess deck of the USS *Paul Revere* at sea. Being the Force Troop, the Navy chiefs singled me out and had a good time ensuring I answered all charges and met all qualifications. I did, and with others, was initiated and gained rites of passage as a Trusted Shellback on September 9, 1970 at longitude 105° 31' east.

My certificate, which as an Army infantryman, I prize to this day, reads "William M. Matz, Jr. has been gathered to our fold and duly initiated as a Trusted Shellback." In 1936, President Franklin Roosevelt, after answering a series of charges, was initiated a Shellback aboard USS *Indianapolis* and honored with a Trusted Shellback certificate. Nine years later, *Indianapolis* was sunk by Japanese torpedoes. It remains the worst sea disaster in U.S. naval history.[22]

The Shellback ceremony began mainly as a morale lifter and for camaraderie among the ship's crew. I am fortunate to have experienced that proud naval tradition. From that day on, I felt much closer to both the staff and the ship's crew. One sailor remarked, "Sir, you now have salt in your blood." I don't know about salt in my blood, but I sure had saltwater all over my body and a salty taste in my mouth for several days.

After crossing the Equator, we headed north and arrived at Da Nang Harbor on October 5 in company with USS *Okinawa* (LPH 3). The Vietnamese government had planned a Vietnamese award ceremony to recognize the many accomplishments of U.S. Navy personnel while on duty in support of the Republic of Vietnam. The ceremony was held on *Okinawa's* hanger deck while she was anchored in the harbor. It was an impressive ceremony with brightly colored signal flags hung the length of the hangar deck.

The U.S. Seventh Fleet's band provided music, and *Okinawa* mustered a smart honor guard. Several dignitaries participated, including the prime minister of the Republic of Vietnam, Tran Thien Khiam,[23] who represented President Thieu.[24] Most of the personnel who were recognized, received their Vietnamese awards in absentia. It was an occasion for me to wear my Army service dress white uniform for the first time. I wore it on several other occasions while serving with the Navy.[25]

While in Vietnamese waters, we received intelligence updates on the latest enemy locations and suspected movements which resulted in no lucrative targets along the Vietnamese coast. Disappointed by this, we sailed for Subic Bay, arriving on October 9. We were chased out of Subic Bay three days later as Typhoon Joan was headed our way. In an October 13 letter to my parents, I wrote, "... we were getting ready to sortie to the open sea to evade Typhoon Joan."

Paul Revere rolled in some moderately heavy seas until October 15 when we headed for Manila, arriving the next day. After a brief port call at the capital city, we arrived back at Subic. As Admiral Rubel's "All Hands" letter of November 15 noted, "*Paul Revere* was underway at 2 AM on 20 October to evade Typhoon Kate." The two typhoons, Joan and Kate, caused a long-awaited combined Philippine-U.S. amphibious exercise to be canceled.[26]

Exercise *Fortress Light* had been planned for several months. It was an across-the-beach/over-the-horizon operation which included Army and Marine units. The site of the exercise was to be Mindoro, an island in the middle of the Philippine archipelago. Typhoon Joan had kicked up rough seas in the Mindoro area, but we had hoped that they would subside in time for D-Day. But Kate passed very close to Mindoro, eliminating any hope that the exercise could be held safely.[27] It was canceled, and I was very disappointed. Joan and Kate were spoilers—they did us in.

I was one of two CTF-76 principal planners for the exercise and worked every aspect of its planning since joining the staff in late June. I wrote several annexes to the plan, coordinated the naval gunfire support, and briefed the admiral and SLF commanders on a regular basis. The entire ARG had been on station. Infantry troops in full battle gear were in the ships' well decks awaiting the assault, and the Marine helicopter squadron was ready to launch. The entire ARG/SLF was leaning forward. It never happened. We stood down.

As plan-wrecking as the typhoons were, we suffered none in comparison to the Filipinos whose homes were destroyed, and whose relatives were killed and injured, considering 600 people were killed and an estimated 80,000 left homeless in the Mindoro Island area. Philippine and American military forces airlifted life-sustaining supplies to isolated communities. For four and a half days we remained on-station providing the command and control and support base for feeding thousands of distressed typhoon victims and treating many of the seriously injured in our amphibious ship sick bays. We were denied a huge amphibious training exercise, but, instead, executed a magnificently successful real-life humanitarian relief.[28]

Typhoons in the Pacific are a nemesis for the U.S. Navy, such as the terrible Typhoon Cobra that hit Okinawa in December 1944 and wrought havoc on the Navy.[29] We spent much of the eight-month deployment riding out and evading storms. While steaming in the South China Sea an unusual opportunity presented itself.

Commander Harwood asked me if I would be interested in becoming an underway officer of the deck (OOD). The OOD is a watch-standing position in a ship's crew stationed on the bridge while underway. He is designated by the commanding officer to be in charge of the ship, including its safe and proper operation during that time. After he briefly explained the duties and requirements, I said, "Yes, I would like to do that."

After a series of rigorous training sessions with the ship's crew, and after close supervision by the most experienced underway OODs, I was certified as a CTF-76 underway officer of the deck. I took the responsibility very seriously and when on watch, never left the bridge, not even to relieve myself. I did my best to hide my anxieties. My watch mates, the enlisted lookouts, talkers, and helmsmen, were very helpful, and my underway watch periods were primarily during daylight hours. Some sailors were never comfortable with an Army guy as OOD while underway!

COMPHIBPAC Navy News Release Number 19-72, dated June 14, 1972 and titled "Ranger Major Runs Navy Ship" read, "Major Matz is one of the select few officers of the Army to ever have earned Navy certification as an Underway Officer of the Deck. Officers so designated must be competent mariners for they are responsible for the ship's operation and safety while underway." The article went on to say, "It is most unusual for an Army officer to hold such qualification."

An *Army Times* article, dated July 5, 1972, titled, "Ranger Major Takes to the Sea" elaborated on my serving "in" the Navy and about "my life being a sea-going affair." Suffice to say, the articles at the time, not only made for interesting reading by Army and Navy readership, but also resulted in several inquiries from close friends wanting to know how a guy who never did well in land navigation exercises could now be responsible for the safe navigation of a U.S. Navy warship at sea. I told them to call the Navy officials who certified me.

On July 1, 1970, one week after I reported for duty with Amphibious Group Three, Admiral Elmo "Bud" Zumwalt, Jr., became chief of naval operations (CNO). Zumwalt's command in Vietnam wasn't a "blue water" force; it was a "brown water" unit. He commanded the Navy's Mekong Delta forces including the flotilla of swift boats and my former unit, TF-117, the joint Army-Navy MRF. Zumwalt reformed Navy personnel policies to improve enlisted life and to help the Navy's plummeting recruiting and retention rates.

His policies were quickly disseminated in Navy-wide communications known as "Z-grams." He authorized beards, sideburns, and mustaches and introduced beer-dispensing machines in enlisted barracks. He authorized enlisted sailors to keep civilian clothing aboard ship for wearing on liberty. He directed commanding officers to assign by-name sponsors for newly arriving personnel. I was glad to see his sponsorship "Z-gram." It was sorely needed. Although I received a formal letter of welcome from the group chief of staff, there was no sponsor assigned to assist me and my family in establishing ourselves in not only a new location, but also with a new service—the Navy.

I was with the Navy at sea when the "Z-grams" hit the fleet, and I didn't notice any adverse impact on the readiness on board the ships upon which we were embarked or within the group staff. The Navy chiefs were not happy, the officers were ambivalent, and the enlisted sailors were ecstatic! For the junior sailors, "Zoomie" was their guy. He knew the primary force-multiplier of the Navy was sailors, not ships, and he sought to improve their quality of life.

Our deployment ended in January 1971, and we returned to San Diego as COMPHIBGRUEASTPAC. In late March, the staff flew to Rodman Naval Station in Panama to join our new flagship, USS *Blue Ridge* (LCC 19). It was built and configured from the keel up to be a command-and-control ship with heavy reliance on computer systems and high-end communications. Originally, its mission was to command wide-scale amphibious operations. Fifty-two years later, as I write this story, it remains in service as flagship of the Seventh Fleet and the oldest deployed warship of the U.S. Navy.

On April 9, with commander Amphibious Group Three and staff embarked, *Blue Ridge* sailed into its homeport of San Diego. It was a beautiful sight. The ship's company in their whites were manning the rails, a centuries-old practice for rendering honors aboard naval vessels. The San Diego City harbor fire boats sprayed their huge water cannons into the air as they escorted our new flagship to its berth at Naval Station, San Diego.

Linda and our three children joined other families in a pier-side welcome and reception. It was a sight to behold! After eight months of steaming in the Western Pacific, the 30-year-old *El Dorado* and 25-year-old *Paul Revere* would both be decommissioned a few years later, and we had the newest ship in the Navy as our command ship. Not only did *Blue Ridge* have the latest, most sophisticated command-and-control suites, it had a large deck which gave me plenty of room to conduct PT sessions for the staff which the admiral decided to continue while in EASTPAC.

Navy Special Warfare

With the relatively low level of enemy activity in the coastal regions and riverine and coastal operations winding down in Vietnam, there were fewer requirements for my afloat Army billet in WESTPAC. On the other hand, the recent organization of the Naval Inshore Warfare Command, Pacific, as a major subordinate command of commander Amphibious Force U.S. Pacific Fleet, coupled with anticipated future joint Army-Navy amphibious and special warfare operations exercises, determined the continuing need for an Army infantry major, but with the revised billet title: Force Army officer/Assistant special operations officer. The billet would be on the staff, commander Amphibious Force, Pacific (COMPHIBPAC) in Coronado, California.

The Navy wanted the incumbent to have amphibious/riverine experience and to be Ranger qualified. I fit the bill perfectly. I was requested by name by Vice-Admiral Nels Johnson, commander Amphibious Force, Pacific. The Army approved, and I began my second year with the Navy on June 30, 1971, as COMPHIBPAC's assistant special operations officer.

My nickname "Force Troop" stayed with me. Emerging from their Vietnam War experiences, all services were formalizing and expanding their special operations capabilities. The Navy was forming its Special Warfare Groups, consisting of UDT, SEAL (Sea, Air, and Land) and Beach Jumper Units (BJU). My primary job was to assist the new units in their organization, development, and training programs to achieve continuity of purpose in conducting joint and combined amphibious, riverine, and special operations.

I was happy with the assignment. It was an easy drive to Coronado Naval Base where our headquarters and most of the Navy Special Warfare units were based. It also afforded me the opportunity to begin work on a master's degree. Infantry branch wanted me to get a master's degree, and with the Navy's support, I enrolled in the University of San Diego and became a candidate for a Master of Arts degree in Political Science.

As a full-time student in the university's evening and weekend program, I had a full plate balancing my Navy workload and the university's degree program requirements, while also trying to help Linda by being an involved dad to our three toddlers. We tried to set aside every Saturday night for ourselves to go to dinner. If I wasn't locked in my study cubicle at the university's library researching/writing my thesis, I joined Linda and the kids for a Sunday outing at either Sea World or one of the Mission Bay beaches. Otherwise, I routinely spent every day and night at work or at the university. There was no free time. It was one of the busiest times of my life! Linda was equally busy taking care of three active kids aged four and under.

My immediate boss was Captain John E. O'Drain, COMPHIBPAC's special operations officer. O'Drain was a graduate of the University of California, Berkeley, and served with Underwater Demolition Team Five (UDT-5) in operations against North Korean forces during the Korean War. A big, burly Irishman, he was fun to work for and full of Korean War frogman stories. He made the routine, mundane staff work fun and was the kind of boss you always went the extra mile for. He made sure I knew the UDT and SEAL officers, and I was always invited to their official and social functions. Come mid-morning, he would say, "Bill, let's get the hell out of here, call the gym and reserve a racquetball court." He wasn't a good racquetball player, and once he found out I was worse than him, he always wanted to play.

Some mornings, I took PT on the Coronado strand with either the UDT teams or SEAL Team One. They knew about my polio, and the fact that one leg was shorter and weaker than the other made no difference to them. They cut me no slack on their morning runs. Immediately following the run, the "seals" and "frogs," as they were called, continued their morning workout by donning rubber swim fins and wading into the surf for a long-distance ocean swim. That was my cue to stand at the water's edge and wave goodbye while I turned and headed to the locker room for a shower.

My leg at that point was rubbery, aching, and needed rest. Some mornings I barely made it to the locker room. I made all the runs and accompanying dry-land exercises but knew my limitations, and a half-mile swim after a run on the strand wasn't my cup of tea. Many a sunny morning, I was sweaty, sandy, salty, and up close with the Navy special warfare guys, and they were every bit a tough, proud lot.

I was involved in every aspect of the command's development of plans and policy for the conduct of special warfare, amphibious and riverine training and operations. Later, I wrote a decision paper on amphibious and naval special warfare training for a course requirement while a student at the U.S. Army Command and General Staff College (CGSC). O'Drain left in March of 1972 to assume command of Naval Special Warfare Group One, consisting of all SEAL and UDT teams and Beach Jumper Units in the Pacific. I was sorry to see him leave, but getting that command was a feather in his cap!

On the distaff (the term commonly used by U.S. military services to describe wives/ladies' functions) side, Linda had some interesting experiences with the Navy wives. At her first Naval Amphibious Base coffee, the admiral's wife, who wasn't known for her sweetness or tact, asked the wives to tell where their husbands worked and what they did. The Navy and Marine wives began effortlessly spieling off Navy acronyms in describing their husbands' unit and work duties, like they were second nature. Not being familiar with Navy terms, Linda replied that her husband was in the Army and worked in a headquarters on Coronado.

That didn't satisfy the admiral's wife who asked her to clarify her answer and be more specific. Linda apologized and said it was no excuse, but she had been in San Diego less than a month and had spent her time moving into quarters and taking care of three children, ages one, two, and four and had not yet learned the Navy acronyms, unit locations, etc. She was uneasy and embarrassed as she felt the stare of disbelief from the admiral's wife, and 25 unspoken expressions of heartfelt pity from the others. Many wives came up to her afterwards and offered to help her get acclimated to the Navy and the many new and confusing acronyms. Indeed, it was the subject of conversation at dinner that night.

In April, I received orders to attend the CGSC at Fort Leavenworth, KS, for the year 1972–3. My two years with the Navy were ending. In a flyer entitled "Force Troop Review," COMPHIBPAC announced the Summer Force Troop Pass in Review ceremony to bid farewell to Major Bill Matz, U.S. Army, COMPHIBPAC's Force Troop. We assembled at the Mexican Village restaurant in Coronado for Mexican fare and drink.

Officers brought their canteens to the no-host bar. There were flags and music. I was able to say goodbye to my COMPHIBPAC and Naval Special Warfare Group (NAVSPECWARGRU) friends as well as to many of my former shipmates from my WESTPAC deployment who also attended the farewell lunch. The bosun piped me aboard and again when I departed. I'm not sure whether they were happy or sad that I was leaving. But as the Force Troop, I bid them farewell with a heavy heart, wonderful memories and a deep fondness and respect for my seafaring naval counterparts. I couldn't have had a better joint assignment!

After a year serving with the Army-Navy Mobile Riverine Force in Vietnam in 1967–68 and two years with the Navy's Pacific amphibious force, both the Army and I were beginning to wonder what branch of service I was in.

CHAPTER SIX

Post-Vietnam War Years

It was a rainy day in mid-June when we loaded our three kids, ages five, three, and two into our Chrysler sedan in San Diego and headed north to pick up U.S. Highway 66—popularly known as Route 66. That day was the only time we remember that it rained while we lived in sunny San Diego. The famous and historic highway embodied a complex, rich history, a remnant of America's past stretching 2,400 miles across two-thirds of the continent.

We traveled from just east of Los Angeles to Chicago, headed to Linda's parents' house in Trenton, New Jersey, for a couple of weeks' vacation before reporting to Fort Leavenworth. Our three-year-old daughter, Heather, was to be a flower girl in our cousin Nancy Smith's wedding on July 1.

Being the first time we drove across the country from west to east, we wanted to visit some national parks along the way. We stopped for a one-night visit in Las Vegas. We passed through the Painted Desert and detoured north to visit the Grand Canyon.

Upon leaving the Grand Canyon, we were traveling south on an Arizona two-lane state highway when I hit a horse mule (a large male mule often called a "john mule"). I was coming over the crest of a hill when out of nowhere it appeared before me standing in the middle of the road. It was huge. Going approximately 70 miles per hour, I couldn't brake in time, or avoid it. I smashed into it head-on, flipped it into the air and through my rearview mirror, I saw it crash-land on the highway behind us. I never heard such a loud thud!

Steam was spurting out of the radiator, and the front hood was crumpled and had broken the windshield. Linda and the kids screamed. It was a loud chorus that I'll never forget. They exited the car, and Linda got the kids off the highway and onto the side of the road. She worried about rattlesnakes that might be in the area, so she made a commotion to scare them away. The mule was a crumpled heap in the middle of the road, dead as a doornail.

We were in an area that bisected the Hualapai and Hopi Indian Reservations on a red-hot day. There was no fencing, and the mules grazed freely on the Indian reservations. An old, broken-down pick-up truck with four Indian men soon arrived. They didn't speak the best English but told me it was their mule, and they were angry as hell to say the least, shouting and demanding money, wanting cash on the spot. I was trying to converse with them when in the distance, I saw an Arizona State Highway Patrol car approaching. There was no traffic on the highway going in either direction.

The state trooper's arrival at that moment was, indeed, fortunate and a pleasant surprise. He immediately took charge of the situation and looking at Linda and the kids, asked, "How are the kids doing?" In his presence, the Indians quieted down and became less hostile and demanding. He had the Indians drag the carcass from the center of the road onto the shoulder, and he radioed a nearby desert gas station for a tow truck. It arrived in due time.

The state trooper wrote up an accident report and handed me a copy. I told him I would pay for the dead mule. "No need to do that. These mules roam the desert and reservations, and don't belong to anyone," he said. I was relieved. "You sure picked a big one to hit, but you are not at fault, so don't feel bad," he assured me. He told me that they were horse mules, and the state allowed them to roam freely. I had not heard the term "horse mule" before that.

The driver of the tow truck must have been in his 80s, tall, wiry, and wearing a big cowboy hat. After he hitched up our damaged Chrysler, all six of us squeezed into the front seat of the truck's cab for the 75-mile drive south to Flagstaff, the nearest town with a repair shop. During the entire ride, the driver grumbled about the Indians.

Our car needed a new radiator, belts, hoses, windshield, a tire, front hood, grill, headlamp, and fender. Parts were not readily available in that part of Arizona, so we had to spend five days in a motel in Flagstaff awaiting new parts. The hood, fender, and grill never arrived, so the auto body shop mechanics repaired the body as best they could. They ended up having to tie down the front hood as the fastener and latch assembly never arrived, and the damaged parts couldn't be repaired.

Once repaired, as best as possible, and minimally passing the road and safety inspection, we were on our way again. I drove the rest of the way to New Jersey very cautiously, stopping often to ensure the wire and rope tie-downs were secure. Our car looked like the Beverly Hillbillies' wacky vehicle lashed together with wire and straps. Like theirs, it was eye-catching, and we had more than a few gawkers as we passed other cars and pedestrians along the way.

Days later, we arrived safely, just in time for Heather to be part of Nancy's wedding. Our last stop before reaching Linda's parents' home was a visit with a college fraternity brother, Barry Kain, and his wife, Nancy, in Elizabethtown, Pennsylvania.

CGSC was a graduate-level school for United States Army and sister service officers. It was a 38-week-long course that educated leaders for full-spectrum joint and multinational operations. During the school year, I also completed my thesis for the Master of Arts degree that I had begun a year earlier at the University of San Diego.

It was, indeed, a busy and challenging academic year. Not only was I attending class every day for the CGSC course, but I was also researching and writing my thesis in the evenings and on weekends, while taking the remaining required political science courses at Kansas State University in Manhattan.

When I completed a chapter, I sent the draft manuscript to my thesis advisor at USD, Dr. A. Paul Theil, PhD, for his review/critique. He was excellent and most helpful, but he cut me no slack. His remarks were concise and to the point: "Go over each sentence. Ask yourself, 'Can I say it in another way and with fewer words?' Is this paragraph important to the overall story line?" he wrote in one of his letters of critique on January 15, 1973. He was a stickler for syntax, punctuation, and clarity of thought. Above all, he demanded in-depth research. He wanted "new sources" to corroborate my "findings."

I was never so stressed academically as when I was researching and writing my thesis. I returned to San Diego in May 1973 for the oral defense of my thesis before a panel of experts assembled by Dr. Thiel. Among them was Dr. Gilbert L. Oddo, PhD, director of USD's Graduate Division. They put me through the wringer with their questions and challenges of my opinion and judgment, and the conclusions I reached about my topic, "Operation *Vulture*: America's Plan to Intervene in Indochina in 1954."

I am indebted to Dr. Thiel for his relentless, persistent drive in making me repeatedly dig deeper into the subject, until he believed I was convinced, beyond any doubt, of my own conclusions. That was an academic drill I had never experienced before, but glad when it was over.

The Master of Arts in Political Science was conferred on me by USD on June 1, 1973. My graduation from CGSC with 1,096 other officers was June 8. I graduated with "Commandant's List" honors. General Michael S. Davison, Commander, U.S. Army Europe, gave the graduation address. A huge sense of relief that "Bill had finished his master's degree" permeated the Matz household. We all celebrated, and I never undertook another major writing project until this book.

The year wasn't without incidents. In June, as Linda was scrubbing the quarters preparing for the Post Housing Office "white glove" inspection required upon vacating our government housing unit, our children were enjoying their last days together with their friends. Our six-year-old son, Billy, and two of his buddies were driven home that day in the back of a MP vehicle, lights flashing. While Linda thought they were playing in the small playground on our street, they went on an adventure hike and strayed into an off-limits field adjacent to the United States Penitentiary grounds that bordered the housing area.

The field contained a small herd of bulls, which at the time, were part of the principal security belt around the penitentiary. The MP who escorted our son to the door told my wife, "Your kid came very close to being chased by a bull and possibly seriously hurt." She thanked the kind and vigilant MP. Young Billy was then put to work helping Mom clean the quarters and pack our remaining goods. The MP reported the incident at his end-of-shift debriefing. The next day, in the Post Bulletin, all CGSC students and their spouses were reminded of their responsibility as parents to "watch their children and always ensure their whereabouts," citing the Matz incident the day before.

After CGSC, I was assigned to the Pentagon, Washington, DC, where I spent the next four years, reporting for duty on July 16. My welcome letter of May 25, 1973, from Major General Charles J. Simmons, read: "Dear Major Matz, on behalf of Lieutenant General Donald H. Cowles, Deputy Chief of Staff for Military Operations and Plans, it is a pleasure to welcome you to the Army General Staff and the Office of the Deputy Chief of Staff for Military Operations and Plans (ODCSOPS)."

My assignment was in the International Affairs Directorate, whose mission and responsibilities were broad in scope, spanning a wide range of responsibilities regarding national security and international political-military activities. General Abrams was chief of staff of the Army at the time.

We bought our first home in the Orange Hunt Subdivision in West Springfield, Virginia. I drove 16 miles each way to and from the Pentagon in my VW Bug, learning how to navigate the Interstate 395 traffic. Battling the traffic was, indeed, the most unpleasant aspect of the tour. Our kids joined the Orange Hunt Swim Team, which was the beginning of what would be long, successful, careers for them in competitive swimming in high school and college. While caring for the kids and managing a very tight budget, Linda sold Avon products door-to-door in the adjacent neighborhoods.

There was no cost-of-living allowance at that time, and the DC area was expensive, so selling Avon products helped with the expenses (she received several awards for top salesperson). The job allowed her flexibility while helping the kids in their many activities. My work hours at the Pentagon were 0700 to 1900, so I couldn't offer much assistance during the week.

Being a young action officer (AO) on the Army staff was a humbling experience. It was all about authoring succinct papers, making "suspenses," and developing and arguing the DCSOPS position. The ODCSOPS Policy Guide for AOs stated: "Get the DCSOPS position in papers and make the 'opposition' happy at the same time."

I was selected for early (below the zone) promotion to lieutenant colonel and was promoted by Major General Wallace H. Nutting, the directorate director, on July 1, 1976. By that time, the International Affairs Directorate had been renamed Strategy, Plans and Policy Directorate, and I moved from being an AO to being the Directorate executive officer.

A month later, I received a letter from the chief, Lieutenant Colonels Division, Military Personnel Center. "This is to inform you that you have been activated from the list of alternate command designees to take command of the 3rd Battalion, 187th Infantry, 101st Airborne Division (Air Assault) at Fort Campbell, Kentucky, in July 1977." I couldn't have been happier! I celebrated the occasion with a few beers that night. I wanted so badly to put the Pentagon in my rearview mirror and get back with troops. Although it would be another 10 months before I took command, there was a bright light at the end of the tunnel!

Commanding the Rakkasans

I assumed command of the 3rd Battalion, 187th Infantry, on July 8, 1977, on Rakkasan Field, Fort Campbell. Originally organized as glider infantry in World War II, the 187th fought in the Korean War as part of an airborne infantry regimental combat team (RCT). Later, the 187th fought in the Vietnam and Persian Gulf Wars. Throughout its history, the 187th has served with distinction in combat, particularly during the Korean War when then Brigadier General William C. Westmoreland commanded the RCT.

During occupation duty in Japan following World War II, the Japanese gave the 187th its nickname, "Rakkasans," which translates to "falling umbrellas" or "parachute". In Vietnam, the battalion received a Presidential Unit Citation (PUC) for their participation in the bloody, 10-day battle of Dong Ap Bia Mountain (aka Hamburger Hill) in 1969. The 3/187 emerged from the Vietnam War as the Army's most decorated airborne battalion. It was an honor to be selected to command this storied battalion!

Shortly after I assumed command, a farewell reception was held for the division's outgoing assistant division commander (operations), Brigadier General Weldon F. Honeycutt. He was a protégé of Westmoreland and had been his aide-de-camp and had commanded the 3/187 during the Battle of Hamburger Hill. The troops nicknamed him "Tiger." A tough, aggressive infantry soldier, who spent most of his Army career with the 187th and the 101st Airborne, he pulled me aside and said, "Matz, you've got the best battalion in the Army. This is my battalion, and you better not screw it up—I'll be watching closely."

"Yes, sir," I responded. Tiger meant it, and his veiled warning that day stayed with me through my 20 months of command.

The Rakkasans were one of three battalions in the division's 3rd Brigade. CSM Alfred B. Walker was my command sergeant major. He had been with the Rakkasans for years and knew the battalion well. I was told I could select my own CSM, but I kept Walker. I liked him, and his heart and soul were with the Rakkasans. Colonel James Thompson, who assumed command a month earlier, was the brigade commander. He was a tall, athletically built North Carolinian who had been a scholarship quarterback on the University of Florida football team in the mid-1950s. He was all business and demanded absolute excellence in everything. He later commanded the 101st. Tolerance wasn't a strong trait of his. I read him quickly, adapted to his demanding, impatient leadership style, kept my distance and commanded my battalion.

The 101st, whose nickname was the "Screaming Eagles," was part of XVIII Airborne Corps, America's Strategic Army Corps, and its mission was "to deploy rapidly worldwide with priority to the NATO area." During the year prior, the personnel strength of the 101st declined appreciably causing the division's readiness condition to decline to C-2 level in July 1977, the month I assumed command. A unit's overall go-to-war readiness status was measured by a "C" rating, which ranged from C-1 (best) to C-5 (worst). Strategic Army Corps units were required to maintain C-1 readiness posture.

Strength and MOS (Military Occupational Specialty) shortages handicapped both training and maintenance and caused deterioration in wartime readiness. The commanding general Major General John Wickham's emphasis was on retention and one year later the Screaming Eagles moved from last in FORSCOM (Forces Command) among the divisions in reenlistment competitive standings to first in rankings for both first term and career reenlistment rates.

The program didn't sacrifice quality in the pursuit of record achievements. Quality indicators throughout the year remained high. Each month, every battalion commander briefed the division commander on his reenlistment rates. My "pucker" factor went up as I stood before the division and brigade commander and briefed my battalion statistics. The overall excellence of the program was recognized by the FORSCOM commander, General Robert M. Shoemaker, in a letter dated November 13, 1978.

The individual unit winners of the competition were 3rd Brigade for brigade-size units and 3/187 for the battalion level competition. My company commanders and NCOs were very proud of that achievement, as retaining quality soldiers was a major goal among Army units at the time. After Vietnam, the Army had to rebuild, and retaining quality personnel was paramount. A commander's performance was largely determined on how well his unit did on reenlistment ... and was one of Wickham's highest priorities. A commander dared not take his eye off reenlistment! The 3/187 continued to lead the division in battalion-level reenlistment during my time as commander.

During the period 1977–78, the All-Volunteer Army was under attack by some congressional critics. There was pressure from some quarters to open additional slots to be filled by female soldiers without hampering combat efficiency. 1978 saw a definite change in the utilization of female soldiers within the 101st, and women began to be assigned to infantry brigade headquarters. The first two women assigned to a brigade headquarters element in the 101st were assigned to the 3rd Brigade in March 1978. The FY 1978 Division Historical Supplement, dated July 23, 1979, stated, "The number of women assigned to infantry brigade and division artillery rose to 61 by the end of the year."

I found that the change in assignment policies for women in divisions created a need for a greater understanding of the role of women soldiers. The concept required some changes in policies and long-held attitudes, particularly among infantry soldiers. I was among those skeptics who held the belief that women don't belong in the infantry, and I firmly believe that today. While women perform admirably in the military services, "common sense" tells you there are "unintended consequences" of women serving in infantry platoons and companies.

The fiscal year 1978 was an important year for the division in several other ways. A particularly important event was the recognition of the Army Air Assault Badge by the Department of the Army. Wickham had campaigned hard for it. The recognition provided authority for recipients to wear the badge as a permanent part of their uniform rather than only when assigned to the 101st, placing the badge alongside such distinguished skill awards as the Parachute Badge. Since, at that time the 101st was the only unit authorized to award the badge, it stood as an Army-wide symbol of the dedication and unique skills of the 101st Airborne Division (Air Assault). Within days of assuming command, I enrolled in the Air Assault Course at Fort Campbell.

The rigorous, fast-paced training was known by some as the "11 toughest days" in the United States Army. Following a four-year desk job in the Pentagon, I wasn't in the best physical condition. I was hoping to have some time to get into shape before enrolling in the demanding course. Thompson had other ideas and wanted me to get the Air Assault Badge ASAP. "You can't command an air assault battalion if you're not air assault qualified," he said. "Bill, you've been through Airborne and Ranger School, you'll make it." He had just completed the course himself.

Air Assault School qualified soldiers to conduct airmobile and air assault helicopter operations, sling-load operations, and it taught rappelling from helicopters and fast-rope techniques. On the morning of graduation, each participant had to complete a 12-mile rucksack march in under three hours before receiving the coveted air assault wings.

The rucksack march was tough on my polio leg, but I made it. It was the first time since being in the Army that I can remember my leg feeling so weak and tired, hurting more than when I finished the final forced march in Ranger School. I was 24 years old then and 39 when I completed Air Assault School. Fifteen years made a difference, and I attribute the weakness of my leg and extreme tiredness partially to my older age. What I didn't know was that my anterior horn cells that fire the muscles in my leg were beginning to deteriorate due to the effects of polio.

A major training innovation began in 1977, called Emergency Deployment Readiness Exercise (EDRE). The EDRE concept was introduced by XVIII Airborne Corps and formed an important part of the division training program. Lieutenant General Volney Warner, the corps commander, saw the EDRE as his "number one training priority." The concept called for "no-notice" rapid deployment of a company or a battalion.

Wheels-up deployment time was 18 hours from alert notification, and the destination varied, depending on the mission selected. There were many scenarios and the Rakkasans were tested several times. A strong point of the EDRE concept was the requirement for continued interface between Army and Air Force units. That close Army-Air Force interoperability proved invaluable during actual hostilities in the years to come.

On Friday, September 15, 1978, General Warner visited Fort Campbell and joined the Rakkasans for a battalion run. "Certainly enjoyed the run with you and your unit last Friday morning. Keep up the good work!" he wrote in a letter dated September 20. He had a gifted way of sizing up a unit's morale, esprit and readiness after spending only a few minutes with the commander and running with the troops. I had the pleasure of joining him and others for his final parachute jump on Sicily DZ before he retired at Fort Bragg in August 1981.

During October 1977, United Artists Films released the film *A Bridge Too Far* based on the novel of the same name by Cornelius Ryan. This was the third major film to focus on the World War II history of the Screaming Eagles. Division leadership encouraged all commanders and noncommissioned officers to see the film and eventually had it shown on post. *Battleground*, an MGM release, told the story of the 101st at Bastogne, Belgium, while the film *The Longest Day* (also from a book by Cornelius Ryan) covered the airborne operation at Normandy on D-Day 1944.

During those years, emphasis was given to active Army units supporting and training Army Reserve, National Guard and ROTC units. Coordination among all these components was collectively called the "Total Army," and that concept prevailed in the post-Vietnam Army. During the period June 10–25, 1978, the Rakkasans deployed to Fort Drum, New York, to train elements of the 28th Infantry Division, Pennsylvania Army National Guard and the 187th Infantry Brigade (Separate), U.S. Army Reserve.

The Rakkasan soldiers and commanders showed enthusiasm and expertise in training the reserve units. I enjoyed the two weeks working with the Pennsylvania guardsmen. Being a native Pennsylvanian made it extra special for me. The 28th ID officers liked the fact a "hometown" boy trained their units and soldiers. They treated me to more Yuengling and Rolling Rock beer than I could ever have imagined and gave me a case of each to take with me back to Fort Campbell.

Major General Fletcher C. Booker, 28th ID commander, wrote to General John N. Brandenburg, who was commanding general of the 101st, on June 24, 1978: "The division as a whole advanced further during this annual training period than in any other AT I've attended in my 37 years of commissioned service." He went on to praise the enthusiasm of every Rakkasan "private." The officers and men of the battalion felt good about the deployment and had given 101 percent effort in training our Army reserve partners!

At that time, a battalion command tour was 18 months, and I was scheduled to leave command on January 16, 1978. Colonel Thompson wanted me to stay longer to lead the 1,000-man task force from the brigade in a Battalion Combat Training (BCT) rotation followed by Joint Readiness Training Exercise (JRTX) *Jack Frost* to be conducted in Alaska from January 15–February 9, 1979. Brandenburg agreed, and a Request for Extension of Command Tour was made through FORSCOM, Headquarters to Commander U.S. Army Military Personnel Center.

The request was approved, and I took the task force, spearheaded by 3/187, to Alaska for JRTX *Jack Frost*. Included in the task force were aviation, field artillery and engineer units. My XO, Major Roy M. Capozzi, did an outstanding job planning for the administrative and logistical challenges we would face.

Task Force Rakkasan consisted of almost 1,100 men, deployed to augment the Alaskan U.S. Army forces in a joint Arctic-weather field exercise. The exercise was conducted to enhance combat readiness in extreme climates and included Arctic survival skills and mountaineering techniques unavailable to soldiers at Fort Campbell.

The JRTX began with preparatory instruction on Arctic survival. We were issued a complete set of Arctic field uniforms, including white thermal boots and other equipment, and we received instruction on the severity and danger of the Arctic climate. It took a couple of fittings of my special polio arch into the thermal boot before I felt comfortable enough to walk in it.

The host unit at Fort Wainwright, the Army's Arctic Brigade, the 172nd Light Infantry Brigade, and the Army's Northern Warfare Training Center (NWTC) at Fort Greeley, provided instruction. Arctic weather skills, familiarization with the Akhio sled, instruction on the wearing and use of snowshoes and skis completed the preparatory training. We then moved to the field to exercise forces in attack and defense of key Alaskan installations and conduct airborne/air assault infantry operations in Arctic winter conditions. The 3/187, during a force-on-force training exercise, spearheaded the "aggressor" force, which included an infantry battalion from the 9th ID and 3rd Battalion, 7th Marine Regiment, in their mission to seize and hold the Alaskan pipeline. The friendly force was composed of the Alaskan 172nd Light Infantry Brigade.

I had never been exposed to such harsh and dangerous weather conditions as on *Jack Frost*. We operated in the field for 12 consecutive days in temperatures of minus 20 degrees Fahrenheit. We were in 3–4 feet of snow and endured gusting winds up to 30 mph. Rakkasan soldiers learned what was required not only to survive in such frigid temperatures but also to operate tactically against a live aggressor force. It pushed us to extremes we had never experienced before.

Before leaving for the exercise, I had received my warning order: "I don't want to hear of any cases of frostbite. Watch your soldiers closely. I'll be up to visit you. Good luck," was the farewell message from my steely-eyed brigade commander, Colonel Thompson, as we departed Fort Campbell for Fort Wainwright. Despite our best efforts and leaders constantly checking their soldiers, seven Rakkasans got frostbite. We evacuated them quickly by helicopter to Bassett Army Hospital on Fort

Wainwright, which was well staffed with medical specialists experienced in Arctic weather injuries. The cases were first- and second-degree frostbite, frostbitten toes, fingers, and noses. A couple were serious, but only one injury resulted in permanent tissue loss.

Black soldiers were significantly more susceptible to frostbite than white soldiers. My battalion surgeon, Captain Ricardo Davila, in his February 8, 1979 medical after-action report, attributed the "relatively few cases of frostbite to early identification of 'susceptibles' and strong command emphasis placed upon prevention." I never stopped worrying about a soldier getting frostbite. It was our worst enemy!

Arctic operations have inherent risks that are often difficult to control. In an article addressing injuries among soldiers, the *Military Medicine Journal*, Vol. 162, December 1997, stated, "As the wind-chill factor decreases to less than minus 20 degrees F., the risk of frostbite goes up exponentially, and even small errors of judgment can quickly cause injury."

On the last night of the force-on-force exercise, we made a night attack. While consolidating on the objective, a ground-burst flash artillery simulator detonated inches from me. The carbon-rich particles blew into the freezing night air with a horrific loud explosion accompanied by a bright flash and cloud of white smoke.

"What in the hell was that?" I screamed.

The blast rocked my feet and legs. I couldn't hear or see anything for a few moments. Major Dave Ohle, my S3 officer, was close by and led me away from the blast area. With my ears ringing and blurry vision, grabbing onto his gloved hand and over-white garment was a lifesaver. The sulfuric smell alone was enough to cause gagging and sickness. Dave was unflappable, an infantry soldier and veteran of two combat tours in the Vietnam War. He was eventually promoted to lieutenant general and became the Army's deputy chief of staff for personnel.

Fort Campbell also hosted a unit exchange exercise with the British Army's First Battalion of the Parachute Regiment (1 Para). Colonel Thompson called me into his office and said, "General Brandenburg and I want you and the Rakkasans to be the host battalion for the Brits."

Exercise *Hickory Stick* brought 650 British paratroopers to train with the 101st. 1 Para had a history of common ties with both the 82nd and 101st, dating back to World War II. Working with the British soldiers was a hoot. The challenge came in understanding their English Cockney accent: the Cockney speakers had a distinctive dialect and accent and occasionally used rhyming slang. My troops enjoyed it as they tried to understand their British counterparts.

Lieutenant Colonel David M. G. Charles, their commanding officer and I hit it off right from the go. We bonded well. He and his troops wanted to mostly do live-fire exercises, as live ammunition was scarce in England. I spent many hours with the British soldiers on the small-unit maneuver ranges going through platoon and company live-fire exercises.

The Brits were not bashful and spent their free time scouring the Tennessee-Kentucky countryside for pubs and girls. The local Clarksville and Hopkinsville people loved the British and opened their homes to them. When it came time to return to England, some didn't want to leave. Lieutenant Colonel Charles, his commanders and NCOs had a time rounding them up for their return flight to England. As I recall, 1 Para returned to England minus at least one soldier. Days later, we found he had eloped with a young lady from Tennessee. I never did learn if he ever returned to England, or if he is still AWOL somewhere in America.

On November 24, my dad called to tell me my mother died. She had been suffering for over a year and died from Creutzfeldt-Jakob disease—an exceedingly rare disease that progresses rapidly and results in degenerative brain disorder and death. Losing my mother was a terrible blow to me. I handled it well at the time, but years after her loss I continue to feel sadness, grief, and sometimes guilt.

When I remember all she did for me, coping with my polio years of rehabilitation and the many sacrifices she made for me, I have remorse and regret. I am sorry I didn't do more for her and spend more time showing her how much I loved and appreciated her. I was always too occupied, too busy with my career. Losing my mother has been one of the most difficult experiences of my life. If I could ever relive a single moment in my life, it would be to tell my mother how much she meant to me. I still carry with me a scar of regret and guilt.

My boss, Colonel Thompson, and his wife Pat were most gracious and helpful in offering to take care of our two daughters, ages eight and nine, while Linda and I flew home to Pennsylvania for the funeral. Our son, Bill, stayed with the family of a friend from his swim team. Army families are a wonderful, caring, and resilient group, and always there for you in time of need!

Linda had just turned 33 when I assumed command of the Rakkasans in July 1977. Although we had been married 11 years, it was the first time she experienced being the commander's wife and living on a division-size post with hundreds of companies, battalions, and brigades. Much is expected of a commander's wife in the Army, particularly if the command is in a high-tempo division like the 101st where units are constantly deploying. The younger wives look to the commander's wife for leadership and support, especially when their husbands are deployed. In those years, it was expected that the commander's wife would be there for them, like a mother is for her child.

At my change of command ceremony, Linda confessed to the brigade commander's wife that she had never been with a unit before, let alone being the leader of the wives' activities in the battalion. Pat Thompson, a lovely southern lady from Georgia, who looked like Nancy Reagan, quickly responded that she need not worry—all she had to do was "care."

Two days after my change of command ceremony, a wives' coffee was held to bid farewell to Mrs. Honeycutt, the wife of the ADC for Operations, who was departing.

Pat called Linda and mentioned that her battalion wives had not responded to attend the farewell and suggested she give them a call to remind them of the event. Having hardly had time to even meet the wives, Linda made the calls.

While in the receiving line to convey her farewell wishes, she overheard someone asking a Rakkasan wife if they were pleased with their new commander. The wife responded, "I don't know about him, but his wife had the nerve to call me and suggest I attend this event. I should give her my babysitting bill for the month." It is Army tradition to hold farewell events for departing leaders and their wives. June and July had been busy with many changes of command requiring dozens of socials and ceremonies to attend. The next time Pat called to ask Linda to remind her battalion ladies of an event, she refused. From then on, Pat referred to her as "the damn Yankee" in her good-natured way of teasing.

In February 1979, I relinquished command of 3/187 to Lieutenant Colonel Edward S. Broderick. I spent my last four months in the 101st as the division's deputy chief of staff. I had been selected to attend the U.S. Army War College (AWC) in Carlisle, Pennsylvania, and had a reporting date of August 3.

We were thrilled that we would spend a year at historic Carlisle Barracks, the U.S. Army's first educational institution. A friend in the current War College class wrote, "Dear Bill & Linda, Congratulations on the news of your assignment to the hotbed of tranquility."

Indeed, it was a great year, a relaxing year quite different from the rigorous, fast-paced life in an operational unit. It gave me an opportunity to deepen my understanding of both self and profession, as well as broaden my perspectives, think anew, question and challenge. It also gave me a chance to rest my polio leg from the fast-paced physicality of an operational infantry unit. The commandant, Major General Dewitt C. Smith, and his wife and the faculty made it a warm and pleasurable experience.

I co-chaired the college's Light Infantry Study, a Department of the Army project being conducted at the war college. It contributed heavily to the development of Army force structure in the 1980s. Among my extra-curricular activities, I was the catcher for our AWC softball team that won the championship at the All-Senior Service School Championship Games in May. Our kids swam competitively on the Carlisle YMCA aquatic team. Linda substitute taught in the Carlisle and Boiling Springs school districts. Indeed, it was a fulfilling year for all.

The overall purpose of the college, as stated by the institution's founder, Secretary of War Elihu Root,[1] "not to promote war, but to preserve peace" stands the test of time. On June 9, I graduated with 228 classmates, including 16 international officers. Friendships and associations made with fellow officers were the most rewarding and lasting part of the year. The relationships served me and the Army well in my following years of service, and I still correspond with a few of my AWC classmates. Despite the "tranquil" and leisurely academic pace, I was more than ready to get back with troops, and the challenges and excitement of an operational unit. I missed it more each day and was longing for the day I would return.

Return to the 82nd Airborne Division

Assignments were announced in April. I would return to the 82nd Airborne as the G3 (operations officer). I was elated! There were probably 10 other lieutenant colonels in our class who wanted that job. Returning to Fort Bragg after 15 years and serving with the "All-American" Division was a dream come true. I couldn't wait to don the airborne maroon beret, emblematic of paratroop forces worldwide. In longhand at the bottom of his welcome letter, Major General Guy S. "Sandy" Meloy III, commanding general, 82nd Airborne Division, wrote, "Bill, other than battalion command, you're getting the best Lieutenant Colonel job in the Army—we are delighted you're coming." General Meloy had been a paratrooper most of his Army career and had commanded two battalions in Vietnam.

We arrived at Fort Bragg in mid-July 1980. The G3 section was well staffed with outstanding officers and NCOs. They brought me up to speed quickly on the pressing issues, and I made it a priority to meet and visit the brigade, division artillery, and support command commanders and their S3s during my first week.

I soon found out the CG's priority. General Meloy had been working on a major revision to the division's Airborne Standard Operating Procedure (ASOP), and he wanted me as a set of fresh eyes to review and edit the revision prior to its publication. He told me it was my priority. "G3, I have been working on this ASOP for well over a year. Many paratroopers have contributed to it. Work with the Chief on the final review. I want it published ASAP," he said as he handed me the 3-inch-thick draft manuscript.

The front and back sides of each page were filled with drawings, diagrams, tables, and narrative. Throughout were statements in bold print on how to beat "Murphy's Law" (if anything can go wrong, it will). It was a rich compendium of airborne knowledge and gave you everything you needed to plan, organize and conduct safe and efficient parachute jumps and airborne operations. The ASOP was the division's Bible.

The "Chief" the CG referred to was the division chief of staff, Colonel Arthur C. Stang. He and I spent night and day poring over every page, every diagram, and every table for the next two weeks. Every edit, every change, had to be approved by Meloy. This was his work. His cover letter forwarding the ASOP to every officer, NCO and jumpmaster read in part, "I expect each of you to know and follow every procedure described in this ASOP. No exceptions and no short cuts."

On September 1, 1980, the ASOP was published and distributed. The next day, September 2, Colonel Stang dropped dead of cardio-respiratory arrest. He had been running on Ardennes Street with a unit during morning PT and was giving the troopers a well-deserved "atta-boy" when he collapsed. I ran with the G3 section on Ardennes that same morning and passed his unit moments before he fell. His death was a shock and a terrible loss to the division, especially the CG. In his eulogy at the memorial service, General Meloy said, "I knew, admired, and respected Art Stang for many years, he was my Chief of Staff, my right arm. He was my friend."

Colonel Raphael Hallada, an artilleryman, replaced Colonel Stang as chief of staff. I worked closely with him in reviewing Volume II of the ASOP which covered joint airborne operations planning. Meloy felt the Army's academic institutions, especially Fort Benning, were not properly teaching airborne operations and that their doctrine was outdated. He was right! Unlike Volume I, which was a revision to an existing document, Volume II was a completely new document. With its publishing on January 30, 1981, Meloy's two-year effort to prepare, in his words, "the most complete source of airborne knowledge in the Army—there is nothing to compare with it in existence today," was finished.

He was extremely proud of this accomplishment and heralded it as his primary achievement as CG, 82nd Airborne. He barked when something wasn't right and demanded that the ASOP be followed to a T when conducting parachute operations. God help the commander or jumpmaster who didn't! But under his tough, crusty veneer was a softer, down-to-earth Texan. General Meloy had empathy for others in all stations of life, was deeply loyal to his friends and enjoyed a joke on himself, being the first to laugh. I was fortunate to have been the division's operations officer for eight months under his command.

On February 6, 1981, Major General James J. Lindsay assumed command of the 82nd, his ninth assignment with the All-Americans! He and General Meloy had been infantry brigade commanders together in the division several years earlier. General Lindsay would serve later as commanding general XVIII Airborne Corps, and then as the first commander in chief of the United States Special Operations Command, retiring as a four-star general. I first met Lindsay during the Cuban Missile Crisis in 1962, when he was a seasoned captain working in the Division G3 section, and I was a brand-new second lieutenant.

Equally savvy when it came to airborne operations, he was a rugged, physically active commander. He was rarely in his office. Colonel Mike Plummer, the new division chief of staff, and I suggested he get out of the office and maximize his time with the units. He did PT and ran with a different unit every morning, jumped with a different company or battalion every week and was always at training. He learned which commanders and units were conducting PT and those who were not. Meloy had been nursing a longstanding injury from a parachute jump and wasn't able to do PT for much of his last year in command.

It was good for the troopers to see their commanding general hooking up for a parachute jump, running with units on the street, and in the field with them at training. General Lindsay knew every inch of every DZ and every training area at Fort Bragg. With the new burst of "get out of the office" physical energy from the CG, the division's emphasis turned to physical training and fly-away exercises (training exercises conducted at locations other than Fort Bragg). I emphasized their need during my first G3 briefing with him, particularly the need to revitalize our EDRE program.

On June 29, 1981, the CG sent a letter, subject: Physical Fitness, to all 82nd Airborne Division leaders. He wrote, "unit physical fitness programs don't challenge the Troopers." He continued, "A daily diet of the airborne shuffle won't improve the condition of your Troopers." The letter challenged leaders to improve their physical fitness training programs. Within weeks, there was noted improvement in both the physical condition and esprit of troopers and units. And the chief of staff, Colonel Plummer, was a physical fitness enthusiast and ensured all staff sections had good PT programs. He was the only person I ever met who could drink beer all night and run a marathon the next day. Mike was, indeed, hard-core and a distance runner.

General Volney Warner, whom I had worked with in 1978 at Fort Campbell, retired from the Army on July 31, 1981, as commander in chief, U.S. Readiness Command. He requested his retirement ceremony be conducted on Sicily DZ at Fort Bragg. On that day, the G3 section arranged for his final parachute jump. I was privileged to be among those he invited to hook up one last time with him in his "final exit." In an August 5, 1981 note he wrote, "To hook up one last time with friends was great. Form is everything when making your final exit. Many thanks for accompanying me on mine." Warner embodied the airborne spirit in a humble, modest, way that made me grateful for the opportunity to have served with America's best—the 82nd trooper!

The following year was like a marathon—preparing for and executing a very ambitious training and exercise schedule. There was no such thing as the 82nd ever being too ready! My philosophy was that we must train and operate on the basis that we are not ready enough. Division and corps EDREs were key to that philosophy. The program had suffered somewhat due to too many less important competing requirements. It needed revitalization. General Lindsay set the goal of a minimum of one battalion-size EDRE every quarter.

We also took a hard look at what exercises we participated in. XVIII Airborne Corps, FORSCOM, and REDCOM (Readiness Command) all wanted to troop list us in their exercises. Some provided little payback in terms of both training and readiness. We pushed back when we saw no value. In my G3 End of Tour Report, dated May 25, 1982, I wrote, "Division must be careful not to get locked into participating in exercises that provide little value."

I pushed for more live-fire training and more range time. I recalled my time as a battalion commander in the 101st two years earlier and our emphasis on live-fire training. We were constantly doing platoon and company live-fire exercises and scoring them. Soldiers could fire blank ammunition all day in "cold" training exercises, but until they moved and shot in "hot" live-fire exercises and experienced firsthand the loud and cracking sound of a bullet, the muzzle flash, the sulfuric smell of wafting gunpowder smoke, then their training wasn't complete. Indeed, like the uptick in unit PT, the training and exercise operational tempo (OPTEMPO) picked-up considerably. My head spins today just thinking about how busy we were and how we had to constantly balance our many requirements and events while always knowing a call could come at any time to fly away on a real-world contingency.

The entire G3 section was busier than ever supporting all facets of the division's training and operations. I don't remember ever being busier as a staff officer and enjoying it so much. Not all the G3 section agreed with me. "Sir, we are doing too much, and the PT is killing me," one of my majors said one morning. As someone nursing a weak leg from polio, I wasn't a bit sympathetic. The G3 section needed to set the pace and establish the standard for the division staff in terms of work ethic, attitude, and physical training. And we did. And my polio leg paid a price from the wear and tear and overuse as a result of this high OPTEMPO.

One morning I got a call from Brigadier General Bobby Porter, the assistant division commander for Support (ADC-S). Porter would later command the 82nd. "G3, come see me." Upon arrival, he asked, "Are you aware of the division policy to maintain in reserve a battalion task-force worth of parachutes for a real-world contingency?" I stood listening in front of his desk. He had just reviewed the "rigger's" report on the status of parachutes in the division and found we had far exceeded the number allotted for training jumps to the extent that we had used parachutes from the real-world contingency stocks. This was unacceptable!

We were about 300 parachutes short of our required contingency stock. "At the rate you're going, G3, we'll be at zero balance in a few more days," he said in his wry, low-key, almost sarcastic manner. He was mad! "Fix it," he said, "and while you're at it, goddammit learn to count."

The "riggers," airborne troopers specially trained to pack and maintain parachutes, worked overtime for a few days packing parachutes and I had to cancel some training jumps. After that ass-chewing, I paid close attention and had a parachute status report given to me every week. As the All-American G3, I had egg on my face, and learned a good lesson!

Airborne Exercises

Exercise *Bright Star* was a combined and joint training exercise led by United States and Egyptian forces each year in Egypt. An outcome of the improved U.S.-Egyptian relationship that contributed to the 1978 Camp David Accords, its purpose was to "strengthen ties between Egyptian Armed Forces and United States Readiness Command's (REDCOM) newly established Rapid Deployment Joint Task Force

(RDJTF), and to demonstrate and enhance the ability of American forces to reinforce their allies in the Middle East in the event of War."

The 82nd spearheaded the *Bright Star* 82 exercise conducted November 9–24 1981. We planned for it for weeks. The November 15 edition of the *New York Times,* in an article entitled "800 U.S. Paratroopers Open War Games in Egypt," billed it as the "largest American military exercise in the Middle East since World War II."[2] *Bright Star* 82 was the most ambitious U.S. demonstration yet of the RDJTF's ability to deploy its strike force in the volatile Middle East.

What made it even more newsworthy was the fact the operation was conducted as Egypt officially ended its mourning period for President Muhammad Anwar el-Sadat,[3] who was assassinated days earlier at a military parade in Cairo. The White House and REDCOM were quick to insist that the exercise had been planned months before Sadat's assassination, despite statements by Secretary of State Alexander M. Haig, that it had been "greatly expanded because of Sadat's killing a month earlier."[4] The deployment of over 4,000 U.S. troops dwarfed the exercise held in Egypt the previous year, which involved only 1,400 U.S. troops and minimal equipment.

The Air Force dramatically demonstrated its ability to put the "rapid" in the Rapid Deployment Force (RDF) by airlifting 360,000 pounds of equipment and 839 paratroopers from the U.S. to Egypt in 16 hours. Our battalion task force boarded six C-141 Starlifter aircraft at Pope AFB in North Carolina, flew non-stop, requiring inflight formation refueling by KC-135 Stratotankers, and dropped our troopers and their equipment in a vast desert wasteland west of Cairo, Egypt.[5]

The parachute assault was preceded by six A-10 Thunderbolt IIs, called Wart Hogs, making passes over a designated 3-mile-long zone. After the A-10s came the parachute drop of our heavy equipment, including vehicles and artillery pieces. Three minutes later, troopers from the 2nd Battalion, 504th Airborne Infantry (the "devils in baggy pants"), bailed out, and the sky was suddenly filled with paratroopers drifting toward earth.

The drop placed platoons and companies near their equipment. All the troopers landed in their designated DZs. We employed what is called a "pop and run" operation, where paratroopers snapped off their parachutes and left them on the ground as they assembled quickly and moved toward their objectives in the desert sands.

Sixteen hours sitting in canvas, side-facing seats in the cargo compartment of a C-141 is not the most pleasant way to fly. I was in the lead aircraft with the CG. Our pilot was Major General Thomas M. Sadler, commander, 21st Air Force flying the lead aircraft. The Air Force had a lot at stake with the non-stop transcontinental flight that would drop 800 paratroopers at a precise point in the Egyptian desert hours after takeoff. Sadler, whose combat missions spanned three wars, wanted to make sure the airdrop went well. And it did!

We inflight rigged, donning parachutes and equipment over the Mediterranean Sea, dropped to 800 feet altitude and flew the prescribed route to the DZ. The lead aircraft arrived at the leading edge of the DZ at precisely the right moment. The green light came on, and the CG jumped. I followed him out the door, and moments later we were on the DZ.[6] "You won't see a better drop than that!" Major Dennis Guyett of the Air Force said afterward.[7]

After 16 hours of formation flying, traversing parts of three continents, an ocean and sea, we arrived in the objective area, on time at the precise location ready to fight. Winds were negligible, and there were only a few minor injuries from the jump, a testament to the superb condition and attitude of the All-American trooper.

After checking our compass, we proceeded east on foot across the desert and miles later arrived at the Egyptian Army camp at Cairo West Air Base. After all those hours of pre-flight staging at Pope AFB and flying across the Atlantic, I couldn't wait to jump and leave that airplane. The warm sunshine of the desert felt good, and my polio leg held up well as we set a good pace on foot toward the air base.

The airborne assault was followed by two weeks of combined and joint training with the Egyptian airborne units and with a mechanized battalion task force from the U.S. 24th ID which had arrived by sea at the port of Alexandria days before the exercise began. At night, I scraped the reddish-yellow sand from my skin and field uniform and welcomed a warm shower if I could get one. At the end of the exercise, most troopers got a pass to Cairo to see something more than sand and blue sky. "I'm enjoying it," said Private First Class Reuben Lantigua, a paratrooper from Long Island City. "All my life I wanted to see Egypt," he told a reporter.[8]

Bright Star exercises with Egypt continue today, occurring about every two years and including other African and Middle Eastern countries. There are always lessons learned when conducting an exercise of that scale and magnitude. *Bright Star* 82 paved the way for future joint and combined rapid deployment exercises and set a high standard of performance for Army and Air Force RDF units.

On December 5, 1981, during a night jump on a division EDRE with a unit from the 2nd Brigade, I fractured my right calcaneus (heel) bone and partially cracked the left calcaneus bone. It was a moonless night when we parachuted into a farmer's field in Georgia near Fort Stewart, impossible to see a thing. I landed hard with both heels hitting the top ridge of a plowed furrow.

I felt pain immediately and knew something was wrong. The CG was two jumpers ahead of me. "G3 is that you?" came a call in the dark. I told him I had a hard landing, but I didn't want him to know I was in pain as I gathered my parachute. We walked off the DZ. I was walking on my tiptoes, as even the slightest weight on my heels produced pain. I felt like a ballet dancer, dancing on my toes trying to keep up with General Lindsay. I was too macho to admit I was hurt.

By daybreak, I was in such pain that I finally gave in. "Sir, my heels are killing me; I've got to see a doc," I said to the CG. I was flown by helicopter to Womack Army Hospital, Fort Bragg. Dr. Jackson, the on-call orthopedic doctor, took X-rays and admitted me to the hospital immediately. I spent three days in the hospital with both legs extended in skin traction. The next day, as the doctor was checking me, one of the residency doctors accompanying him on his rounds asked, "Are you the Bill Matz that went to Lansdowne-Aldan High School?"

It was Captain William Meade. Willie and I were on the high school football team together and he recognized my name. What a small world! I was released from the hospital on December 8 and walked with a mid-calf boot cast on my right leg until it healed. The left leg didn't require a cast. I was worried as the worst break was to my right heel, which was my polio leg. Both heel bones recovered well with no residual complications.

There was no reduction in either the number or intensity of fly-away exercises for the division. The new year, 1982, had the division troop listed for participation in several major REDCOM and RDJTF exercises. Although we were at peace, it was a busy Christmas period for the G3 section. I spent half of Christmas Day at the office much to Linda's chagrin!

After participating in REDCOM Command Post Exercise (CPX) *Gallant Knight* 82 in January–February at Fort Bragg, the division set its sights on *Gallant Eagle* 82, a two-week field training exercise (FTX) designed to test the ability of the RDJTF to move quickly and operate in a high desert terrain against a heavy force.

We had developed and had been training on a new airborne anti-armor defense concept and wanted to test it in a desert environment against a live opposing force. As part of the exercise held at the National Training Center, Fort Irwin, California, we executed a mass tactical airborne operation at daybreak on March 30, beginning the first phase of *Gallant Eagle* 82.

Heavy equipment loads were dropped minutes before 1,780 paratroopers parachuted onto five DZs designated Rock, Stone, Gold, Silver, and Nelson. The DZs were separated by miles of rough desert terrain, and each was 2 kilometers wide by 6 kilometers long. The sun had just risen when the light turned green, and the jumpmaster's resounding shout, "Go!" pierced the wind and noise. I was in the lead aircraft with the CG. He was the number one jumper in the left door; I was number one jumper in the right door. We jumped on DZ Nelson.[9]

We were delivered by 16 C-141B aircraft after an overnight transcontinental flight from Pope AFB and by 12 C-130 aircraft from an intermediate staging base (ISB) at March AFB, California. Like the *Bright Star* deployment four months earlier, I was glad to exit the aircraft. The sky was filled with paratroopers as we descended to the desert floor below. The airborne operation took 40 minutes, from 0530 to 0610. It was the largest military airdrop at the time since World War II and would prove to be one of the most disastrous!

The airborne operation ultimately resulted in at least 158 injuries, including six deaths. One death was from a parachute malfunction. I landed hard on the rough, rocky, desert floor. I absorbed the initial impact of landing on my left leg, protecting my weaker leg from the hard impact of the ground (as I always did). Other than a few bruises and a minor abrasion to my right hand, I was fine. Many other troopers were not so fortunate. On the DZs powerful updrafts blew dozens of paratroopers off course and slammed them into the ground. One trooper crashed into a vehicle and was killed.

The wind dragged other troopers, sometimes head over heels, across the hard rocky terrain and prickly sagebrush when they were unable to pop their safety catches to release their chutes. Specialist 4 Daniel Maynard, 24, of New York City, who suffered a fractured pelvis, said, "I hit the ground, rolled about three times and started to pass out." In addition, 54 percent of the casualties and all the deaths occurred on DZ Silver where the wind gusts were the worst. Four troopers died on DZ Silver soon after impact due to head trauma and massive internal injuries. Two others died later of their injuries in nearby hospitals.

Of the 158 troopers treated on the day of the operation, 22 were treated and returned to duty, four were dead on arrival and 132 were admitted to Army and local civilian hospitals.[10] The division, corps, and other Army unit medical personnel did an outstanding job managing such a mass casualty situation. Not only did they triage, treat, and evacuate the many casualties from the five DZs, but they also provided accurate and timely information to the CG, the field commanders and the media.

The *Military Medicine Journal* stated, "The type of injuries supports the finding that the paratroopers made extremely hard landings and that extensive dragging occurred after landing due to higher than detected wind velocity ... 52 injuries (33 percent) were disabling to the extent that the individual required more than 7 days for recovery."[11]

Numbers showed that 8.9 percent of the jumpers were hurt, while an injury rate of 1 percent is considered normal in an exercise of that size. We were mystified. How could it have happened? Major John Dye, the division spokesman, when interviewed, said that two DZ safety officers had measured the wind at ground level and found it to be under the maximum 15 miles-per-hour allowable for training jumps before they gave the go-ahead to jump. Dye said the division averaged about 110,000 parachute jumps a year, "and we've never come into a situation like this before. Winds are our worst enemy."[12]

Some theorized that winds had careened off a nearby range of low mountains and swept back across the large open area of the DZs, creating crosscurrents, updrafts, and unusual turbulence. People who lived in the desert spoke of the phenomenon. We had never seen it before. One grim lesson learned was that more complete wind measurements would be taken for future mass tactical jumps before the go-ahead signal was given.

On the last day, before we left Fort Irwin for our return to Fort Bragg, Mike Plummer and I visited our remaining troopers still recovering at Loma Linda University Medical Center in San Bernardino.

Easter Sunday with parents and sisters, Becky (next to me) and Diane, c. 1949. (Private collection)

My mother was very proud of her two-year-old son's Honorable Mention award in the national contest. (Private collection)

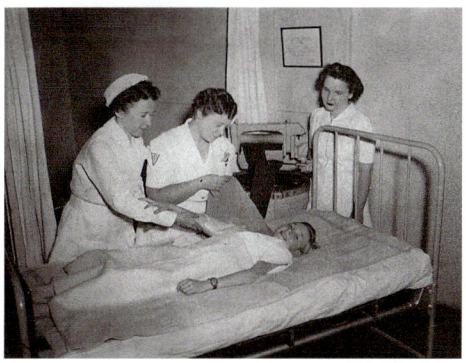

Receiving Sister Kenny moist hot pack therapy for polio at Home of the Merciful Savior for Crippled Children Hospital in 1944. (Courtesy of Home of the Merciful Savior for Crippled Children)

Fellow polio patients in hospital play yard, 1944–45. (Courtesy of the Home of the Merciful Savior for Crippled Children Hospital)

Riding my tricycle wearing orthopedic shoes with ankle-foot brace in 4th of July parade c. 1948. (Private collection)

High School football team seniors. I'm in the second row, on the left (Jersey #40). (LAHS)

Our 1960 Gettysburg College lacrosse team. I am seated fourth from the left in the second row (Jersey #76). (Gettysburg College Spectrum)

Newly commissioned Second Lieutenant Bill Matz with my proud parents and sister Becky, at Gettysburg College ROTC commissioning ceremony in 1961. (Private collection)

Standing in our Bay Garvey prior to its maiden voyage in Great Egg Harbor Bay in June 1961. Left to right: my business partner and fraternity brother, Joe Baily; myself and Joe's brother, Tom. My first attempt at entrepreneurship failed! (Atlantic City Press)

Attending Airborne (Jump) School in 1962. (82nd Airborne Division)

With my squad of West Pointers in Ranger School at Florida Ranger Camp. I'm in the center, kneeling. (USAIS)

My platoon leaders when I commanded B Company, 2/8th Cavalry on the DMZ in Korea in 1965. Left to right: Bob Balderson, Bob Lauridson, myself, Jim Tarkenton, and Bill Duncan. Two years later, Tarkenton was killed in combat in Vietnam and Balderson, Duncan and I were wounded. (Private collection)

With the reflagging of the First Cavalry Division to the 2nd Infantry Division in 1965 in Korea, my company became B Company, 2/23rd Infantry. (Private collection)

Our wedding in July 1966. (James E. Morgan)

Lieutenant Bill Matz, 82nd Airborne Division, 1964. Linda showed this photo to our baby son every night while I was in Vietnam. He knew his dad when I came home! (Private collection)

Armed with an AR-15, manning a patrol base in Long Dinh District, Vietnam, Christmas Day 1967. Lieutenant Jim Hasselmann, my field artillery forward observer, is to my left. Pant legs are unbloused to allow drying and fresh air circulation. (Private collection)

With one of my Vietnamese *chieu hoi* Tiger Scouts. Ten months earlier he was fighting with a Viet Cong (VC) unit. (Private collection)

3/47th Infantry commanders, Dong Tam, Vietnam, 1967. Left to right: Captain Ray Sanders, A Company; Captain Craig Boice, B Company; Lieutenant Colonel George Bland, battalion commander; Captain S. Santiago, Headquarters Company; Captain Al Sabitsch, D Company; and myself, C Company. (Private collection)

General Creighton Abrams, commander US Military Assistance Command, Vietnam, presenting me with the Distinguished Service Cross, September 1968. (US Army photograph, 9th Infantry Division).

Receiving the Silver Star from the 9th Infantry Division commanding general, Major General George O'Connor in 1967. (US Army photograph, 9th Infantry Division)

A captured VC weapons/ammunition cache in the Delta, 1967. (9th Infantry Division)

Charlie Company infantrymen moving through a rice paddy in 1968. Note the heavy nipa palm along the banks. (9th Infantry Division)

Me and my RTO crouching for cover during battle in Vinh Long in February 1968. My RTO was killed three days after this photo was taken. (Photo by 9th ID Signal Battalion Combat Photography Team)

One of my tunnel rats, PFC Felix Rios-Ramos, moving through waist-deep water. (*Octofoil* Magazine Jan–Feb 1968)

Charlie Company infantrymen crossing a stream in the Mekong Delta. (9th Infantry Division)

Charlie Company infantrymen crossing shoulder-high water in the Mekong Delta. (9th Infantry Division)

9th Division infantrymen clearing out enemy in My Tho during the Tet Offensive in 1968. (Photo by 9th Infantry Division Combat Photography Team)

9th Division infantrymen assisting wounded comrade in My Tho during the Tet Offensive 1968. (Photo by 9th Infantry Division Combat Photography Team)

Enjoying a cold 3.2 percent beer with my Lieutenant platoon leaders after an operation in Dinh Tuong Province in 1968. Left to right: Keith Meitz, Bill Waugh, myself, Dick Evans, and FNU Harrison. Evans and I were wounded several days later during Tet. (Private collection)

Reuniting with family upon my return home from Vietnam, October 1968. (Private collection)

As underway officer of the deck, manning phones on the bridge of USS *Paul Revere*, in WESTPAC 1970. (US Navy)

Rear Admiral Rubel presenting me the Navy Commendation Medal for meritorious service while assigned with Amphibious Ready Group Three during deployment to WESTPAC in 1970–71. (US Navy)

Middlebury College ROTC Detachment officers in 1969–70. Left to right seated: Captain Paul Pietrzak, Captain Larry Spohn; standing: Lieutenant Colonel Jim Hefti, Major John Daily, myself, Major Herb Koenigsbauer. (Private collection)

Lieutenant Colonel Jim and Sally Hefti present us with the traditional baby cup at the birth of our daughter, Rebecca, in January 1970. This was our second baby cup while at Middlebury and Hefti warned "we are running out of cups." (Private collection)

On the beach with the kids, Billy, Heather, and Rebecca in Ocean City, NJ, 1972. (Private collection)

Our Chrysler with a crumpled front end after hitting the horse mule in Arizona. The windshield and engine parts were replaced, but the hood, grill, and other body parts had to be tied together as we limped home to New Jersey, 1973. (Private collection)

Family while attending Command and General Staff College, Fort Leavenworth, KS, 1973. (CGSC Yearbook)

Major General Wallace Nutting and Linda pin on silver oak leaves at my promotion to lieutenant colonel in the Pentagon in 1976. (ODCSOPS)

Third Brigade, 101st Airborne Division commanders. Left to right: Lieutenant Colonel Chuck Westpheling (1/503 Infantry), myself (3/187 Infantry), Lieutenant Colonel Dan Campbell (2/503 Infantry) and Colonel Jim Thompson, brigade commander, 1978. (Private collection)

3rd Battalion, 187th Infantry (Rakkasans) passing in review during the 101st Airborne Division Annual Review in 1978. (101st Airborne Division)

3rd Battalion, 187th Infantry officers, Fort Campbell, 1978. Rakkasans all! (Private collection)

General Robert Shoemaker, commander, US Forces Command, presents the award for the highest reenlistment rate among combat battalions in Forces Command in 1978 to the Rakkasans. Joining me in accepting the award are Sergeant Grassi, battalion reenlistment NCO, and Command Sergeant Major Alfred Walker, battalion CSM. (101st Airborne Division)

Establishing communications after the parachute assault into the Sahara Desert in Egypt during Operation *Bright Star* in 1981 with the 82nd Airborne Division. (Private collection)

At center calling cadence for the division staff as Major General Jim Lindsay leads the 82nd Airborne Division in the Annual Division Review, May 1982 at Fort Bragg. (82nd Airborne Division)

82nd Airborne Division principal staff officers at holiday event, 1982. Left to right: myself, G3; Division comptroller; Lieutenant Colonel George Landis, G4; Lieutenant Colonel Jim Shepherd, G5; Colonel Mike Plummer, chief of staff; Lieutenant Colonel Jim Guest, G1; and Lieutenant Colonel Tony Lackey, G2. (Private collection)

Our family en route to Korea in August 1982 with our German Shepherd, Sariah. (Private collection)

CSM Buck Taylor passing the brigade colors to me as I assume command of the 4th Training Brigade at Fort Knox in July 1983. (Fort Knox PAO)

After presenting Secretary Weinberger with the Random House Dictionary upon his departure as secretary of defense in November 1987, he quipped, "I'll now have to pay more attention to what Random House says." (Courtesy of Office of Secretary of Defense)

Our family join Secretary Carlucci in his Pentagon office prior to hosting my promotion and farewell ceremony. Linda's father, John Heal, joins us, 1988. (Courtesy of Office of Secretary of Defense)

President Reagan wishing Linda and me a safe trip to Fort Ord, California. "You are beating me out there," he said with five months remaining in his presidency, 1988. (Courtesy of the White House)

Secretary Carlucci and Linda pinning on the star at my promotion to brigadier general in the Pentagon in 1988. (Courtesy of Office of Secretary of Defense)

Climbing the last hill on a morning run with the 7th Infantry Division Support Command on the Monterey Peninsula. Colonel Randy Poore, Support Command commander, is on the right. Note the muscle atrophy and thinness to my polio leg (right leg). (7th Infantry Division)

Major General Cavezza, Command Sergeant Major Mock, and I applaud troops as they finish a 4-mile division run in 1989. (7th Infantry Division)

Speaking with Panamanian villagers during Operation *Just Cause*, 1990. (7th Infantry Division)

Noriega's Learjet at Patilla Field after Navy SEALS disabled it by rocket fire during Operation *Just Cause*. I'm at the right as 7th Infantry Division troops prepare it for evacuation. (7th Infantry Division)

Lightfighters patrolling Panama City during Operation *Just Cause*. (7th Infantry Division)

Receiving line at the aloha welcome ceremony upon arriving Fort Shafter, Hawaii in 1990. General Fred Weyand, former US Army chief of staff, is greeting us. (Courtesy of USARPAC)

US Army Pacific commander Lieutenant General Mick Kicklighter and me at my aloha welcome ceremony in Hawaii in 1990. (Courtesy of USARPAC)

Trooping the line with the US Army Pacific CSM Gary Carpenter at the aloha welcome ceremony in 1990. (Courtesy of USARPAC)

Running with the 25th Infantry Division in 1990. Major General Fred Gorden, the division commander, is on the left in a white T-shirt. The PT runs were becoming more difficult! Note my thin, atrophied polio (right) leg. (25th Infantry Division)

Adorned with leis, Linda and I pose for a last shot with Lieutenant General Johnnie Corns and his wife, Carol, before leaving Hawaii, 1991. (Courtesy of USARPAC)

Greeting the Chairman of the Joint Chiefs of Staff, General John Shalikashvili, on his visit to Fort Lewis in 1994. (Courtesy of Fort Lewis)

First Corps Commander, Lieutenant General Glenn Marsh bidding us farewell at my retirement from the Army ceremony at Fort Lewis in 1995. (Courtesy of Fort Lewis)

Al-Qaeda detonated the truck bomb at the base of these two apartment buildings at approximately 11:20 p.m., May 12, 2003. All eight Americans killed were asleep in these buildings. (Vinnell-Arabia)

Briefing Secretary of State Colin Powell and US Ambassador Robert Jordan during their visit to the devastated Vinnell compound the morning after the Al-Qaeda attack in 2003. (Vinnell-Arabia)

With the SANG program's Saudi senior managers. (Vinnell-Arabia)

The leadership team on the Vinnell-Arabia/Northrop Grumman SANG modernization program. The six others to my left in desert tan uniform are all retired Army colonels. (Vinnell-Arabia)

Linda and me with President George W. Bush at a White House reception in 2006. (Courtesy of the White House)

President Obama thanking NAUS for unwavering support of the Veterans Advance Funding Bill prior to his signing the bill in a White House ceremony in October 2009. (Courtesy of the White House)

NAUS at the 2009 Veterans Day Breakfast at the White House hosted by President and Mrs. Obama. We were at the table with the Japanese American Veterans Association president, Robert Nakamoto, and executive director, Terry Shima. (Courtesy of the White House)

With Secretary of Veterans Affairs, General Ric Shinsecki, keynote speaker at our NAUS Annual Meeting in 2010. With us is our chairman of the board, Vice Admiral Jim Zimble, US Navy (Ret.). (Courtesy of NAUS)

Fiji fraternity brothers and wives gather at our house in Great Falls for a reunion in 2010. (Private collection)

With my fraternity brothers at our Gettysburg College 50th reunion in 2011. Left to right: Barry, Mel, myself, Mike, Ken, Skip, Dick, Bruce, Chuck, Fred, and Ron. (Private collection)

Pledging allegiance to the flag with ROTC Cadet Benjamin Flanders at the Dedication of the Gettysburg College Veterans Memorial on October 22, 2011. (Gettysburg College)

Linda and me with our seven grandsons at Great Falls Park in Virginia in 2007. (Private collection)

The Veterans Disability Benefits Commission. Back row, left to right: Ray Wilburn, executive director; commissioners Rick Surratt, Jim Livingston, Ken Jordan, Jennifer Carroll, Don Cassiday, Nick Bacon, and John Grady. Front row, left to right: Butch Joeckel, Larry Brown, Joe Wynn, James Terry Scott (chair), Dennis McGinn, and Bill Matz, 2007. (VDBC photo)

Visiting Senator Lindsay Graham in 2005 to thank him for introducing TRICARE Reserve Select legislation. (NAUS)

NAUS legislative director Rick Jones and I meet with Senator John McCain to discuss NAUS concerns on the FY 2011 National Defense Authorization Act. (NAUS)

Secretary of Veterans Affairs, Lieutenant General Jim Peake, retired, joins NAUS in placing wreaths on the graves at Arlington National Cemetery as part of the Wreaths Across America program in 2008. Joining us at left are Chad and Darleen Chadwick. (NAUS)

With Representative Joe Wilson, SC-02-R, at a meeting to strategize our fight to stop the DOD's attempt to increase TRICARE fees. (NAUS)

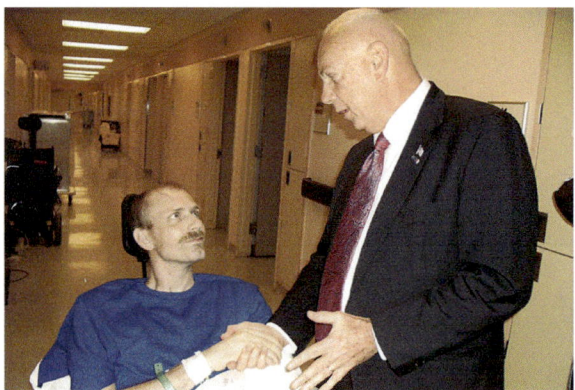

Visiting wounded warrior SPC George Wilmot, USA, at Walter Reed Army Medical Center as he recuperates from wounds received in Afghanistan. (NAUS)

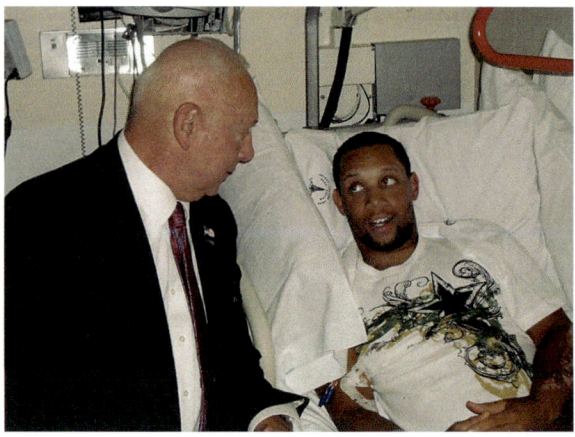

Visiting a wounded warrior at Walter Reed Army Medical Center as he recuperates from wounds received in the Middle East. (NAUS)

Left to right: Vice Admiral Jim Zimble USN (Ret.), myself and Fred Fielding at my retirement dinner from NAUS in 2011. Fred reminded me this was my third retirement since I retired from the Army! (Courtesy of NAUS)

I was honored to receive the University of San Diego Hughes Achievement Award in 2016. (University of San Diego)

Our 50th wedding anniversary in 2016. A memorable event spent with family and friends. (Private collection)

ABMC cemetery staff assist me in placing a wreath in commemoration of the 2,289 war dead interred at the Aisne-Marne American Cemetery in France on Memorial Day 2018. (Image courtesy of ABMC/Julien Nguyen-Kim)

Rendering honors with President Trump and the first lady at the 75th anniversary of D-Day ceremony at the American Cemetery in Normandy, France, 2019. One hundred and seventy-one World War II veterans attended. (Courtesy of ABMC)

President Trump presenting ABMC with an encased American flag during the 100th anniversary of World War I ceremony at the Suresnes American Cemetery in France, 2018. (Courtesy of ABMC)

Linda and me with ABMC Chairman David Urban and his wife Kellie at the 75th anniversary of D-Day ceremony at the American Cemetery in Normandy, France, 2019. (Private collection)

Speaking at the 75th Anniversary of the Battle of the Bulge Ceremony at the American Cemetery in Luxembourg, 2019. (Courtesy of ABMC)

Engaging with the Commander-in-Chief at the Army–Navy football game, December 2018. The Army won 17–10!

With Prime Minister of Luxembourg, Xavier Bettel; HRH the Grand Duke Henri of Luxembourg; the king and queen of Belgium, King Philippe and Queen Mathilde; US Ambassador to Luxembourg J. Randolph Evans at the 75th anniversary of the Battle of the Bulge ceremony, 2019. (Courtesy of ABMC)

Major General Brian Winski, commander, 101st Airborne Division, pins a 101st lapel pin on me at the 75th anniversary of the Battle of the Bulge luncheon while US Ambassador to Belgium, Ronald Gidwitz looks on. (Courtesy of ABMC)

At General George Patton's gravesite, Luxembourg American Cemetery with David Urban, Chairman of ABMC; Mark Esper, Secretary of Defense, and General Mark Milley, Chairman of the Joint Chiefs of Staff, 1919. (ABMC)

Thanksgiving with family in Great Falls, VA, 2019. (Private collection)

Linda and me at our grandson Ben's wedding, 2021. (Private collection)

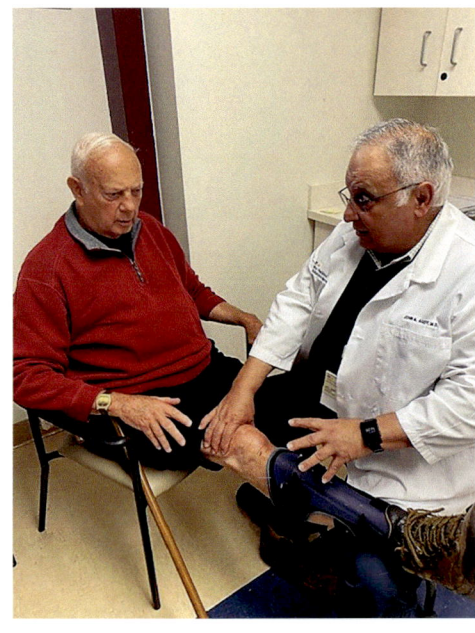

A check-up visit with Dr. John Aseff at the National Rehabilitation Hospital's Post-Polio Clinic, 2022. (Courtesy of NRH)

My time as the All-American G3 was drawing to an end. Upon return, we conducted our after-action reviews and made the necessary changes to the division's ASOP. General Lindsay pinned on my Master Parachute badge at a brief ceremony at division headquarters.

On May 28, we had the annual 82nd Airborne Division All-American Review on Pike Field. It was a great day! It began by honoring hundreds of former paratroopers proudly passing in review as our national and division colors majestically waved over the division formation of thousands of troopers. General Lindsay presented awards to the Trooper of the Year, and four other Trooper of the Year finalists.[13]

Then, at the playing of *Sound Off*, 12,000 paratroopers stepped off smartly to begin the pass in review. It was an awesome sight and a moment I will never forget. Marching abreast in formation with my fellow division staff officers as the CG led the division in the pass in review, I was counting a quiet cadence with the sound of the drumbeat so we could all march in step and was having a hell of a time trying to keep the G1, Lieutenant Colonel Jim Guest, in step. I am not sure Jim ever got in step as we passed the reviewing stand, came to "eyes-right" and saluted the reviewing officer. The G4, Lieutenant Colonel George Landis, and I kidded him about that for years.

Leaving the review that day, I couldn't help but reflect on the previous two years and all that had transpired. The opportunity I had been given to serve as the 82nd Airborne Division G3 under the Army's two finest airborne commanders, Generals Meloy and Lindsay, was the priceless opportunity of a lifetime. It was a milestone for me, and today, 40 years later, it remains among my most cherished memories. The challenge, experience and learning were unparalleled. There would be other interesting assignments and challenges awaiting me, but I'll always have a sentimental weakness and special soft spot in my heart for the 82nd Airborne Division and Fort Bragg!

CHAPTER SEVEN

Return to Korea and the Pentagon

Prior to departing Fort Bragg, I was selected for promotion to colonel. I received orders in March assigning me to the J3/G3 staff of U.S. Forces Korea (USFK)/Eighth U.S. Army (EUSA). My report date was July 10, 1982. The assignment would be much less taxing to my polio leg which was crying out for a rest after two years of arduous airborne duty. I was thankful for that, as I was always cognizant of my weak leg and how far I could push until it gave out—something I always dreaded and wouldn't let happen no matter the pain and challenge.

Linda and the kids were looking forward to going to Korea. I was looking forward to my return to "Land of the Morning Calm" as South Korea was referred to. I was interested in seeing how the country and the disposition of U.S. forces had changed since I last served there as a captain in 1965–66. In his letter of April 30, 1982, Major General William C. Moore, assistant chief of staff, J3/G3, welcomed me to the "J3/G3 team" and told me I would be assigned as chief, Force Development Division, USFK/EUSA, on Yongsan Army Post, Seoul. "This will be a bit different than the 82nd but will be strikingly similar to the ODCSOPS syndrome!" he wrote.

I first met General Moore when he and I were serving in the 82nd in 1962. I was a second lieutenant; he was a major and battle group S3. We were preparing for the invasion of Cuba during the Missile Crisis. He was a West Point graduate of the Class of 1952 and an infantry combat veteran of the Korean and Vietnam Wars. A bright, no-nonsense leader, he was telling me what to expect. I was glad that I would be working for him. General Robert W. Sennewald was commander USFK/EUSA.

Arranging the travel and flying from Philadelphia to Seoul proved interesting. The assignment was a 24-month accompanied tour with my family. My order read, "Concurrent travel for wife and three kids is authorized." However, my assignment order didn't include our German Shepherd, Soriah. We had to get special authorization to take her. We had gotten her from a German Shepherd breeding farm in North Carolina while we were with the 82nd.

Soriah was a pure-bred and well trained at the breeding farm. She was highly intelligent and had a friendly temperament. However, she chewed my leather boots, which I wasn't happy about (being hard to come by since they were two different sizes and fitted with a special arch to accommodate my polio defect). She was a member of the family. After weeks of persistent coordination with the Army, U.S. Customs, and the Japanese and Korean Custom authorities, we received authority for her to travel with us. She traveled in an approved heavy-duty dog crate in the baggage compartment of the plane.

My travel order authorized leave in Hawaii and Japan en route to Korea, so we spent a few days vacationing in Hawaii. Both Honolulu and Tokyo had strict quarantine laws, and we had to leave her in the airport quarantine station at each airport. She didn't want to leave the kids and growled loudly at the quarantine station attendant in Hawaii. He was scared to death, and he told me if the dog didn't behave, he would return her to the mainland. Our daughters worked with Soriah and calmed her down. Upon arrival at Kimpo Airport in Korea, she went into a two-week quarantine.

I learned later, the Army was discouraging personnel assigned on accompanied tours to Asian countries from bringing large dogs with them. That fact had never been brought to my attention. Headquarters, Department of the Army, dispatched a message to EUSA on June 16, stating, "Colonel Matz, accompanied by his wife and three children will be arriving Kimpo 1110 hours, July 9, 1982, on NW Orient Flight 191 from Tokyo. The family will be bringing a German Shepherd dog which will require immediate quarantine processing upon arrival." Despite the minor flap, our family of six arrived safely ready to begin our tour in Korea.

The first few days in Korea were spent in a small hotel outside Yongsan Post as our quarters would not be ready for several days. We were asleep in our room on the top floor the first night when a civil defense drill was called by the Korean authorities. Because of heightened concerns about the threat posed by their communist North Korean neighbor, drills in South Korea had become routine. There was no warning.

The drill simulated an air and artillery attack. We awoke to sirens blasting, and there was a total blackout. We evacuated the hotel, using the stairs to a shelter outside. The loud broadcasts from the street and from within the hotel were all in Korean. We couldn't understand a word. No one was speaking English. We quickly grabbed our valuables and made it down the stairwell. It was pandemonium, unrestrained disorder. We were the only English speakers in the hotel. Most of the Koreans seemed to know where to go so we followed them to a shelter. Sometime later an "all clear" signal was given, and we returned to our room. We didn't sleep the rest of the night. Linda and the kids, as always, took it in stride. What a welcome to Korea!

Once we got settled into our quarters on Yongsan Post and the kids were enrolled in the Department of Defense Dependent School (DODDS), Linda took a job teaching home economics, typing and health at the Seoul International School in downtown Seoul. It was a coeducational American curriculum, and though the teaching staff and students were primarily American, there were approximately 38 other nationalities represented in the student body. There were challenges teaching at an international school overseas which Linda met head on and with love for her students.

Our Korean houseboy, Mr. Chu (a wonderful man who became part of our family) was frightened of Soriah, and it took weeks before he felt safe and comfortable with her. They eventually became best friends, and he walked her on a leash throughout the embassy housing area on Yongsan Post where we lived. Mr. So, the Force Development Division driver, never did get over his fear of her.

As chief, Force Development, I was the command's principal staff advisor on all matters pertaining to force structure, modernization, tactical stationing and unit status reporting which measured a unit's readiness. It required several trips to the Pentagon to participate in force development and modernization meetings. The flights back and forth between Washington, DC and Seoul were long and tiring, which I never got used to. We were transitioning the Army's 2nd ID to a new structure, which required significant resources and manpower. We were also fielding new equipment and weapons systems to Eighth Army units throughout the Korean Peninsula.

Those areas were vital to the modernization of the Army in which I had little prior experience but quickly learned their importance. I had some bright officers working for me, and we were able to obtain from Department of the Army the resourcing and manpower needed to keep the command's modernization programs on schedule. On September 30, 1982, General Moore promoted me and three other USFK/EUSA officers to colonel.

Linda and General Moore pinned on my eagle insignia, and our kids were present. "A great day for 'Colonel' Bill Matz," he wrote on a photo of the promotion. "Congratulations—I was honored to have officiated." In my brief remarks, I recalled an earlier promotion in Korea. In 1965, my battalion commander promoted me to captain on the rear ramp of an M113 armored personnel carrier while on a training exercise in the DMZ.

On March 2, I got a phone call from my sister, Becky, crying and very upset. Our father had just died of a heart attack. It was sudden and unexpected. Weeks earlier, he purchased an airline ticket to fly to Korea to visit with us in late March. He was excited and wrote how he couldn't wait to see us; we also looked forward to it. I flew home for the funeral and to help my sister as executor of his estate. Linda stayed in Korea with the kids. To this day, she is upset that she didn't return with me for the funeral and burial. We missed my father. He and my mother were always the first to visit us during our many moves in the Army. The kids were upset as their Grandpop was a real companion to them, always taking them swimming and to good restaurants for a treat. When we were home on vacation, this gifted salesman took Billy (Grandpop called him "the Champ") on his business calls where his customers knew all about his wonderful grandchildren. He always had the best nicknames for the girls as well. There was a wonderful bond between my father and our kids.

Linda wrote in our 1983 Christmas letter, "It does not seem the same without him popping in on us. He was a wonderful father and grandfather for the Matz family." Like all fathers, he was proud of his son and especially proud that I was an officer in the Army. And I will never forget how he insisted I get rid of the leg brace and crutch while a young boy. He pushed me to my limits, rarely giving me any slack. He set a high standard for me as only a father can do. He came to Hawaii in July to spend the last couple days with us before we continued to Korea. I sensed then that he was feeling sad, melancholy that we were leaving. He was living alone, and I regret I wasn't there with him when he died.

General Moore was an outstanding writer and took pride in the many papers and documents he authored as the USFK/EUSA Operations chief. We worked on a force modernization paper for General Sennewald's decision. Moore reviewed it, made a few changes, and told me to "walk it around the headquarters" and get concurrence from the other staff agency principals.

One general officer on the staff, who delighted in nitpicking, made several changes to the paper. Moore was furious. "Bill, go back and tell him he doesn't have enough stars to change my paper." I did, but not using those exact words, and the paper went forward without any changes from that general. I often think of that incident when I am sitting in a board meeting watching my colleagues waste time changing "happy" to "glad."

In January 1983, I received a warning order from the Army's MILPERCEN (Military Personnel Center) that I would be activated off the colonel command alternate list for a brigade command beginning the summer of 1983. There was no mention of which brigade it would be. That meant I would not complete my two-year accompanied tour with EUSA. Sometime later, I was notified that I would take command of the 4th Training Brigade at Fort Knox, Kentucky, in August. I accepted this with mixed feelings. I wanted so much to command an infantry brigade in an infantry or airborne division. I had never heard of the 4th Training Brigade, and had never been to Fort Knox, but knew it as the Armor School and Training Center. I kept asking, "Why would the Army send me there to command?"

The command was an Initial Entry Training (IET) Brigade consisting of four basic training (BT) battalions, one advanced individual training (AIT) battalion, and the Fort Knox Reception Station. The BT and AIT battalions trained 24,000 soldiers annually and the reception station processed and reassigned 35,000 new accessions annually. The brigade also conducted the annual one-of-a-kind ROTC Basic Camp training 4,800 ROTC cadets each summer. The program of instruction (POI) for the BT battalions was infantry oriented, and the Infantry School and Center at Fort Benning had a proponency for the 10-week basic training course.

The previous two commanders had been armor branch officers, and Major General Frederic J. Brown III, the Armor Center commanding general, wanted an infantry colonel to command the 4th Training Brigade. It made immanent sense. Also, the Army was adding emphasis to the importance of TRADOC (Training and Doctrine Command) commands and assigning officers selected by Department of the Army central selection boards to command training battalions and brigades.

General Maxwell R. Thurman, the Army's deputy chief of staff for personnel at the time and later the TRADOC commander, was personally involved in selecting the colonels to command TRADOC brigades. I was told later that when Thurman was presented the recommended slate of colonels for brigade command for his approval, he changed me, who at that time was slated to command a FORSCOM infantry brigade, to command the 4th Training Brigade.

After I got over my initial combination of shock and disappointment that I would not command an infantry brigade, I settled down and began to think "basic training." I became excited about the new and different challenges I would have as a training brigade commander, and I was looking forward to it.

Prior to leaving Korea, Lieutenant General Choi Moon Kyu, deputy chief of staff for operations, Republic of Korea Army (ROKA), arranged for me and Linda to visit the ROKA installations along the DMZ. I wanted to see the camps (Camp Wilbur and Camp Young) I had lived in and the guard posts that I manned when I was with the 1st Cavalry and 2nd ID in 1965–66. It was quite different. The dusty, dirt paved MSR (main supply route) that went north from Seoul to the DMZ was now a modern highway.

My former company and battalion base camps had different names and were occupied by ROKA troops guarding the DMZ. We went down into a tunnel that had been used by North Korean infiltrators. Strands of fencing and barbed wire marked minefields. One guard post inside the DMZ that my company manned in 1966 was still there and occupied by very alert ROKA troops. You could see clearly across the DMZ and well into North Korea. It was springtime. The scenery was beautiful, yet an eerie feeling permeated the air.

We heard broadcasts coming from the North coaxing those in South Korea to come across the border. We visited the site where in 1966, North Korean infiltrators ambushed an eight-man 2nd ID patrol from my brigade, killing seven American soldiers and one South Korean soldier. The entire trip to the DMZ brought a wave of nostalgia that swept over me, bringing back long-gone moments, evoked by the desire to return to a time and place of my earlier soldiering. It was a memorable visit!

We left Korea with mixed emotions. Our three kids were in their early teens. Heather was on the summer swim, cross country, and soccer teams for her school. Becky was also swimming and playing soccer while doing professional modeling for the Nike Company and traveling the peninsula with the Nike-sponsored Korean professional baseball team. Billy's record-breaking time the year before broke the North Carolina state record for the boys' 13–14 age group in the 100-meter breaststroke and placed him in that age group's national Top 10 Times in America—a major accomplishment!

They were enjoying Korea but longing for a more challenging and competitive environment in athletics. The Koreans were beginning to train their athletes for the Summer Olympics that would be held in Seoul in 1988, and the Korean swim clubs would not allow American kids to train with them. Our son needed coaching and more challenging workouts.

I was enjoying my job with USFK/EUSA. I always respected the ROKA officers and soldiers. They were tough, disciplined, and well trained. They fought tenaciously against the communist forces during the Korean War, and they supported us with two infantry divisions during the Vietnam War. They are our friends and best ally in Asia. I hated to say goodbye to my Korean Army counterparts. We left Korea for the United States in late July. We didn't have the irritating hassle and inconvenience of quarantine and customs requirements for Soriah we experienced a year earlier, coming to Korea and the flight home was long, but pleasant. I was looking forward to my command at Fort Knox.

Commanding a Basic Training Brigade

On August 26, 1983, the brigade's colors were passed, and I assumed command of the 4th Training Brigade (TB) from the outgoing commander, Colonel Donald L. Smart, on Brooks Field. General Brown was a stickler for detail who insisted on exactness and maintaining an exceedingly high standard in everything done. I was impressed with the pristine condition of the field and grounds and the flawless conduct of the ceremony.

Fort Knox was a school and training center. Thousands of young soldiers, officers, and ROTC cadets receive their basic and advanced training there each year. It was the first stop in their Army careers. It was important that we set the highest standard, one they and their families would remember as they progressed through the Army.

The change of command ceremony concluded with the playing of *Garryowen* and the *Armor Song*. *Garryowen* was the United States Army 7th Cavalry Regiment official march tune, and with the *Armor Song*, was played at every ceremony at Fort Knox. "Bill, Welcome to Fort Knox, you better get used to the playing of those songs. They are our hymns! You are going to hear them often," said Colonel Roy C. Price, chief of staff of the Armor Center. He handed me a copy of the ceremony program with its bright yellow cover. Yellow was the official color of the armor branch. My cultural indoctrination into the armor community had begun!

My first meeting was with the battalion CSMs and first sergeants. The brigade CSM, Alan M. "Buck" Taylor, a rugged West Virginian, who served during the Korean War, and was a twice-wounded combat veteran in Vietnam, set up the meeting. Buck was stubborn, but always looked out for his brigade's best interest. I liked that. He and the post CSM often tangled over issues and General Brown and I would have to step in. Seven CSMs and 27 first sergeants attended. I wanted to meet the senior NCOs as soon as possible, and they wanted to hear from the new brigade commander. NCOs were charged with the care, training, education, and readiness of every soldier. In basic training units, that takes on an added dimension.

NCOs are the drill sergeants responsible for skillfully executing the mission of transforming civilians into competent, confident soldiers. That includes the responsibility for teaching young soldiers the fundamentals of military life, Army core values, esprit de corps, leadership, military bearing, customs, and courtesies. The session with the senior NCOs who supervised and who oversaw the drill sergeants, gave me a feel for how well the brigade was accomplishing its mission.

They spoke; I listened and learned. No one took notes. Everyone was free to speak. In those two hours, I learned a wealth of information about my new command. The CSMs and the first sergeants appreciated it. In the following days, I visited the battalions and Reception Station, meeting with the officers and drill sergeants.

I noted that the current BT POI wasn't challenging. It didn't reflect current TRADOC policy and was completely void of air land battle doctrine, an aggressive new warfighting doctrine that required fundamental changes to our approach to basic training. With General Brown's support, I actively sought involvement with the TRADOC/Infantry School BT POI Task Force, and through that medium we contributed significantly to the development of a new, more robust, and challenging POI.

The 4th Training Brigade's initiative to test and integrate many new training modules even before the TRADOC validation began at training centers at Forts Dix, Jackson, and Benning, paid high dividends for both the graduating soldiers and for the training cadre. "We need to toughen their training," I told General Brown. I convinced him to let me begin bayonet/pugil training in our basic training courses. The development, testing and implementation of a seven-hour bayonet/pugil module is tangible evidence of our toughening the POI. The basic trainees enjoyed it.

We also constructed a new and more challenging live-fire infiltration course that challenged young trainees. We lowered the height that live rounds were shot over a crawling trainee, and we added more demolition pits and barbed wire obstacles. It required substantial amounts of 7.62 mm machine gun ammunition. We were finally resourced for it, and it became part of the BT POI. By August 1, 1985, we were conducting a new bayonet assault course and hand-to-hand combat training in newly constructed combatant pits.

Indeed, basic training conducted by the 4th Training Brigade was becoming tougher and more challenging. In the spring of 1985, we lost a trainee to a lightning strike. After pausing for a typical Kentucky pop-up late-afternoon thunderstorm, the trainees resumed their lesson on field camouflage. One soldier was chopping vegetation with his steel entrenching tool (shovel), when lightning struck his shovel and passed through his body. CPR was administered immediately, and he was rushed to the hospital. My XO, Major John A. Buckley, having been notified, arrived at the hospital just as the ambulance from the field arrived.

Observing the efforts in the hospital emergency room, he quickly realized that nothing could be done for the soldier. A review of the incident indicated the cadre did everything right, having the soldiers ground their steel gear and lie down away from it until the storm passed. It was literally a bolt out of the blue.

My AIT battalion that trained soldiers to maintain the Army's primary front-line fighting vehicles didn't include spending a single night in the field. I didn't like that and wanted to get them out of their comfortable barracks and into the field. With TRADOC support we implemented a three-day/night bivouac and a five-day/night Field Training Exercise (FTX) for the AIT POI. Soldiers experienced maintaining vehicles in simulated combat in the field. They found changing a truck transmission in the field was quite different and much more challenging than doing it in a clean, comfortable, maintenance building. The commanders and noncommissioned officers readily supported the changes that toughened training and made it more realistic for soldiers. I was cautioned more than once to "ease up Matz" but I kept the pedal on the gas while managing not to upset my boss.

One morning my phone rang. It was General Brown's aide-de-camp. Brown had received a call from Lieutenant General Charles W. Bagnal, deputy commander for training at TRADOC. Bagnal wanted to visit Fort Knox and get briefed on our BT and AIT initiatives, and General Brown asked me to develop an itinerary for him. Bagnal spent two days, March 20–21, 1984, with the brigade. He visited basic rifle marksmanship training and walked a live-fire lane with a BT company that was conducting a close combat movement exercise. He spent the night observing the newly implemented AIT and BT field training exercises. I briefed him on the status of our pugil and bayonet training modules and on the live-fire infiltration course we were constructing.

In a close-out seminar session with my officers and NCOs, he said, "I am particularly pleased with the development and initiative of a three-day bivouac and five-day FTX for your AIT soldiers." It was a good visit. My drill sergeants appreciated being able to brief the general on improvements to the POI that they had personally worked on. Giving the NCOs and company commanders credit for and ownership of the improvements to training sparked their morale to new heights! On March 6–7, General Brown sent me to the annual Infantry Conference at Fort Benning to brief the attendees on "BT/AIT training enhancements." Benning was the proponent for the BT POI, and the Infantry Center commander wanted to update commanders on new and proposed changes being developed by the 4th Training Brigade.

An extremely important aspect of IET is the PT program. The drill sergeants took pride in their company and platoon PT programs. I ran with a different company each morning and made it a point to talk with the trainees and drill sergeants afterward. We had 17 Master Fitness Trainers in the brigade, and I was pushing hard to get a minimum of one per company. We instituted quarterly brigade runs. It took a while to get permission to do that. Post authorities were concerned that a large formation of soldiers running early in the morning would congest traffic. We finally pushed through the bureaucracy, receiving permission after we laid out a running route that didn't interfere with the early morning traffic.

Brigade runs did little for the building of endurance, but they did much to embrace unit esprit, morale, cohesion, and teamwork on a grand scale. The brigade runs also gave me an opportunity to address all soldiers and cadre at a single formation. General Brown and Brigadier General Gordon R. Sullivan, commandant U.S. Army Armor School (Sullivan would later serve as the 32nd chief of staff of the Army) were always invited to run with us, and they often did. I wanted to expose our young trainees to general officers and wanted them to see firsthand generals participating in unit runs and training.

I can still see both generals on separate occasions, running back and forth along the five-battalion formation trying to spend running time with each unit. The largest brigade run consisted of 4,100 soldiers and cadre. "Great running with all these young trainees—let me know when the next run is," General Sullivan said, as he departed the brigade area.

The quality of personnel in the brigade was outstanding. I was fortunate when General Brown agreed to release his very capable secretary of the general staff (SGS), Major John A. Buckley, to be my brigade XO when Lieutenant Colonel Roger Casalengo departed. John was a superb XO and had the complete support of our very capable brigade staff, led by Captain Tom Turning and Major Roy Altobello. He kept everything running smoothly, which allowed me to get out and participate in training with the drill sergeants. The young captains I was getting from the armor and infantry advance classes were among the best I had known. In the March/April 1985 timeframe, I had more captains knocking on the door than I had vacant positions.

The unpopular stigma of commanding a TRADOC basic training unit was going by the wayside. I caveat that by saying most captains still preferred commanding FORSCOM operational units. And they should have. I appreciated General Brown as chief of armor helping my armor captains who desired a second command after their BT/AIT command to get an armor or cavalry troop command. The fact that my company commanders knew that possibility existed was a positive incentive for them. My drill sergeants were top notch. They were smart, in good physical condition and demonstrated flexibility in adapting their demeanor and leadership style when necessary. And the first sergeants and CSMs were also top notch and dedicated to maintaining high standards, ensuring we were sending the best-trained soldiers to the force.

Each summer, the brigade supported the Army's ROTC Basic Camp conducted at Fort Knox. All four BT battalions stood down from training IET soldiers and converted to billeting and training over 4,000 ROTC cadets, which posed challenges, as our cadre had to adjust to implementing a completely different program of instruction. We also had to adjust our billeting and other facilities to accommodate both men and women cadets. Our BT/AIT mission only involved male trainees. I worked closely with Brigadier General Tom Lightner, commanding general, Second ROTC Region, and his staff, in organizing and conducting the ROTC Basic Camp.

Our family enjoyed Fort Knox. Linda wrote in our 1983 Christmas letter, "Fort Knox is like small town, USA." It had outstanding facilities and programs for all family members. It was even one of the very few Army posts to have its own high school. Every Friday night in the fall, the entire post turned out to support their Fort Knox Kentucky State Champion Football Team and magnificent marching band.

All three of our kids swam for the high school swim team, and Bill and Heather also swam for the Gold Vault Swim Club (a USA Swimming Club) that worked out three hours a day, year-round. Heather also ran on the high school cross-country team and was selected for the Kentucky High School All State "A" Team. Becky swam for the high school swim team and was also a school cadet and manager of the baseball team. Linda taught basic skills to soldiers who needed to get a high school diploma. We were all busy enjoying what we were doing and would compare notes at the dinner table each night!

During swimming practice his senior year, our son suffered a sudden rupture of veins and arteries in the brain called an arteriovenous malformation (AVM) causing bleeding into the brain and spinal cord. He was evacuated from Ireland Army Hospital by ambulance to the Kosair Children's Hospital 45 miles away in Louisville. The doctor at Kosair told us, "This is a serious bleed, and your son might not make it through the night." Linda and I held each other and prayed as we never did before as we sat with Bill in intensive care.

He made it through that night and the next, and the next, as they treated the hemorrhage by inserting a shunt into his frontal lobe to relieve the swelling and pressure. Dr. Henry D. Garretson,[1] a renowned neurosurgeon at Louisville Hospital, who helped pioneer the AVM treatment, was out of town but was on the phone instructing resident physicians on emergency treatment.

Unfortunately, the attending physician mistakenly damaged the fourth cranial nerve affecting the eye while performing the delicate procedure. Bill was in a coma and unconscious for several days, and we later discovered that his peripheral vision was permanently damaged. His high school and swim club friends and the entire Fort Knox community were extremely supportive. While Linda and I spent days traveling 45 miles back and forth between Fort Knox and the hospital in Louisville, the Fort Knox community took charge of caring for our two daughters and helping Linda's parents who flew in from New Jersey.

On November 28, Bill had brain surgery to remove the AVM. By that date, he had made a miraculous recovery from the trauma and was strong enough to undergo infratentorial craniotomy surgery. The nurse warned us that with that surgery, his personality could change, and it did ... he came out of the surgery joking with the nurses. We thanked the Lord that he was okay and out of danger!

His competitive spirit led him onto the comeback trail, and Dr. Garretson gave him permission to swim in the four-state All-Star swimming meet in Indianapolis, Indiana, on February 10. He had been recruited by Jack Ryan, the men's swimming coach at West Point and was accepted pending the results of a final neurological examination. He was looking forward to being a West Point cadet and an Army officer.

The first thing he said when emerging from the coma in intensive care was, "I guess I can't go to West Point now." My wife, the eternal optimist, told him he could do anything he put his mind to. A neurosurgery intern pulled her aside and said, "You should not give him false hope." Linda replied, "He needs a goal, something to work towards to get better." We always remain positive no matter how dire the circumstance!

Fast forward to July 1985, two days before R-Day (Reception Day—the day cadet candidates report to West Point), I took him to Walter Reed Army Medical Center for his final neurological examination. After a two-hour-long examination, including visual and cognitive skill tests, the head of neurosurgery determined he was "not commissionable." Permanent memory loss and damage to his peripheral vision resulted in that determination. West Point was out. Bill and I were upset. Linda was relieved as she believed the academic and physical rigor of the academy would be too taxing on him as he still had the AVM (it was slowly growing back and would require further surgery).

The West Point coach, Jack Ryan, called his good friend, Peter Brown, the swimming coach at Pennsylvania State University. Coach Brown called and said, "Bring him up." He needed a first-year breast stroker who could also swim the individual medley to round out his freshman recruiting class. Although disappointed that he couldn't attend West Point, Billy got a partial scholarship and swam for the Penn State Nittany Lions for four years.

Billy was recipient of the 37th Annual Game Guy Award presented by the Kentucky High School Athletic Association. The award was given each year to a Kentucky high school athlete who had overcome some physical handicap. I was never prouder of him than on his graduation from high school when he received his award. He overcame a terrible physical setback and went on to swim competitively at the NCAA Division I level in college.

During the challenging time of our son's bleed and his following surgery and recovery, we couldn't have been part of a finer, more compassionate, and caring community. All of Fort Knox prayed for his recovery. General Sullivan's wife, Gaye, was the first to knock on our door and offer her time to look after our daughters while

Linda and I spent days at the hospital in Louisville. My XO's wife, Mary Buckley, who was president of the Gold Vault Swim Club, rallied the coaches, swimmers, and their parents who provided untold support. The Buckleys were a blessing during this time of need! There were times during his eight-day intensive care stay that I gripped his hand and would not leave him. I had a tough time coping with my son's situation. Seeing him lie there in intensive care hooked up to life-support machines tore me apart inside. Linda coped with Bill's setback much better than I did. She was much stronger than I was!

The Browns established a wonderful command and family climate. When they entertained at their quarters, it was done with exquisite taste. At a formal dinner party, Linda and I were seated apart at their long dining room table with the other brigade commanders and their wives when General Brown proposed a toast. We raised our long-stemmed crystal wine glasses, and as Linda set her glass back down on the table it shattered. All eyes gazed first at the broken glass, then at her. She looked up with tears rolling down her cheeks. So embarrassed and so sorry, she helped Mrs. Brown clean up, while apologizing profusely for the broken glass. She offered to buy a new glass. Mrs. Brown politely said, "That was a family heirloom that can't be replaced." Linda was crushed and stayed awake that night trying to figure how she could possibly make it up to her.

Two months later General Brown hosted a stag reception in his back yard for a group of international officers visiting the Armor Center. I leaned on one of the wide-armed classic Adirondack chairs that graced his yard and broke it. The chair crumpled, and the arm fell onto the ground. General Brown, with an off-handed glib, witty, comment spontaneously said, "Thank God there is only one infantry colonel on post, they break everything."

It brought more than just a quiet chuckle from the group of mostly armor officers from other countries. I am sure at that very moment he remembered his wife Ann's beautiful crystal glass that Linda broke two months earlier. Word got around post, "If you invite the Matzes, it's best you remove your valuables." I did, however, drive a bright armor-yellow-colored VW bug that took a little pressure off me among the armor community.

Linda tried to attend as many basic trainee luncheons and graduations as she could. We both enjoyed attending the graduations and meeting the parents of soldiers who just successfully completed basic training. Proud mothers and fathers wanted their photos taken with their son and his drill sergeant. The graduations were happy moments for all concerned. I felt good about being a basic training commander and that I had given my all to ensure the young volunteer soldiers were well trained. I remember these moments well and cherish the memories.

On Thanksgiving, I had the CSM pick two basic trainees to join us for dinner. The company first sergeants and drill sergeants didn't think that was a good idea, but we did it anyway. Once they got over their nervousness at eating dinner at the brigade commander's house, they seemed to relax and enjoy their meal. Our children made them feel very much at home. One trainee told us at the dinner table that he was glad to get away from his drill sergeant. I understood that. I learned more about my drill sergeants than I could have ever imagined. It was difficult getting them to leave when the drill sergeant came to pick them up. I don't think my drill sergeants ever forgave me for that!

As my brigade command time was ending, my CSM, Walter Tobias, and I wanted to get our NCO and officer cadre together for a social event. We decided to have a formal Dining-In, one of the United States Army's most valued customs. It was a function that provided an occasion for the officers and NCOs to gather in an atmosphere of camaraderie, good fellowship, fun and social rapport. My XO, the Dining-In's punch master, oversaw the planning and ensured it conformed to the traditional Regimental Mess. The theme of the Dining-In was the 4th Training Brigade's integration into the Army's Regimental System.

I invited General Sullivan to be our guest speaker. I told him I wanted the Dining-In to also have a "teaching" moment. An avid and well-read historian himself, he spoke about camaraderie and trust among officers and their importance in combat. He cited General William Tecumseh Sherman's letter of March 10, 1864, to General Ulysses S. Grant, summing up their long, bloody, and successful western campaigns during the Civil War.

Without a note, he recited almost verbatim Sherman's words, "'I knew wherever I was that you thought of me, and if I got in a tight place, you would come if alive." "Powerful," Sullivan said, "powerful." He was telling us that Grant's and Sherman's way to victory was "built on the mutual trust that their friendship inspired." He was conveying to us why trust is so important in the Army. By building trust with subordinates, peers, and superiors and respecting each other, Army leaders create strong working relationships that win battles. Indeed, it was a teaching moment!

He also reminded all of us besides teaching basic soldier skills, we were responsible for developing soldiers of good character. He quoted the scene from the movie *An Officer and a Gentleman* where the drill instructor (played by Louis Gossett, Jr.) said to the officer candidate (played by Richard Gere): "I'm not talking about flying; I'm talking about character!"

I will never forget the lessons from that Dining-In at Fort Knox! The next day, I received a handwritten note from General Sullivan, "Dear Bill—Thank you and your officers for a great time last night. You are in command of a splendid group, and this, combined with your very important mission, should be a justifiable source of pride! Cordially, Gordon."

In April, I received orders assigning me to the Office of the Deputy Chief of Staff for Military Operations and Plans (ODCSOPS), Army staff, which meant a return to the Pentagon and reengaging with the Washington, DC, traffic on Interstate 395 every day. That was something I wasn't looking forward to.

On July 26, 1985, I passed the 4th Training Brigade colors to Colonel John N. Sloan. Moments before the ceremony was to begin, it started raining lightly. Everyone was seated. All six battalions and the band were in formation standing on the final ready line at parade rest. There was no thunder or lightning. I wanted to continue with the ceremony outside, but I was overruled, so we moved inside. My brigade XO was commander of troops. He did an outstanding job adjusting to the weather and calmly moving everyone inside at the last moment.

Like the change of command ceremony two years earlier, *Garryowen* and other armor march songs were played. The program, however, was different. This time it was printed in infantry blue instead of the standard bright armor yellow. I felt I had to leave a mark whereby the fine people of the Armor Center and my many armor officer colleagues would remember me as their "infantry" brother in arms! We departed Fort Knox that day for Washington, DC, with Linda driving the Mercury station wagon and me driving my yellow VW bug.

I am grateful for the two years I spent at the Home of Armor and Cavalry in 1983–85. Professionally, it was among the best of my many tours of duty with the Army. Linda and the kids remember it as their "best and most fun-filled post" for the family. The community, spirit, leadership, and mission were unbeatable, and I left with a deep appreciation and understanding for the importance of initial entry training. The habits, standards, and values learned in basic training stay with a soldier for life!

Back to the Pentagon

I reported to the Pentagon in August 1985. Lieutenant General Carl E. Vuono had just become the deputy chief of staff for operations and plans. He later became the 31st chief of staff of the Army. In a brief office call, he welcomed me and told me that I would be the deputy in the Training Directorate, one of five directorates in ODCSOPS. He said my new boss, Major General Tom Tait, would be arriving shortly from Germany and would be the director.

General Vuono was the Army's consummate trainer, and he wanted to ensure the Training Directorate had a credible balance of officers with heavy and light force experience when developing training policy and resourcing training programs. General Vuono knew more about the Army's training needs than anyone else, and he spearheaded the institution of new and dynamic individual and collective training programs which continue in various forms throughout the Army today. As I left his office, he looked at me and said, "Start thinking about a light force combat training center." Major General Norman H. Schwarzkopf, who later was the United States Central Command (CENTCOM) commander during Operation *Desert Storm*, was Vuono's assistant.

General Tait was a giant in the armor/cavalry community. He was every bit of 6 feet 6 inches tall and loved soldiers. He liked to be referred to as "Viking 6," his former unit call sign. He brought with him a wealth of heavy force experience, particularly in cavalry and armored cavalry units, and knew the Army's training needs in Germany. He was larger than life physically and mentally, with a booming voice, and he could intimidate you if you let him. I liked his power-down approach to leadership and his rough sense of humor. In our first meeting he told me, "Tell me everything you know about our 'light' forces and their training needs. I'm depending on you." In his role as Army inspector of training, he was often away from the Pentagon visiting units, leaving me in charge. This was much to his liking as he was like a "bull in a China shop" and like me, didn't enjoy Pentagon duty.

As deputy, I was responsible for supervising the daily operations of the Directorate. The Training Directorate developed policy and programs and prioritized the resourcing for all Army training, including initial entry and unit training, leader development, and overseeing the integration of modern technology, ammunition, and the flying hour program into the "Total Army."

The training priority for the Army at that time was the enhancement and development of combat maneuver training centers. The National Training Center (NTC) was created in California in 1981 to train Armored Brigade Combat Teams to fight and win in Europe. A similar maneuver training center was needed in Europe to train the many U.S. Army mechanized and armor units stationed there.

More important was the need for a similar combat maneuver training center to train the Army's light forces, one that would provide challenging force-on-force training for battalion- and brigade-size task forces from the new light, airborne and air assault divisions. As Vuono alerted earlier, "Start thinking about it." That new combat training center became my priority as I oversaw the DCSOPS training task force that planned, developed, and resourced it. I became a strong voice for this when I chaired the Army's light force study at the war college.

We spent months planning the new training center, including visits to Fort Chaffee as part of the site selection process. After obtaining General Wickham and Secretary of the Army John O. Marsh's final approval in 1987, the new center was established at Fort Chaffee, Arkansas. It was named the Joint Readiness Training Center (JRTC). In 1993, JRTC moved from Chaffee to Fort Polk, Louisiana, where it trains the Army's active and reserve brigade combat teams and is the most realistic and relevant unit training in the Army.

Working closely with U.S. Seventh Army in Europe, we also established the Combat Maneuver Training Center (CMTC) in Hohenfels, Germany in 1987. CMTC, since renamed the Joint Multinational Readiness Center (JMRC), now trains units from our European allies, including the Ukrainian military prior to their 2022 war with Russia. Jobs in the Pentagon are often not satisfying, but I felt good about being part of the Army team that brought the superb unit training centers into formal status.

Late one afternoon, I received a call from Colonel Ronald H. Griffith, XO to the DCSOPS. Ron and I served together in ODCSOPS in the early 1970s and were Army War College classmates. "Bill, come right up, General Schwarzkopf wants to see you, and he's not happy." My antenna went up. I hurriedly climbed the two flights of stairs to the DCSOPS's office suite on the third floor of the Pentagon's E-ring.

I was racking my brain, thinking, "What could this be about?" Ron was waiting for me. "Bill, the Bear's steaming mad about something." Ron opened the door slowly to Schwarzkopf's office: "Sir, Colonel Matz is here." I entered his office and Ron shut the door behind me.

Only Schwarzkopf and I were in his office. He was seated at his desk in the center of the room with hands folded.

"Sir, you wanted to see me?" I asked, standing in front of his desk. Before I could say another word, he said loudly, "Don't ever come late to my meeting again. That was embarrassing for me this morning." When I tried to explain what happened, he leaned forward, blood rushing to his head, and abruptly interrupted, "I don't want to hear that. You were late, don't let it happen again—there's no further discussion!"

"Yes sir!" I saluted and left. At that moment, what else could I do!

He had chaired an important meeting with Army staff and secretariat principals that morning. General Tait was on a training visit, and I represented the Training Directorate and arrived late because Schwarzkopf's office gave me the wrong conference room. By the time I sorted it out and got to the right room, his meeting had started. I quietly took my place at the meeting table. After I left Schwarzkopf's office, I told Ron exactly what had happened and why I was late. He shared that with General Schwarzkopf.

When I got home that evening I told Linda, "Today I got the worst ass-chewing of my life." My 16-mile-drive home that night gave me time to think about the over-the-top scolding I received because I was a few minutes late to a meeting. He reacted grossly out of proportion to the situation. I heard from some who served with him and knew him well, that he was prone to fits of temper and anger.

I have sometimes wondered what would have happened or how he would have reacted had I not shown up at all, or if the 6-foot 6-inch tall General Tait had shown up late! In spite of my chagrin, the end result was positive. I was never late to another meeting with Schwarzkopf, and he never mentioned it again. In fact, just a few weeks later, as my intermediate rater on a Complete the Record Officer Efficiency Report (OER) for the brigadier general selection board preparing to convene, he wrote one of the strongest, most professional endorsements I ever received.

In the summer of 1986, General Vuono received his fourth star and left to command the U.S. Army Training and Doctrine Command. Lieutenant General Robert W. RisCassi became the DCSOPS. He had been commander of the Combined Arms Center at Fort Leavenworth, and he brought the latest training ideas from the field. A clear thinker and visionary, who always saw through the clutter and minutiae, he was one of the Army's most capable and brightest leaders. He later served as vice chief of staff of the Army and commander in chief of United Nations Command in Korea, retiring as a four-star general.

In May 1987, I was nearing the end of my second year in the Training Directorate when I was told to report to General Arthur E. Brown, who had just been appointed vice chief of staff of the Army. He told me the chief of staff selected me to be the Army's nominee for the position of executive secretary in the Office of the Secretary of Defense (OSD) working for Casper W. Weinberger.

That was a highly desired colonel's position in the OSD that all services sought to fill. Each service nominated a 06-level officer. An Air Force colonel was leaving the position. General Brown told me the Army wanted the position badly. "This is important to the Army, get your uniform ready and ensure you wear all your service ribbons and awards. You will be interviewing with the secretary of defense. Do good, we need this position," he said.

Two days later, I was interviewed by the deputy SECDEF, Will Taft, and Air Force Major General Gordon E. Fornell, senior military assistant to the SECDEF. The next day, I met with Secretary Weinberger. He had met with one of the other service nominees earlier. General Fornell, a decorated combat pilot of the Vietnam War, showed me into the secretary's office, a huge, tastefully decorated room. His office wasn't just a place to work, but a museum as well.

Dominating it was a 9-foot mahogany desk used by General John "Black Jack" Pershing after World War I. Weinberger had been well briefed on my assignments and qualifications. "I'm impressed, colonel, with your war record and the assignments you've had. I see you are in the infantry," he said, rattling off some of my assignments. He was extraordinarily polite and courteous and remembered everything he read and heard. He told me he had read several books about the infantry and that he had enlisted in the infantry in 1941 and served with the 41st Infantry Division in New Guinea during World War II.

He talked about going on patrol as a rifleman against Japanese forces. In his book, *Fighting for Peace* published in 1990, he wrote the best men he knew "during the war were infantry soldiers and that serving with the infantry was the only right and honorable way to serve."[2] Until that point, he did all the talking. I said very little. After he finished talking about his service in the Army, he said, "I am going to bring you into the department to serve as my Executive Secretary. How do you feel about that?"

"Sir, I'm very honored, and I'll do a good job. But I really want to get back to troops."

"That day will come," he said. "But right now, I need you here."

We shook hands and I left his office. As I returned to my office, I thought how genuine and engaging he was, truly respecting the infantry soldier, and I had a warm spot in my heart for him. I was pleased to be working for a secretary that had served in the Army in combat and held the infantry in such high esteem. While working for him he often harkened back to his days as an infantryman in the jungles of New Guinea, sharing with me his fondest memories of that time. A couple of those stories can't be shared in this book.

General RisCassi hosted a farewell luncheon in one of the Pentagon's dining rooms for me and several other officers who were leaving DCSOPS. He thanked us and wished us well. Most were returning to troop units and happy to leave the Pentagon. General RisCassi remarked "and Colonel Matz is moving down the hallway and will remain in the Pentagon."

As executive secretary of the Department of Defense, I led the secretariat staff and was responsible for the management, control and review of all action and information correspondence for the secretary and deputy secretary. I was also the OSD point of contact between DOD and the White House, National Security Council (NSC), State Department, Central Intelligence Agency (CIA) and other government agencies. The secretariat staff also prepared the SECDEF's annual report to Congress, his report on Soviet military power, and his weekly report to the president.

Secretary Weinberger, referred to as "Cap" by his close friends, had spent most of his undergraduate time at Harvard College writing editorials for the *Harvard Crimson*, the daily college newspaper. In law school, he was editor of the *Harvard Law Review*, a journal of legal scholarship. He was an excellent writer and expressed his views on paper better than anyone I knew. He prided himself on his rhetorical skills. That posed a challenge for me and my secretariat staff, as he generated volumes of correspondence and reports that needed to be sent out in perfect form. Cap was a communicator. He loved sending and receiving letters/notes.

Our job included drafting the SECDEF's correspondence to include personal letters ensuring they were grammatically and factually correct. Fortunately, the Army, Navy, Air Force and Marines sent bright, outstanding officers to serve on the secretariat staff. Most were excellent writers. One of the best was a Navy SEAL who drafted most of Secretary Weinberger's personal correspondence. At times, he would challenge us on the use of correct syntax—the order of words and phrases to form proper sentences. "Is that sentence a simple sentence, a compound sentence, or a complex sentence?" he would ask. I learned, "All rules of syntax are grammar rules, but not all grammar rules are syntax rules." Indeed, Cap Weinberger was a skilled writer and grammarian of the first order. He often challenged what we drafted and delighted in teaching us about the English language.

During the several months I was executive secretary for Weinberger, two events stand out among many the DOD dealt with requiring a full-court press. Both involved Iran. The first was the Iran–Contra affair, where the administration secretly sold arms to Iran and used some of the proceeds to fund the Contras, a right-wing rebel group in Nicaragua. The Congress was investigating events occurring two years earlier. Cap was scheduled to appear before a combined committee co-chaired by Senator Daniel Inouye and Representative Lee Hamilton on July 31 and August 3.[3]

He didn't take that lightly and spent hours preparing for the hearing; my office assisted. We gathered documents, notes, transcripts of earlier meetings and other materials for his review. We were constantly providing papers and information he requested. Will Taft, Rich Armitage and other key people in the department grilled him daily in preparation for the hearing. I was a "backbencher" in his office during some of these prep sessions. It was an extremely busy time, and I felt a certain tension throughout the office.

Iran–Contra was a serious matter, and the hearings were covered extensively by the press. We wanted our boss to do well at the hearing. Weinberger did his homework and was well prepared. It was a difficult period for Cap Weinberger when he learned his department transferred weapons parts to the CIA to send to Iran. He wrote in his book *Fighting for Peace*, "I seriously contemplated resignation."[4]

The second event requiring a full-court press was the Persian Gulf tanker escort enterprise. In 1987, Kuwait asked for U.S. assistance for its oil tankers passing through the Persian Gulf. Iran was openly attacking shipping. Congress had some doubts about the U.S. Navy providing escorts for Kuwaiti tankers and wanted more influence over the Navy's escort operation. Some even discussed defunding the escort operation.[5] The secretary of the navy (SECNAV), James Webb, also had misgivings and expressed them in a memo to Weinberger in July.

Webb was concerned that there was no clear end date to the operation and that it would not have the support of the American people.[6] I remember the day Cap learned of Webb's objection. It surprised him that his SECNAV didn't support the maritime operation; he was disconcerted and not happy. Weinberger did well convincing Congress and addressing Webb's concerns.

The event proving just how dangerous the region had become was the USS *Stark* incident, when it was hit by an Iraqi F-14 firing two Exocet missiles into the ship, killing 37 sailors on May 17, 1987.[7] Making matters worse was when an Iranian airliner was mistakenly shot down by the USS *Vincennes*,[8] as the U.S. Navy was on high alert after the *Stark* incident. During that period, the Iran-Iraq War raged from 1980 to 1988 creating additional political and military tensions throughout the Middle East. This was, indeed, a busy time, and stubborn Cap was relentless in driving the administration's position to support the Kuwaiti request. He closed ranks on the issue and no one, absolutely no one, was going to change his position!

Supported by Secretary of State George Schultz, Weinberger argued strongly to support Kuwait. President Ronald Reagan agreed, and the first escort operation was July 21. Weinberger, in his book, argued it was the right thing to do, "making sure that nonbelligerent and crucial commerce could and would flow freely in the open international waters of the Gulf without being subjected to mining or other attacks by Iran."[9]

With increasing Chinese-made Silkworm missile attacks by Iran on our ships, Cap told the U.S. theater commander, General George Crist, USMC, "You can shoot first and not wait until fired upon."[10] What a bold and courageous directive altering the Rules of Engagement! He knew that, but he also knew what made sense and what didn't. He wanted his sailors and ship captains to know the SECDEF supported them. He didn't want to lose another sailor in Arabian waters. On August 15, 1988, the Iranians gave up and agreed to accept a cease-fire. The United States Navy's escort of the tankers was an unqualified success. Cap won the day!

Weinberger was fiercely loyal to his friend, Ronald Reagan. During the time I was his executive secretary, his positions and policies supported the president and were always in the best interest of America. When he believed he was right, he would never give in. He walked into my office on a couple occasions reciting one of his favorite quotes of Winston Churchill, "Never give in—never, never, never, in nothing great or small ..." He knew the quote by heart, and he lived by it. We heard it often!

On July 20, I received a call from Colonel Ron Griffith requesting my presence. He was still XO to DCSOPS and just received a prepositioned copy of the brigadier general promotion list and said, "Make sure you come to work tomorrow." That was late afternoon. The next morning the list was officially announced and Weinberger and Fornell walked into my office to congratulate me and handed me the department's official marked copy of the list, which I still have. Stamped in red are the words "SECDEF HAS SEEN." I swelled with joy at that happy moment. It was a humbling moment as both extended their hand in congratulations ... with Cap giving me a wink of the eye as he left. I said nothing to Linda the night before about my brief meeting with Ron Griffith, but after the secretary and Fornell left my office, I called and gave her the good news.

On November 5, Weinberger announced he would resign as SECDEF. On November 23, Frank Carlucci, National Security Advisor, was sworn in as the new SECDEF. On November 12 a farewell reception was held at the Anderson House in Washington, DC, in honor of Secretary Weinberger.

Days later, on behalf of the executive secretary staff, I presented him with a copy of the 4-inch-thick newly published *Random House Dictionary of the English Language*. My staff thought it a most appropriate gift. We all signed it. It was a warm, pleasurable, moment for the former infantryman and editor of the *Harvard Law Review*. He was delighted and couldn't thank us enough! Standing, leafing through the pages, he so eloquently quipped as to how he and Random House "didn't always see eye to eye," and he would now "pay closer attention to what they had to say." I never saw Weinberger so at ease and in such a good mood as when he received the dictionary. Indeed, a light relaxing moment! Going through life, there are moments indelibly seared in your brain, and this was one of those moments for me.

On November 17, an Armed Forces review and awards ceremony was held on the Pentagon parade field in his honor. It was hosted by Admiral William J. Crowe, Jr., chairman, JCS. The distinguished guest was the president of the United States (POTUS). Cap reviewed the troops for the last time. President Reagan in his remarks said, "To anyone who calls for even the slightest slacking off in commitment to a strong and ready national defense, I'll have two words: Cap Weinberger."

The *Washington Post* the next day reported, "Both Weinberger and Reagan wiped tears from their eyes as the military band played the hymns of the four services."[11]

Secretary Carlucci reported for duty on Monday, November 23. Carlucci was no stranger to the Pentagon. He had been Weinberger's deputy secretary from 1981–83. He also held office with Weinberger in the Nixon and Ford administrations. He was very competent, spoke softly, and had a serious demeanor. Coming from the NSC, he was knowledgeable about the defense and security issues the administration was grappling with.

Carlucci approached matters with a lighter, more conciliatory tone than Weinberger. He took the extra step needed to placate his opposition. He let Congress know he was willing to negotiate on contentious matters. He whispered to me once about the importance of compromise. He got immediately involved in the Persian Gulf escort affair and, being a former naval officer, held a hard line and supported our Navy escorting the tankers. In my view, he was the perfect replacement and served as SECDEF until the end of the Reagan administration. General Colin Powell, his deputy at the NSC, became the national security advisor.

On July 22, Secretary Carlucci hosted a promotion and farewell ceremony for me in the Pentagon. I had received orders to be an assistant division commander in the 7th Infantry Division at Fort Ord, California, weeks before, and my promotion date to brigadier general coincided with my departure. It was a beautiful ceremony followed by a lovely reception. The secretary and Linda pinned my star on. When it came time for me to say a few words, I welled up with emotion and pride. Our three kids, Linda's dad, brother and family, my sister and her family, as well as several cousins, attended.

Secretary Carlucci insisted I invite as many family and friends as I wanted. "I'm Italian and family is everything. Invite them all. This is a big day for you and them," he said to me. He enjoyed the ceremony and was smiling and the most relaxed I had ever seen him. Standing at the podium, he remarked as to how the executive secretary wrote all his correspondence. "He probably wrote this," he joked, referring to his prepared remarks. He was gracious and spent time with our friends and family. By the end of the day, he was worn out having had his photo taken several times with each family member and friend.

On July 26, Linda and I had a courtesy call with President Reagan in the Oval Office. It was arranged by the White House staff with whom I had worked closely as the OSD point of contact with the White House. The president said to us, "I'm upset with you—you are beating me to California." He was in a good mood and in his last few months as president and was looking forward to returning to his ranch in Santa Barbara. Indeed, we were honored and thrilled that the Commander in Chief would take time in his crowded schedule to say thank you and farewell. I was taken back by his witty quips and humor about leaving DC and returning to the ranch. Our brief visit gave him pause to realize there was light at the end of the tunnel for him, also! The aide twice had to interrupt us to keep the president on schedule.

When we left in July 1988, after three years in the Pentagon, our two daughters had graduated from Lake Braddock High School. Heather and Billy were attending Penn State University, and Becky was beginning college at Philadelphia College of Textile and Science (renamed Thomas Jefferson University). Linda would leave her many friends at Moran, Stahl and Boyer, where she had been a relocation specialist for the last two years. We would be "empty nesters" for the first time. Linda was looking forward to life on the beautiful Monterey Peninsula. I was looking forward to returning to troops and serving with the 7th ID.

Newly promoted and assuming my first general officer assignment, little did I know that the following year I would be involved in a major event that would test the U.S. military's ability to deploy rapidly to eliminate a regional security threat!

CHAPTER EIGHT

Deposing a Dictator—Panama Invasion

It was a beautiful day in early August 1988, when we arrived at the main gate of Fort Ord, California. My wife, daughter, and I had been on the road for six days driving from Virginia. I had never been to Fort Ord or California's Monterey Peninsula. The large sign at the entrance that read, "7th Infantry Division (Light) Home of the Lightfighters," was a welcome sight after three years in the Pentagon. Lieutenant Ted Wilson and a Specialist 4 Yeager met us at the gate. They would be my aide-de-camp and driver my first few months. "Welcome, to the 7th ID!" they barked out. Then they escorted us to our quarters. I was back with troops and couldn't be happier.

Having just been promoted, I wanted to be an assistant division commander. I had been told initially I was returning to the 82nd Airborne where I had served two previous assignments. But it is not uncommon for division and corps commanders to trade among brigadier generals, who have only limited influence on their assignments—and ultimately, with the approval of the General Officer Management Office (GOMO), I landed at the 7th ID, instead. It didn't matter to me, and in hindsight, I believe someone above was looking out for me knowing that I had a weak, injury-prone leg from polio. Sustained parachuting and hard landings with the 82nd would've been detrimental at that point in my life.

In July 1983, General John Wickham, my division commander when I commanded an air assault battalion in the 101st Airborne in the late 1970s, became the new Army chief of staff. Within three months of his appointment, he announced his decision to create light infantry divisions (LIDs) to address contingencies and growing threats requiring more rapidly deployable forces to operate in restrictive environments. That was a major change of direction for the Army. The decision was met with some resistance both inside and outside the Army. Lieutenant Colonel Timothy Wray, in an essay written for the National War College, stated, "The decision took the Army largely by surprise ... no 'consensus building' had been done to prepare the Army community to accept the new force structure and its strategic role."

The "heavy force" community saw light divisions as a "serious threat to their own preeminence within the Army's mission hierarchy." Power, prestige, resources, and numbers of command positions were at stake. Outside the Army, the Marine Corps and civilian defense contractors were in opposition. The Marines saw it as encroachment on their turf. The contractors saw light divisions as a shift away from what had been their lifeblood, manufacturing and selling expensive sophisticated equipment, weapons, and vehicles.[1]

In 1983–85, I was commanding the 4th Training Brigade at the Armor Center at Fort Knox when Wickham made the announcement. I was the only Infantry branch colonel on a post filled with scores of armor branch colonels and hundreds of other armor branch officers. They kiddingly asked if I had anything to do with the decision.

"You infantry guys are trying to put us out of business. The real threat is an attack through the Fulda Gap. Tell me how a light division is going to stop a Soviet armored unit." They would tease and banter with me with comments like that and questions every day. I listened and rebutted, "You can always branch transfer to the infantry!" None ever accepted the offer.

Wickham found strong support in General William DePuy, who had been highly influential in the development of Army doctrine in the 1970s, and John O. Marsh, who was secretary of the Army.[2] He launched the Army's light divisions for sound strategic reasons. I was on the Army staff during Wickham's last two years as Army chief of staff when he oversaw the fielding of the LIDs and whenever I passed him in the corridor, he would say, "Keep those light divisions in your sights, we're almost there."

The decision to develop more deployable light forces revealed for me the institutional prejudices, interests, and jealousies of various groups, both inside and outside the Army. Some senior Army officers were so close-minded they couldn't relinquish their Cold War thinking, convinced the only threat was the Warsaw Pact requiring heavy forces in Europe. If the Army was going to be a globally strategic force, then structure changes were needed. Moving to a more low-intensity crisis response structure was the answer. It was an interesting time to be on the Army staff.

In my 33 years in the Army, the enhancement of special operating forces, coupled with the creation of the new light infantry divisions, were perhaps the soundest decisions made in ensuring the Army can provide combat capabilities across the entire conflict spectrum. My new outfit, the 7th ID (Light) was one of five new light divisions. Four were in the active force, and one an Army National Guard division. Although the 7th ID was known as the "Bayonet Division" from its World War II and Korean War service, the division's soldiers would become known as the "Lightfighters."

Tactically sound and capable of rapid deployment, the 7th ID could deploy across the operational continuum. With a strength of 10,700 and much lighter equipment, the division could be moved into the objective area by fewer aircraft (500 sorties) than heavy divisions. By comparison, a heavy mechanized division had 17,500 soldiers and required 1,502 sorties.

The 7th ID completed certification as the first LID in 1986 and was well into its training cycle when I arrived at Fort Ord. We concentrated our training on lower-intensity conflict scenarios then occurring in the Caribbean and Central America. In addition to training at Fort Ord, we trained at Fort Hunter Liggett, a 166,000-acre sub-post of Fort Ord, and Camp Roberts, a California National Guard post. Hunter Liggett was 85 miles south of Ord and Roberts was 25 miles southeast of Hunter Liggett.

We couldn't have had better training areas than these for the likely contingencies we trained for. The Army gave the 7th ID priority, equivalent to the 82nd and 101st Airborne Divisions, for assignment of Ranger qualified officers and NCOs. By that time, I had served with every type of infantry division, including mechanized (1st Cavalry Division), standard (2nd and 9th ID's Infantry Divisions), airborne (82nd Airborne Division) and air assault (101st Airborne Division (Air Assault) and now I was with the Army's first newly designed LID. I couldn't have been happier!

This was the circumstance in the 7th ID when I arrived. Four weeks earlier, I received a welcome letter from the division's CG, Major General Carmen "Carm" J. Cavezza. He had just arrived at Fort Ord himself with his wife, Joyce, and was still unpacking. His letter of July 8, 1988 covered the "Lightfighter Spirit" and the "challenges ahead" and closed with "welcome to 'Lightfighter Country' and the 7th Infantry Division (Light)."

I first met Cavezza a year earlier in the Pentagon when he was XO to the secretary of the Army, John O. Marsh, and I was executive secretary to the SECDEF, Casper Weinberger. Our offices were on the 3rd floor, E-Ring. On occasion, Carmen called me to his office to relay a personal request from Mr. Marsh, who was always wanting some "close hold information" about a particular issue involving the OSD or the Army. "Bill, keep me informed, this is very important to the Army and to Mr. Marsh," Carm would say in his low-key, soft-spoken manner. We were both glad to leave the "five-sided puzzle palace," as we jokingly called the Pentagon, and return to a troop unit.

Carm was a superb infantry soldier. A graduate of the Citadel and a 173rd Airborne Brigade wounded veteran of Vietnam, he was the right person to lead the new division as we trained and prepared for the challenges ahead. I was the assistant division commander for support (ADC-S) responsible for the readiness of all units in areas of logistics and deployment. That meant supervising those units and systems that prepared the division for alert and deployment and for sustaining the division when it was deployed.

The 7th ID was part of the Army's XVIII Airborne Corps Rapid Deployment Force (RDF). The Corps, sometimes called "America's Contingency Corps," was the Army's strategic response force with the mission to rapidly deploy anywhere globally to "shape, deter, fight and win." When America needs armed forces in a hurry, the first phone call usually goes to Fort Bragg. Units assigned to the RDF train at a higher intensity than the rest of the military. RDF units are required to be "wheels up" (en route to the objective by aircraft) within 18 hours of notification. As ADC-S responsible for deployment operations, I had my work cut out for me.

The 7th ID's wartime mission read: "When directed, rapidly deploy as a light infantry combined arms force to conduct military operations in support of national objectives." From the mission statement, a unit derives its mission essential task list (METL), the fundamental tasks that units are designed to perform in an operational environment. Unit training is focused on the METL. When General Cavezza and I arrived, the 7th ID was well along in training to its wartime tasks.

The one "essential" task that needed some fine tuning was the task to "alert and rapidly deploy by air." Cavezza had served in the 82nd Airborne, and I had served in both the 82nd and the 101st—the Army's two premier rapid deployment divisions. We knew the 7th ID's ability to perform that important task needed improvement. We worked to improve the division's EDRE program. A good EDRE program was crucial for a unit whose mission is to rapidly deploy by air.

I pestered the hell out of XVIII Airborne Corps' G3 to increase the number and size of EDREs for the 7th ID. They were our wartime contingency headquarters and were responsible for planning and conducting most EDREs. Their priority was always the 82nd and 101st, but we needed those exercises to improve our deployment capability, and the number of company- and battalion-size EDREs began to increase. Soon our EDRE program allowed the division to conduct two "no-notice" rapid deployments simultaneously, one to Korea on Exercise *Team Spirit* and one to the southeastern U.S. on Exercise *Sand Eagle*.

In addition to the 10,700-man 7th ID, there was also a separate 2,500-man Combat Support Brigade consisting of four combat support and combat service support battalions, at Fort Ord. It was called the Bayonet Combat Support Brigade (BCSB), which reported to me. I met with the commander, Colonel Jim Carlson, and had him develop a plan where his brigade would assume responsibility for all deployment operations including out loading divisional units during EDREs and contingency operations. I also tasked him with upgrading the division's alert holding areas and the Departure Airfield Control Group (DACG) facilities. In their current state, those facilities were inadequate to support a rapid deployment force unit in the time required.

Prior to that, the BCSB was only minimally involved in deployment operations. Carlson's four battalions could do a hell of a lot more to ensure the division's timely outload and deployment. They were underutilized and making them a close partner with the 7th ID would free up the 7th ID and their commanders from their stateside outload responsibilities allowing them to concentrate on planning for their wartime mission and tasks in the objective area.

Carlson was a highly competent, can-do commander. An "attention to detail" guy if there ever was one, he took the task on with vigor and his brigade, in a matter of months, upgraded and, where needed, built new alert holding areas for personnel, vehicles and ammunition. We also installed larger weighing scales, fencing and lighting to secure the holding and lock-down areas. Our DACG facilities at our departure airfields were also upgraded and relocated closer to the airfield ramps and installed secure communications equipment. We reconnoitered and secured better highway routes for our vehicle convoys traveling the 152 miles from Fort Ord to Travis AFB.

We developed a close working relationship with the Air Force. The 22nd Air Force was responsible for providing and managing tactical airlift west of the Mississippi and the Pacific, including the 7th ID. Their 60th Military Airlift Wing (MAW) at Travis AFB supported us. I made several trips to Travis briefing the 60th MAW's command group and base support elements on the new light divisions, including the organization, mission, and requirements of the 7th ID.

This helped wing/squadron commanders and their aircraft loadmasters on how to configure their aircraft for transporting 7th ID units. I wanted to ensure we had the base commander's full support for our DACG forward element and that they would provide the facilities and logistical support required. We conducted rehearsals to ensure the new upgrades worked. Cavezza wanted me to "sensitize" them to our 18-hour "wheels up" requirement. I got to know the 60th MAW commander and his staff well and they gave me a BOQ (Bachelor Officer Quarters) room at Travis AFB. It was a good marriage!

Within a few months, we had significantly improved our deployment outload systems and facilities giving us confidence that we could perform our RDF mission within the 18-hour "wheels up" time constraint. "Bill, you and Carlson are driving us crazy with all these new deployment ideas and facilities, besides, you're wearing the Air Force out. You need to lighten up a bit, don't piss 'em off," Cavezza said at a division command and staff meeting. He was half-joking in his quiet unflappable way, while his more serious side knew that the 7th ID as a new RDF unit, needed to wrench up its deployment capability significantly. Carm had a wonderful and witty sense of humor. It wouldn't be too long before the call came that would test the division's ability to rapidly deploy.

In addition to EDRE and deployment training, the division remained busy honing its other warfighting skills and tasks. We successfully completed a Battle Command Training Program (BCTP) exercise designed to train division and brigade commanders and their staffs. It was followed by a *Warfighter* training exercise that focused the 7th ID on its warfighting competencies. Our infantry battalions deployed on a regular basis to the Army's Joint Readiness Training Center (JRTC) in Louisiana and the National Training Center (NTC) in California for one- to two-week "force on force" training exercises. We always had a battalion or two training at Fort Hunter

Liggett or Camp Roberts. To further toughen the soldiers, sometimes our battalions would road march on foot a portion of the 85 miles from Fort Hunter Liggett back to Fort Ord, requiring civilian authorities granting necessary highway/road clearances.

Ensuring the combat readiness of the Division Support Command (DISCOM) and the separate engineer, air defense, military intelligence, and signal battalions and the military police company, was also a responsibility of the ADC-S. They provided vital support to our infantry, field artillery and aviation units. Working with the commanders and soldiers of those supporting units was an enlightening and broadening experience for me. I had always been with infantry units and never paid much attention to the logistics and sustainment side and the crucial support they provided to the combat forces.

I selected a bright quartermaster branch officer as my next aide. Lieutenant Dick Scott, a 1986 USMA graduate, had performed well in DISCOM. Although young and inexperienced, I was struck by his maturity and self-confidence. A kid from New Jersey, there was nothing shy about him! He had a solid reputation as a soldier and his knowledge of the support battalions was invaluable. Besides, he played lacrosse at West Point and that appealed to me having played myself at Gettysburg College. We would sometimes get our old lacrosse sticks out and toss the ball around.

Although we were busy, it wasn't all work. Linda and I enjoyed the camaraderie of the combat support battalion commanders and their wives. The "separates" as they came to be known, were a fun and vivacious group. The commanders and their wives—Mike and Betty Meuleners, Pete and Diane Franklin, Doug and Karen Hayden, Chuck and Aline Hurd, and Tom Armeli—were full of energy and kept me and Linda busy organizing our social functions and events and were very supportive of post activities.

Our division commander's wife, Joyce, devoted a great deal of her time and wonderful homemaking talents to ensuring a vibrant and fun-filled social life. She and Carm hosted what my wife described as the "best and most delicious" dinners and luncheons for the commanders, staff, and their spouses. "Eat it while it's hot … and there is plenty more! Carm, pass the rolls," Joyce would say as she ensured everyone's dinner plate was full. Linda and many of the other spouses followed her lead in hosting numerous social events.

All wanted to reciprocate in kind in appreciation for the Cavezzas' generous and gracious hospitalities. The division was known for its caring and bonding culture both socially and professionally. The wholesome socialization engendered among the officers and their spouses helped take the edge off the rigor and intensity we all felt due to the heightened training and exercise deployment operating tempo. Serving with the Cavezzas, a very effective and caring command team, was a joy!

Cavezza saw to it that the individual Lightfighters maintained a high level of physical fitness. All battalions, and most companies, had their own weightlifting and exercise equipment. My aide ensured we were bench-pressing and curling free weights at least three times a week. During those early morning workouts lifting weights, we spotted for each other. While I would struggle to bench press eight reps at 130 lbs, he would do eight reps at 220 lbs, easily.

Companies and battalions conducted unit runs five days a week on the hilly roads and along the rugged beach paths of Fort Ord. Each morning, we ran with a different battalion. At 0630 you could hear the echo of the cadence and singing as the units ran in formation all the way to the neighboring towns of Marina and Seaside. It was "go, go, go" from dawn to dusk. Many a night we spent in the field with the troops, even when not deployed. Emphasis was placed on training at night. As we received and began training with the newer night vision goggles, soldiers became increasingly confident about their ability to move, shoot, and communicate during darkness.

Moving and operating at night would pay dividends in the months ahead. We wanted to "own the night." Indeed, the training and operating tempo was at a high pace; soldier morale and unit esprit were even higher! I began to feel the pangs of a new culture. The vigorous spirit of the Lightfighter had permeated the once quiet and laid-back Fort Ord. The entire post took on a new sense of urgency. The infusion of a nucleus of NCOs and officers who had served in the Army's other high-spirited RDF units had paid off. We were craving a real-world contingency to test our RDF capabilities.

We monitored the lower intensity threats in Central America posing challenges to the United States. The most significant was evolving in Panama, the roots of which began in 1981 when General Omar Torrijos, the *de facto* ruler of Panama and head of the Panamanian Army, was killed in a fiery plane crash.[3]

At the time, Lieutenant Colonel Manuel Antonio Noriega was Torrijos's intelligence officer and right-hand man, and within two years promoted himself to general and assumed command of the newly formed Panamanian Defense Forces (PDF), thus becoming *de facto* dictator. To this day, some blame Noriega for the crash that killed Torrijos.[4]

Noriega was corrupt and ruthless to the core. Torrijos called him "my gangster."[5] He ruled the country through the military power of the PDF, taking control of the news media and becoming a major conduit for the evolving drug trade in South and Central America.

Despite that, the U.S. maintained a close relationship with Noriega throughout the mid-1980s, driven largely by the U.S. concern about the security of the Panama Canal, ensuring it remained a viable international waterway. According to John Dinges, a Latin American specialist and author, "Noriega was reported to have played a role in the Iran–Contra Affair" that plagued the Reagan administration in the mid-1980s.[6]

In 1985, U.S.–Panamanian relations began a steady decline. As head of a narco-militaristic regime, Noriega and his PDF-led government were increasingly harassing American citizens stationed in the Canal Zone. When President George H. W. Bush declared drugs a "major threat" to America in 1988, "a Florida federal court indicted Noriega for drug trafficking." That infuriated Noriega ... and "relations further deteriorated."[7]

When I was executive secretary for Weinberger and Carlucci in 1987–88, there was hope that a Panamanian solution, such as a *coup d'etat* or national election, would remove Noriega. That was an occasional topic of discussion during the weekly meeting between Weinberger and SECSTATE George Shultz. I helped prepare Weinberger's and Carlucci's read-ahead materials for those meetings. Both secretaries, as well as the White House, were paying very close attention to the unfolding events in Panama.

Weinberger didn't trust Noriega, and always believed the U.S. would eventually clash with this dictator. "I told the president, this guy is not going to go away, we're going to have to go down there and teach him a lesson," Cap would say. He was right! The coups we hoped for never happened, as Noriega was able to crush two coup attempts by his officers to overthrow him in March 1988 and October 1989, having the coup leaders killed or imprisoned. When his candidate was losing the national presidential election in May 1989, he put down the opposition with brute force declaring the election null and void.[8]

Meanwhile, he continued to wage an ever-increasingly violent campaign of harassment and intimidation against U.S. personnel.[9] With tensions rising, U.S. Southern Command in Panama began preparing for military action. Chief among U.S. concerns was the safety of U.S. citizens and the security and uninterrupted operation of the Panama Canal.

Operation *Blue Spoon* was a contingency plan in the *Prayer Book* series of U.S. military plans for Panama. It "envisioned that in the face of a threat to American interests, the U.S. would land military forces at Howard Air Force Base,"[10] which would then move to counter the threat. Due to Noriega's defiance and the ever-increasing tension, it became clear that a methodical, sequential build-up of forces, called for in *Blue Spoon*, ran the risk of Noriega's PDF not only preemptively interdicting and blocking the buildup, but also moving to take American hostages and facilities, most critically, the canal and its supporting facilities. This wasn't acceptable.[11] We always worried about Noriega taking hostages.

Consequently, *Blue Spoon* evolved into a more rapid operation focused on quickly taking down Noriega and his PDF units with overwhelming force. Thus, from June 1989, the U.S. military plan shifted from a slower sequential operation to a rapid, overwhelming, and decisive invasion.[12]

With Noriega overturning the presidential election in May, President Bush announced that the U.S. had seen enough thuggery in Panama. He proclaimed that Noriega "must go."[13] The *Chicago Tribune* reported on May 14, 1989, that Bush told reporters aboard Air Force One, "The will of the people should not be thwarted by this man and a handful of Doberman thugs."

Within the U.S. military command structure, some changes had also been made. General Maxwell Thurman, an aggressive, no-nonsense officer, replaced General Fred Woerner as commander, United States Southern Command.[14] I always saw Thurman as a soldier possessing uncommon vigor and a determination to win. He had a brilliant mind and knew every facet of the Army and its capabilities. He was full of self-confidence, always winning the argument and getting what he wanted and doing it his way. Thurman needed a warfighting operational command to plan and execute the operation.

He requested the Army's XVIII Airborne Corps as the Joint Task Force (JTF) warfighting headquarters that would plan and execute the operation. Lieutenant General Carl W. Stiner was the corps commander. Thurman told the Army chief of staff, "I want Stiner to be my war planner, my warfighter."[15] He was experienced, and Thurman knew and trusted him. Thurman and Stiner knew that they might have to launch operations at very short notice. Execution of rapid operations against multiple targets dispersed throughout the country required a fully staffed operational headquarters that could execute on short notice.[16]

Stiner focused specifically on developing and rehearsing a detailed plan to eliminate Noriega's entire PDF. XVIII Airborne Corps was organized and ready for such a mission. Not only would they plan the operation; they would also command all joint forces during execution. Thurman worked closely with the chairman of the Joint Chiefs, General Colin Powell, to define U.S. goals and objectives for Panama. These would become the basis for all planning. The U.S. objectives, as approved by President Bush, were to:

"1. Create an environment safe for Americans.
2. Ensure the integrity of the Panama Canal.
3. Provide a stable environment for the freely elected Guillermo Endara government.
4. Bring Noriega to justice."[17]

The decisive tactical objective was the destruction of the PDF. With that decision, Stiner and his staff began to flesh out, in detail, the plan for the U.S. invasion of Panama.[18]

After Noriega declared the May national elections null and void, President Bush ordered an additional 1,900 combat troops to Panama to increase security for American personnel and facilities, augmenting the almost 12,000 U.S. troops permanently stationed there. The operation was code-named *Nimrod Dancer*.[19]

As part of the operation, 7th ID deployed elements of its 1st Brigade on an EDRE to Howard AFB in Panama. We knew something was in the wind. We got the alert call at 0800 hours, May 10, and immediately executed deployment outload procedures at Fort Ord and Travis AFB. N-Hour was 1330 hours, May 11, and the first aircraft was "wheels up," departing Travis AFB at 0650 hours, May 12.

Operating out of Fort Sherman, located at the Caribbean (northern) end of the Panama Canal, the 7th ID force became Task Force Atlantic. Atlantic's mission was to "defend the integrity and security of U.S. defense sites and to protect U.S. military and civilian populations," representing a deliberate demonstration of force. From May 11, the 7th ID maintained a brigade in Panama as Task Force Atlantic throughout an indefinite period.

We stayed busy resupplying our forward-deployed brigade. Cavezza and our other assistant division commander, Brigadier General Bob Ord, ADC for Operations, and I logged many air miles taking turns visiting our troops in Panama and participating in XVIII Airborne Corps invasion planning that was secretly taking place at Quarry Heights in Panama. General Ord later became Commander, U.S. Army Pacific.

Operation Nimrod Dancer demonstrated that the 7th ID could deploy within the 18-hour window, arriving in the theater of operations prepared to conduct combat operations, if necessary. The many weeks spent working and training with the Air Force to improve the division's rapid deployment outload systems were paying off. I was feeling more confident every day in our ability to get to the objective quickly and fight and win!

During July to November, violent acts directed toward U.S. soldiers and their families increased. Many people were openly harassed by the PDF and the "Dignity Battalions." These battalions were paramilitary militia units created by Noriega in early 1988 to augment the PDF in defending against possible invasion by the U.S. Their ranks were filled with cronies of his ilk, tough, violent hoodlums. Noriega also used them to suppress domestic political opposition. There were approximately 12 battalions, initially, each with 25–250 male and female volunteers. Five were in Panama City, and the others were dispersed throughout the country.[20]

Noriega's Dignity Battalions brutally suppressed the demonstrations resulting from the May election. They severely beat the three winning presidential/vice-presidential candidates who ran against Noriega's candidate and shot and killed their bodyguards. These were covered in the May 22, 1989, edition of *Time* magazine. This brought worldwide attention to the brutality of the Noriega regime.[21] We in the 7th ID took notice of the increasing violence and began to study the order of battle of the PDF in detail.

In October, 3rd Brigade replaced 1st Brigade as Task Force Atlantic. The 3rd Brigade was commanded by Colonel Keith Kellogg, a hard-charging infantryman and Vietnam veteran. He later commanded the 82nd Airborne and retired as a lieutenant general. In 2018, while serving in the Trump administration's NSC, Vice-President Mike Pence chose Kellogg to serve as his national security advisor.[22]

One morning in early November while cooling down from a long division headquarters run, Cavezza pulled me aside: "Bill, I want you to go to Panama to attend an important planning meeting with Stiner and to visit the 3rd Brigade."

My aide-de-camp and I boarded a resupply aircraft the next night out of Travis AFB for Panama.

On November 3, the JCS approved Stiner's invasion plan. During the planning, the codename for the plan changed from *Blue Spoon* to *Just Cause*, a more appropriate name. The purpose of the meeting I attended in Panama was to brief the major force participants on the new plan and the JCS's decision. Joint Task Force South (JTFSO), with XVIII Airborne Corps as the nucleus, and Stiner commanding, was established as the command that would execute the operation. Everyone was sworn to secrecy. Stiner and Thurman were adamant, there would be no leaks. Secrecy was paramount: God help anyone who mentioned it outside the secure planning facilities!

I also spent a day with 3rd Brigade on a routine armed route reconnaissance patrol in the northern half of the canal zone. As our armed Humvees drove through the outskirts of Colon city, we came face to face with armed PDF troops manning check points along the roads, and we slowed down. They stared us down with angry looks while moving toward our vehicles, clutching and pointing their rifles in a way that signaled we were not welcome. "Yanquis, go home!" shouted one of the PDF soldiers, waving his unholstered pistol at us as we drove by slowly. I thought to myself, "The clock is ticking. How much longer will we put up with this kind of bullying and intimidation?"

The threats and intimidation by the growing numbers of Dignity Battalions, had penetrated PDF forces where soldiers were now brazenly demanding U.S. forces leave Panama. The average Panamanian citizen didn't want Americans to leave. Not once during the reconnaissance patrol did I feel seriously threatened, but I could see and feel firsthand the tension building between U.S. and PDF soldiers as their armed vehicle patrols often crossed each other's paths.

PDF soldiers increasingly walked in front of U.S. vehicles yelling threats while brandishing weapons. It wouldn't take much to provoke either a young PDF soldier or U.S. Lightfighter to level his weapon and fire. Our troops had strict orders not to engage verbally. Restraint became critical to avoid igniting an international incident during that tense time.

I returned to Fort Ord and briefed Cavezza. My exact words to him were, "Sir, I bet we're fighting in Panama before Christmas. Our guys aren't going to take this crap much longer!" I wonder if later Cavezza thought I was a prophet!

After JCS approval of OPORD 1-90 (*Just Cause*) the forces involved began to refine the target lists. TF Atlantic would seize eight targets in the Canal Operating Zone from Gamboa to Colon. Kellogg's brigade, at Fort Sherman, began preparing and rehearsing for those missions.

Unfortunately, the fast-paced OPTEMPO, coupled with the relentless, five-day-a-week arduous PT schedule, was beginning to challenge my polio leg. I was experiencing increased pain and the leg was weakening with each stride. My left leg was compensating even more for the weaker limb. My father's words of 40 years earlier telling me to "suck it up" when I was recovering from paralysis, rang in my ears. His words, coupled with the enthusiasm of hundreds of Lightfighters I ran with daily, kept me going. I couldn't let them down.

Completing the long unit runs and road marches became the rite of passage for acceptance into the Lightfighter culture. Finishing the runs and road marches reinforced the value of the culture. Those who fell out faced scorn and ridicule. Leaders dare not fall out of a unit run! As a brigadier general and an assistant division commander, that couldn't happen. The thought of humiliation and losing the respect of your soldiers was daunting, something I never wanted to experience, so I kept going, damn the weak polio leg!

The 7th ID's 3rd Brigade stepped up their training and the heightened pace pushed JTFSO to a high state of readiness. It also increased the tension between U.S. and PDF troops. Kellogg informed us daily of what was occurring in his AO and of increasing incidents of harassment by the PDF.

On December 15, the Panamanian National Assembly at the behest of Noriega formally declared that a state of war existed between the United States and Panama. On December 16, PDF soldiers stopped U.S. military personnel at gun point and fired upon American military vehicles at highway check points, including four unarmed U.S. officers in a car in downtown Panama City, wounding three. One, Marine 1st Lieutenant Robert Paz, later died of his wounds.[23]

A U.S. Navy lieutenant and his wife who witnessed the incident were arrested and brought to a police station for questioning. The interrogators beat the officer and pointed a gun at his head, while other PDF goons forced his wife to stand against a detention cell wall while they assaulted her; she collapsed.[24] The following day, the PDF handcuffed and beat U.S. Army military police soldiers.

The two weeks before Christmas saw many holiday parties at Fort Ord and Monterey. Linda and I supported the Pajaro Earthquake Christmas Benefit on December 10 at the Monterey Conference Center for the victims of the 7.2 quake that struck the San Francisco/Fort Ord/Monterey corridor on October 17 leaving many families homeless.

It was an impressive soiree with people coming together to help those in need, and it wasn't without celebrities. Clint Eastwood was Master of Ceremonies. He was joined by Doris Day, Chevy Chase and Paul Anka, to name a few. Representative Leon Panetta played the piano. For us, it was the beginning of the Christmas season festivities.

A week later, we were enjoying a lovely holiday reception and dinner at Bob and Gail Ord's quarters on December 17 when the phone rang at 2130 hours. It was General Stiner alerting us to the impending operation in Panama and to be ready to execute *Just Cause* in a matter of hours. Cavezza took the call and pulled Bob and me aside, informed us and swore us to secrecy. The wives and everyone else present were beyond curious as to what the call was about. We said nothing.

Powell had briefed the president that afternoon and told him we needed to launch *Just Cause* now to "minimize the time available for the PDF to seize U.S. citizens" and neutralize the PDF.[25] Cole's account of the meeting as published by the JCS, Joint History Office, states the president needed "no further persuasion." He ordered execution of *Just Cause* with the words: "Okay, let's do it. The hell with it!"[26]

Stiner's call was our warning order, and at 0900 the next morning we initiated our X-Hour crisis-planning sequence. All 7th ID units identified to support Operation *Just Cause* began preparing for deployment. We activated all outload deployment facilities to include our DACGs.

In the interest of maintaining security and keeping the press away, I personally met with the Monterey Regional Airport authorities to alert them we would soon be using the airport to deploy but couldn't say where or when. Travis AFB alone couldn't handle the number of air sorties and personnel, equipment, and ammunition that had to be moved in the short amount of time required. Monterey Airport authorities knew the possibility that we could take over their airport at some point, but never thought it would happen.

They initially balked as many flights into and out of Monterey were scheduled over the Christmas holidays but seeing the handwriting on the wall, the airport authorities joined the Lightfighter team and were most cooperative once the deployment began. Over their strong objection, I activated our DACG facility at Monterey Airport that day. Since some civilian flights were cancelled, I wasn't a popular guy, but time was of the essence, and I wasn't going to wait. 7th ID Operations Center received the JCS execute order the next day at 0700, Tuesday, December 19. "Wheels up" for the first aircraft was 0100, December 20. The call we had been waiting for, the mission we were craving for, finally came. We were ready. D-Day was December 20, with H-Hour at 0100. The invasion of Panama had begun.

We did well until that point maintaining secrecy. The massive movement of U.S. aircraft as they repositioned to airbases in the States, and began arriving at Howard AFB in Panama the afternoon of December 19, compromised strategic surprise. It was obvious that something was about to happen. Cole reported, "At 2200 on the evening of December 19, Dan Rather of CBS News commented: 'U.S. military transport planes have left Fort Bragg. The Pentagon declines to say whether they're bound for Panama ... only that XVIII Airborne Corps has been conducting what the Army calls an emergency readiness measure.'"[27]

A San Francisco TV and radio station reported "troop-carrying aircraft" arriving at Travis AFB and military vehicles leaving Fort Ord for Travis. Our wives at Fort Ord had been watching local and national news reports on TV about the possibility of U.S. forces invading Panama ever since our "warning order" call on December 17, yet, ironically, sworn to secrecy we could say nothing. "Bill, I saw on the news that Army units from California were preparing to go to Panama. That's got to be you," my wife said. She knew it was!

While we were feverishly preparing to meet our H-Hour launch time, and not being able to disclose the details, our wives were learning all about the impending operation from the news broadcasts. I thought they had better information than we had. Linda had long since made reservations for all of us to go skiing at Lake Tahoe over Christmas with our two daughters who would be home from college. The five-day ski trip went as planned, but without me.

I was on the ramp at Travis AFB with my aide-de-camp, when Cavezza, his command sergeant major, and personnel from the Division Tactical Command Post (DTAC) arrived. The damp, gray-colored fog was so thick that night, I couldn't see more than 20 feet in front of me. An Air Force ramp NCO said, "You could cut it with a knife." He was right!

"Bill, where the hell are you—slow down!" yelled Cavezza, as I was in front following the Air Force ramp guide leading them to their airplane on the ready ramp. Theirs would be the first to take off. The fog was so thick that we couldn't see the airplanes all painted gray, let alone the tail numbers, and went to the wrong airplane.

Finally, after we went to two airplanes, we located the right C-141, and Cavezza and his party loaded. Over the din of the four jet engines, I yelled, "Good luck!" His reply was, "Make sure you get all the troops loaded and out of here." The loadmaster on Cavezza's airplane secured the troop door, and the C-141 began taxiing away from its parking space. Minutes later, it took off. We couldn't see anything, but we heard the loud climbing roar of the engines as the C-141 lifted into the sky. My Air Force counterpart and I looked at our watches. It was 0056 hours. We made our "wheels-up" time for the first aircraft with four minutes to spare.

As reported in the July–December 1989 history of the 60th Air Wing, "The dense fog threatened to delay the early morning operations, but a radar-controlled lighting system enabled schedules to be met. Colonel James C. Schaffer, DCO, stated, 'I'd be lying … if I didn't tell you I was worried about the weather.'" I was in awe of the Air Force pilots and ramp personnel and their ability to load and take off under the strain of such adverse weather conditions.

During my *Just Cause* oral history interview on April 30, 1992, I told that story when asked by Dr. Robert Wright of the U.S. Army Center of Military History, "What was the funniest story that happened to you during the deployment?" At the time it wasn't so funny, but later I thought about it and laughed. That incident also portrayed for me the ominous task that laid ahead for me in loading thousands of troops and equipment that night and the next morning. I had a sinking feeling in my stomach.

Thirty-eight additional sorties of C-141s and C-5s departing from Travis AFB at 20-minute intervals and 35 C-141 and C-5 sorties departing from Monterey Regional Airport at 25-minute intervals carried the remainder of the 7th ID invasion force, their equipment and ammunition. The last sortie arrived at Howard AFB, midday December 23.

In addition to the 7th ID force (eight of the nine infantry battalions deployed), the U.S. invasion force consisted of mainly a reinforced brigade of the 82nd Airborne, the 75th Ranger Regiment and other Special Operations forces including Army Special Forces, Navy SEALs, Air Force Special Operations Personnel and Psychological Operations Specialists.

The Panamanian Defense Force numbered about 12,000 troops, including National Guard and police forces. Only 4,000 were combat troops. In addition, there were 12–18 Dignity Battalions. We were never able to correctly determine their number or strength. It was an overwhelming U.S. force in size and capability. Including the U.S. forces already stationed in Panama, the *Just Cause* force exceeded 20,000 troops.

At H-Hour, 0100, TF Atlantic, led by the 7th ID's 3rd Brigade, attacked into the Colon-Gamboa sector and by 1030, reported all missions accomplished. Special Operations units also struck, simultaneously, at several targets. Navy SEALs tragically lost four men killed and eight wounded as they encountered unexpected stiff resistance while attacking to disable Noriega's airplane at Punta Paitilla Airfield.

At 0100 the Ranger Force conducted parachute assaults against multiple targets from Rio Hato to Fort Cimarron. An additional 2,700 paratroopers from the 82nd joined the Rangers in the largest U.S. airborne operation since World War II. Their mission was to isolate Panama City, while heavier assault forces within the city neutralized the PDF headquarters at the Comandancia.

TF Semper Fi, consisting of U.S. Marines, secured the Bridge of the Americas and the area surrounding Howard AFB while TF Bayonet with the 193rd Infantry Brigade captured Fort Amador and destroyed the PDF headquarters and the Comandancia building in Panama City. TF Pacific conducted the last major combat assaults on D-Day and by 2000 hours had secured all their objectives.

It was swift and overwhelming. Throughout all the attacks, many PDF soldiers dispersed rather than surrender. They continued to resist using hit-and-run tactics, sniping and sabotage. They were armed and began operating in small roving bands. Some Dignity Battalions began to loot and intimidate the citizens rather than stand and fight the "Yanquis."[28]

With the PDF largely defeated by December 23, the Dignity Battalions constituted the chief danger. They remained a threat to U.S. personnel. They were not trained, disciplined soldiers; these were common thugs and, in some ways, more dangerous.

Meanwhile, at Fort Ord, small groups of demonstrators from the San Francisco and San Jose areas had converged on the main gate. They were a mix of college-age students and some older citizens carrying placards and make-shift signs. Some had bull horns. They chanted anti-war slogans protesting the legitimacy of the invasion and demanding entry to the post.

Representative Panetta's office called me wanting me to go to the gate "now" and "talk to them." I was 156 miles away at Travis AFB overseeing the deployment of our troops and couldn't leave and I said I would get there when I could. Panetta's caller wasn't happy with my answer and called again, insisting I go to the gate immediately and meet with the protestors. I was beyond angry and gave him a "wait out" and passed the call to our military police company. They handled the situation, dispersing them away from the main gate.

There was no physical violence, and no protesters ever entered the post, but they tried. In each case, they were stopped by our MPs. They were loud and clamorous, but harmless, like most protesters. When I finally arrived two days later, after the last airplanes took off from Travis, a few protesters were still camped out on the sand dunes across the highway from Fort Ord's main gate. When they saw me, one woman yelled, "Bring the troops home now! You have no right to invade another country!" They eventually left.

Meanwhile in Panama, the hunt was on for Noriega. When the first U.S. forces arrived in the early morning hours of December 20, Noriega fled. He abandoned his commanders and soldiers and hid, showing his true colors. He finally sought asylum in the Vatican's Embassy in Panama City which granted him temporary political refuge until he could be transported to either Cuba or Spain.[29] For the next 10 days, U.S. and Panamanian authorities tried to secure Noriega's release. The Nunciature (diplomatic post of the Holy See) area was cordoned off. Finally, with the Vatican's assistance conditional that the U.S. agree not to pursue the death penalty at his trial, Noriega surrendered.

With the capture of Noriega and the defeat of the PDF, major combat operations of *Just Cause* were over and U.S. forces began redeploying back home. The 7th ID forces would remain to assist civil affairs troops in carrying out the next phase of the operation called *Promote Liberty*. That phase included mop-up operations of remaining enemy holdouts and re-establishing law and order. Despite the defeat of Panama's armed forces, the Dignity Battalions and armed criminal elements continued looting and sniping at U.S. and Panamanian citizens. They had to be removed.

Small bands of armed, former die-hard PDF and Dignity Battalion personnel scattered to the countryside, toward both the Columbian and Costa Rican borders, had to be located and neutralized. U.S. Ambassador Arthur H. Davis wanted infantry forces to remain to ensure the Endara government would be safely protected against any reprisals.[30] We had indications that could happen. Those critical post-invasion tasks fell to the 7th ID.

Up until that time, I had remained stateside, balancing my time between our two departure airfields and Fort Ord, ensuring the timely outload and deployment of 7th ID forces and the daily resupply of equipment and ammunition. With the troop deployment and initial resupply missions complete, and knowing I was spoiling to get into the action in Panama, Cavezza called me from Panama, "Bill come on down, I need you here now and want you to take charge of the Army and Marine forces for the Promote Liberty stability operations."

Stiner and his XVIII Airborne Corps element were returning to Fort Bragg. The responsibility for executing the Promote Liberty phase of the operation was turned over to Cavezza as the commander of Joint Task Force Panama. I had been packed and ready to go since before D-Day, and it didn't take me long to get an aircraft out of Travis AFB to Howard AFB. My aide and I were in Panama within hours of Cavezza's call.

Carm would kid me later about my "bitching" and wanting to get out of Fort Ord and Travis AFB and join the action in Panama. I always knew being the ADC for Support, that when the call to deploy came, I would not go with the assault force. My job was to remain behind and get the troops and equipment to the departure airfields and onto the airplanes for their scheduled deployment times. Cavezza had been the ADC (Support) in the 82nd and he knew how I felt.

When I arrived at Howard AFB, Colonel Jim Wright, our DISCOM commander met me, and we flew by chopper to JTF Panama headquarters at Fort Amador. My job getting the troops out of Travis and Monterey was complete. I needed a change and was happy to get to Panama to help with the continuing operations. After a quick turnover briefing, Cavezza said goodbye and returned to Fort Ord, and I took command of the 7th ID's Promote Liberty stability operations.

Promote Liberty presented a different challenge. We had failed to prepare for the transition to stability operations. As Noriega's regime fell, "There was no coherent plan to assist the transition to a new government ... We underestimated the complex threat that emerged as the PDF dissolved."[31] In an interview on April 2, 1990, the spokesperson for the J3, OJCS, admitted it was a weakness in the planning.

Despite this void in planning for a restoration mission, Army civil affairs units, bolstered by the 7th ID and 16th MP Brigade, were up to the task. For the next three weeks, our infantry units swept the distant countryside searching for and rooting out the last pockets of PDF and Dignity Battalion resistance moving at night whenever feasible.

At daybreak, they swept quickly into the towns surprising any PDF or Dignity Battalion holdouts. In one way, it reminded me of the search and destroy operations we employed in the Mekong Delta in Vietnam 20 years earlier, but on a far smaller scale, and without the intense combat of Vietnam. Sporadic fire fights in the country towns lasted but a few minutes before the PDF soldiers dropped their weapons and surrendered. Our units captured many PDF soldiers and Dignity Battalion members, evacuating them back to the prisoner of war/detention centers in Panama City.

Our 2nd Brigade conducted air assault operations west to the Costa Rican border covering over 14,000 square miles while operating in six military zones, clearing almost 100 towns in western Panama. Later, they expanded their AO eastward to the Colombian border. 1st Brigade (9th Regiment) cleared a large portion of Panama City of PDF, snipers, and looters, and re-established normalcy in the streets. I spent almost every day in either a Blackhawk helicopter or in my Humvee moving through the countryside with the 1st and 2nd Brigade troops.

By January 15, our DISCOM shipped over 20,000 captured weapons and assorted ammunition to the U.S. for inspection and inventory. Each weapon and piece of equipment had to be catalogued, inspected, wrapped and crated, which was a tedious task.

Among the many arms we captured were thousands of crated Panamanian T65 bayonets with scabbard. They were made for the T65 assault rifle manufactured in Taiwan for the Republic of China forces. Those type bayonets were used in Haiti, El Salvador and Grenada in the 1980s. I learned later they were going to be distributed over the holidays to the PDF forces. We received permission to allow each of our soldiers to keep one as a souvenir.

We worked closely with the newly created Fuerza Publica (Public Force). The new civil police force was composed largely of former policemen and PDF soldiers vetted by the Endara government. The Fuerza Publica made great strides at restoring basic functions throughout Panama City. I worked with their leader, Roberto Armijo, and with the U.S. Army's 96th Civil Affairs Battalion in implementing the many stability tasks.[32]

While searching the Province of Cocle for pro-Noriega holdouts, my aide and I came across Noriega's lavish beach house at Decameron Beach near Rio Hato about a two-hour drive west of Panama City on the shores of the Pacific Ocean. The 75th Ranger Regiment had been there on D-Day where some of the heaviest fighting took place. Much of the area was shot up badly. It was an eerie feeling knowing that Noriega and his family had probably been there less than three weeks earlier. During his reign, many Panamanians were said to have been murdered on the grounds.

As part of our search, we also visited Noriega's house in the exclusive Altos del Golfo neighborhood just outside Panama City. The lavishly furnished house was pretty much untouched, and Christmas decorations remained in place to include a beautifully decorated Christmas tree with wrapped gifts for his wife and daughters underneath. We took photographs but didn't disturb anything.

By the end of January, civil-military efforts were taking shape, and security was restored to the point that the remaining 7th ID forces could return to Fort Ord. Our rifle companies had located and neutralized the last remnants of the enemy. I'll never forget the attitude and smiles on the faces of the Panamanian people.

U.S. causalities for *Just Cause* included 23 killed, 322 wounded. Two 7th ID soldiers were among the 23 American KIAs. I attended the funeral service and burial for Private First Class Gibbs who was interred at the City of Monterey Cemetery. Four days earlier, on December 26, I visited with his young wife and mother at their quarters on Fort Ord.

Operation *Just Cause* was a total defeat of the enemy. In military terms, it was rapid, decisive and with overwhelming force, brought the total collapse of enemy resistance while minimizing collateral damage. In a statement to the press, General Carl Vuono, chief of staff of the Army, said, "The deployment was done so well, the Congress and the press could find no fault with it."

The 7th ID's performance proved beyond a doubt General Wickham's sound strategic reason for launching the light infantry divisions six years earlier. It demonstrated the need for, and the value of, inserting rapidly deployable light infantry forces to augment airborne, Ranger and Special Operations forces. Indeed, it more than validated the strategic concept that drove the creation of the light divisions.

General Jim Lindsay, commander in chief, United States Special Operations Command, noted in a letter to Defense Secretary Dick Cheney that *"Just Cause* demonstrated that both conventional forces and Special Operations forces had made significant progress toward joint interoperability since the poorly executed 1983 intervention in Grenada."[33] During *Just Cause*, cooperation and coordination among all services was outstanding.[34]

While there are many facets of *Just Cause* that could be written about in more detail, including the successful operations of the other U.S. invasion forces, I dwelt here largely on the 7th ID's role and participation, as this is the unit I served with at the time and soldiering with the Lightfighters was a big part of my life's story. Our intense deployment exercise training conducted with our Air Force counterpart paid off![35]

The tough individual and unit training, and the careful planning, including preparations and rehearsals, all contributed to the success of the operation. The USAF performed a herculean task in getting all personnel, equipment, and ammunition into and out of Panama without missing a beat.

All this pales in comparison to the physically tough, disciplined, well-trained infantryman, whose esprit and courage is the decisive factor in the outcome of any ground operation, whether Army or Marine. The infantry owns the last 100 yards of the battle. Indeed, the most valuable player in *Just Cause* was the infantryman. For the 7th ID, it was the Lightfighter. I'm proud to have served with them!

Upon my return to Fort Ord from Panama, I received assignment orders to Headquarters United States Army, Pacific, Fort Shafter, Hawaii. I wasn't excited, as I didn't want to leave the Lightfighters! We had only a few days to say farewell, clear quarters and pack our household goods. Linda and I departed immediately after our farewell ceremony on February 26, 1990 for Sea-Tac Airport and our flight to Hawaii.

CHAPTER NINE

Post-Cold War Years

Headquarters, U.S. Army, Pacific (USARPAC) was located on Fort Shafter, on Oahu, Hawaii. Built in 1905, it's the oldest military base on Oahu, 5 miles from downtown Honolulu and Waikiki Beach. When I arrived, the command was undergoing a redesignation from U.S. Army Western Command (WESTCOM) to USARPAC. USARPAC is the Army component of the U.S. Pacific Command (PACOM) (renamed the U.S. Indo-Pacific Command in 2018).

Lieutenant General Claude M. "Mick" Kicklighter was the USARPAC commander. I was assigned as deputy commander/chief of staff. As principal assistant to the commander, I had two roles. As deputy commander, I was utilized in my true capacity as the second-in-command. When he discussed duties with me, I told the CG that I saw myself as a key enabler within the command who could provide organizational flexibility in the execution of our diverse mission.

In my role as chief of staff, I supervised the headquarters, USARPAC staff. In both capacities, I interacted routinely with PACOM, the other service component commands, U.S. Army Japan, U.S. Army Alaska and the 25th Infantry Division at Schofield Barracks. I also directly supervised seven brigade-size commands including the U.S. Army Chemical Activity Pacific on Johnston Island in the North Pacific Ocean.

Having served in Vietnam and having had an afloat tour with the U.S. Navy 7th Fleet amphibious forces, two tours in Korea and multiple long training periods in Thailand and the Philippines, I had a good understanding of the Pacific Theater and an appreciation for current and future requirements of the Army's most geographically and widely dispersed command. I also had great respect for the courage and fighting qualities of our allied Asian soldiers.

Linda was excited about going to Hawaii as were our kids, even though they were on the mainland in college or working and couldn't join us. The "aloha" greetings began before we left Fort Ord. It's no secret that "aloha" is one of the most well-known Hawaiian words, used as a greeting but also as a farewell or goodbye. I found it to be a way of life. And the Hawaiian people used it to convey kindness, positive intentions, and respect for others.

On March 8, General Kicklighter welcomed us with a beautiful retreat ceremony on Palm Circle. My predecessor departed in January. Kicklighter had been without a deputy/chief of staff for six weeks, and he made it a point to invite our service counterparts and friends of the command so I could meet them. One tall, lanky gentleman in a white sport jacket and aloha shirt came through the receiving line and said, "Aloha and congratulations, you finally made it. What took you so long? Mick needs some help." It was General (Ret.) Frederick C. Weyand, former chief of staff of the United States Army and a resident of Honolulu. I had heard much about Weyand, a combat veteran of three wars, but never met him. A strong supporter of USARPAC, I would see him often for the next 20 months.

Our new address was 8 Palm Circle on Fort Shafter. Palm Circle consisted of a large grass parade field formed in a roughly oval shape with 15 two-storey, wood-framed officers' quarters lining the north and east sides of the parade field. The elegant homes, built in 1907, were part of the Palm Circle Historic District listed on the National and State Historic Registers of Historic Places. Quarters 8 was featured in the 1953 movie *From Here to Eternity*. Here we were, "empty nesters," living in a 5,700 square foot, five-bedroom home with three expansive *lanais* (roofed, open-sided verandas). Linda welcomed the challenge of furnishing and decorating the beautiful historic house.

We found the quarters to be very practical as we had many visitors, both official and family and friends. The 19x17-foot dining room and the *lanais* were put to good use as we shared with the Kicklighters the many entertainment and representational functions required of the command. Every Pacific Theater ally army general officer, to include the Aussies "Down Under," found a reason to visit USARPAC. We enjoyed hosting them. "To my CGSC classmate, after 18 years, it was so good to see you during my visit to Hawaii," wrote General Kim Jin Young, deputy commander in chief, ROK/US Combined Forces Command.

Linda and I soon discovered we had friends we never knew once they learned we were in Hawaii. Not a week went by that we didn't get a telephone call or letter from a friend or relative saying, "We're coming to visit you in Hawaii." Social obligations with the command were a huge part of the assignment, and General and Mrs. Kicklighter were welcoming and gracious hosts!

Johnston Atoll, an unincorporated territory of the United States, consisted of four small islands in the North Pacific Ocean 750 miles southeast of Hawaii. The U.S. stockpiled chemical weapons there after World War II. In 1985, Congress mandated that all U.S. chemical weapons stockpiles be moved to Johnston Island and destroyed. Chemical weapons, including Sulfur Mustard (HD), Sarin (GB) and VX Nerve Agent were shipped from stockpiles in Germany and the Solomon Islands to the atoll. JACADS (Johnston Atoll Chemical Agent Disposal System), a two-storey, 80,000-square-foot facility, was built on the island. The Army had responsibility for operating the system, which included large burning plants.

USARPAC was designated the Army agent for the disposal and had the mission to oversee the chemical weapons destruction.[1]

That was no small mission! We were behind schedule for our testing phase and needed to ramp up. Within days after my arrival, Kicklighter said, "Get out there to Johnston Island and see what is going on. They need to get moving, we're behind." I flew to Johnston Atoll the next day and met with the commander in charge of the operation, an Army chemical branch colonel who reported directly to me. He was inundated with Environmental Protection Agency (EPA), U.S. Fish and Wildlife Service who managed the atoll's wildlife refuge, and other government agencies bothering him with senseless bureaucratic matters detracting from his mission.

I made immediate, on-the-spot changes which freed him up and reduced demands on him and his command from the "pestering" agencies. "Back off," I told them. "If there is an issue or problem, call me." The minute I set foot on the island I was fitted with a protective mask and given instructions on how to don the mask and what to do in the event there was a chemical gas leak. When I entered the facility to inspect the burning plants, I donned a Level "A" HAZMAT (hazardous materials) protection suit which offered the highest level of protection against chemical and microbiological hazards.

I was totally encapsulated in the chemical and vapor-protective suit from head to toe. The suit was bulky, uncomfortable, and hot, but could mean the difference between life and death and had to be worn. I hated it! The entire island was always in a state of readiness in the event there was a suspected leak, and the JACADS facility alarm sounded. Handling deadly munitions from the time they arrived on the island until they were destroyed in one of the burning plants wasn't a job for the meek.

Upon my return to Fort Shafter, we received word the secretary of the Army, Michael P. Stone, would visit USARPAC on March 20–21. He wanted to visit JACADS disposal facility and receive an update on our progress. Various Congressmen were prodding him with questions about the new facility and asking when testing would begin. His wife, an avid bird and wildlife enthusiast, accompanied him. Kicklighter asked me to escort him to the island for the visit. Linda was asked to go along as company for Mrs. Stone.

Congress had passed a law requiring the Army to complete operational verification testing (OVT) to guarantee each type of munition could be disposed of safely. The secretary wanted to know the status of our preparation to begin testing. Our JACADS personnel gave him a very thorough briefing, after which the secretary and his wife visited the wildlife refuge at the far end of the island. The visit went well. Soon after, the OVT phase began, and the first disposal took place on June 30, 1990.

That day, JACADS became the first U.S. chemical weapons disposal facility, and I was inundated with the media covering the event. The OVT phase lasted until 1993. Full scale operations began, and the last chemical weapon was disposed of in 2020. During my tenure at USARPAC, I spent considerable time overseeing the

JACADS operation. If I never see Johnston Island again or a HAZMAT protection suit, it would be fine with me!

When I arrived, the command was deeply involved in the planning for the future organization of the Army in the Pacific. I chaired the Pacific Vision General Officer Steering Committee, which required overseeing the planning and coordination of actions involving the reorganization. At the same time, I was the USARPAC representative to the Nunn-Warner phase down efforts in the Pacific.[2] The United States began to rethink security in the Pacific. What sort of presence was necessary and affordable? Are current command relationships the best way to carry out contingencies or fight a war? These were questions that were being addressed in Washington, and it was in that setting that we were reorganizing the post-Vietnam War Western Command to the new more appropriate U.S. Army Pacific Command. It was an interesting time. On August 30, 1990, Western Command was officially renamed United States Army Pacific, reestablishing USARPAC as the Army component to CINCPAC. At that time, we also assumed command of U.S. Army Japan.

Whenever I could leave the office, I visited the 25th ID at Schofield Barracks, about a 40-minute drive from Fort Shafter. Major General Fred Gorden was the commander, and he invited me to visit often. The 25th ID had recently been reorganized as a new light infantry division, and Fred asked me to assist them in their planning and organizing for their new rapid deployment mission. He knew of my contingency deployment experience with the XVIII Airborne Corps and with the 82nd Airborne and 7th Infantry Divisions. He was eager to get the help.

I appreciated his request, and I spent time advising his staff and commanders on development of division ready force (DRF) sustainment packages. Their outload facilities and ammunition packages needed improvement. The 45th Support Group worked for me. I added to their mission essential task list the task to outload 25th ID units during a deployment. I called my Air Force counterpart, Major General Mike Kerby, vice commander U.S. Pacific Air Forces at Hickam Air Base. He designated specific "call-forward" areas and an area for the division's DACG for their exclusive use. That improved the division's readiness posture and their ability to execute a no-notice deployment.

In early July 1990, the Army announced the selection of 35 officers for promotion to major general. The names were forwarded to the U.S. Senate for confirmation. I was surprised to see my name on the list. I had only been promoted to brigadier general less than two years earlier. The confirmation process took longer than usual, and I remained in a "promotable" status until I left USARPAC in November 1991. Nevertheless, it was a welcome signal. I was grateful for the Army's confidence in me!

On August 2, 1990, Iraq invaded Kuwait. President George H. W. Bush announced that "the United States considered Iraq's aggression unacceptable." The U.S. response to that was Operation *Desert Shield*, which consisted of the build-up of combat power and the defense of Saudi Arabia. My former units, the 82nd and 101st, began to deploy to Saudi Arabia. In rapid succession, other combat units followed. I watched that every night on TV and was envious of my friends in those units who were deploying to the Gulf. That was an exciting time, and I knew they would soon be in a fight. I wanted to be with them.

In early December, I received a call from Major General Tom Tait, my boss when I was on the Army staff. He had just moved to Fort Lewis to become interim commander of I Corps and Fort Lewis when Lieutenant General Calvin Waller was tapped to go to the Gulf to be Schwarzkopf's deputy. He was "probing," trying to get a feel whether I thought I could get released from USARPAC to come to either Fort Lewis or go to the Gulf. "If you can't go to the Gulf, I can use you here," he said. I had hinted earlier to Kicklighter that I would like to be "on my way to Saudi Arabia as part of the buildup, and could I be released to go?" He politely declined to pursue my desires any further.

U.S. VII Corps was moving from Europe to the Gulf. After Tait's second call, I approached Kicklighter again about going, and again he said "no." At that point, I stopped asking. I wanted so much to be part of the fight that was going to happen in the Gulf. I thought my chances of joining a division or a staff deploying to the Gulf would increase if my boss agreed to release me. He didn't!

General Kicklighter was right. My job was in USARPAC where I was needed and could best serve the Army. The invasion, Operation *Desert Storm* began on January 16, 1991. I was busy overseeing USARPAC support to *Desert Shield/Desert Storm*. That included sending USARPAC units, equipment, and ammunition packages to CENTCOM in the Gulf. We quickly mobilized, trained and deployed the 731st MP Company, a USARPAC reserve unit from Guam. I put them on the airplane at Hickam AFB, and they arrived in Saudi Arabia on February 3, 1991. They processed approximately 20,000 Iraqi prisoners during *Desert Storm*, returning home on May 10, 1991.

My biggest regret while in the Army is that I missed the Gulf War. I trained and prepared my entire Army career for a moment like that, and I missed it! It was a dispiriting experience; some things are hard to get over.

When I arrived at USARPAC, there was no organized PT program for the staff. The staff was doing PT only when they felt like it. That changed, and we began organized PT sessions three mornings a week. The other two mornings, the individual staff sections conducted PT on their own. As our morning PT runs took us into the heavily vegetated, hilly, and steep canyons of Fort Shafter, my polio leg was beginning to tire more quickly, and I had to restrain myself from overdoing it and trying to run the hilly courses.

During my annual physical examination in 1990, Dr. Robert Bausch advised me to ease up on my running to preserve what muscle I had remaining in my leg. The anterior horn cells that were once destroyed by the polio virus and re-innervated were beginning to show serious deterioration. "You are experiencing a progressive weakness in your leg, and you should not over work it," he warned. His examination advice, which he repeated several times, caused me to seriously consider the weakening condition of my leg and make an adjustment.

One interesting duty as deputy commander was serving as chairman of the Oahu Joint Family Housing Policy Advisory Board. The Army was executive agent for all military family housing on Oahu. USARPAC's Oahu Consolidated Family Housing Office (OCFHO) executed the management and funding of several thousand sets of family quarters for all services. That included over 30 sets of general/flag officer quarters. As chairman of the joint policy board, I personally handled all matters pertaining to the general/flag officer quarters.

Kicklighter delegated that authority to me. "Sir, I really don't want to do this," I respectfully told him on a couple occasions, to no avail. I was the person who approved or disapproved a flag officer's request to change, upgrade, or refurbish his quarters, always citing the appropriate clause in the housing regulation that supported the decision.

"Sir, this is not an authorized expenditure, it can't be approved," I once told a four-star general. He wanted to recarpet his entire quarters and install a small putting green on the grounds. He wasn't taking a "no" answer from a one-star general. His request went up the chain. After further, careful review, I delivered the final decision, again denying his request. Although I always had the full support of my boss, dealing personally with flag officers (and their wives) on matters pertaining to their living quarters wasn't my most pleasant duty. As a small consolation, I did get to know many of the flag officers on Oahu that I otherwise would not have. This didn't endear me to some.

One day we got a call from Senator Daniel Inouye's office in Honolulu. The senator was in Hawaii and wanted to visit the Army's Pohakuloa Training Area. It sits on a high plateau near the Hualalai volcanic mountains on the big island of Hawaii. It was the largest U.S. training area in the Pacific consisting of bombing and gunnery ranges and a large maneuver area. The 25th ID and Marines trained there. Some native Hawaiians were threatening to sue both the Army and the state for failing to protect ceded lands at Pohakuloa.

I met the senator at the center's main gate, and the garrison commander and I escorted him through the training area. We answered his questions and when thanking us and shaking hands to say goodbye he said, "You and I both have the Distinguished Service Cross," and then briefly mentioned his time with the 442nd Regimental Combat Team during World War II. I was surprised that he brought that up. It had nothing to do with the matter at hand.

Weeks later, we were told there was no longer a threat of lawsuits, and we could continue normal training at Pohakaloa. I met Senator Inouye years again later in Washington, DC, after I left the Army and was president of a veterans' organization, and my staff and his were working on veterans' legislation. Inouye's DSC had been upgraded to the Medal of Honor.

On July 11, 1991, Lieutenant General Johnnie H. Corns assumed command of USARPAC. General Sullivan, who had just become 32nd chief of staff of the Army only 20 days earlier, presided over the change of command. It was good to have the chief visit USARPAC. The units and salute battery conducted a beautiful change of command and farewell ceremony on Palm Circle parade field to honor General and Mrs. Kicklighter. Linda and I were honored to host the farewell dinner for the Kicklighters who had contributed so much to USARPAC and who would be remembered for their generous and gracious hospitality.

While I was at Harvard attending one of the national security programs at the JFK School of Government, Linda flew back to be with our son, Bill, during his scheduled surgery at the University of Virginia Medical Center in Charlottesville. The AVM which ruptured causing his brain hemorrhage in 1985 had grown back, and he needed surgery. We were fortunate to get Dr. Ladislau Steiner, of the Karalinska Institute in Stockholm, Sweden, who was practicing at UVA, to perform the operation. Internationally, he was known as the "Emissary of Gamma Knife Surgery." The treatment was a success. The malformation was obliterated, and our son lived a normal life ever since. Linda joined me in Cambridge for the graduation ceremony at Harvard.

While empty nesters in Hawaii, our daughters who were in college in Pennsylvania joined us during the summer and worked in Hawaii. Heather graduated from Penn State and became a fitness trainer at the Honolulu Athletic Club. Using her experience competing as a member of the women's swim and track teams at Penn State, she transitioned into the grueling sport of triathlon, an endurance multisport race consisting of swimming, cycling, and running over various distances. Having competed in 22 Ironman races, including 11 Ironman World Championships in Kona, Hawaii, she continued to train and compete in the triathlon, while also training others to be triathletes.

We had many family members and friends visit us in Hawaii. When we were not hosting them, we were busy attending and hosting "official" command functions. We escaped to Waikiki Beach at Fort DeRussy or played a round of golf at the Par 3 Course at Hickam AFB whenever we could with our Palm Circle neighbors, the Farrises, Fields, and Childs. Somehow, with all those activities, Linda managed to teach English at the Pacific International Language Institute in Honolulu. Indeed, we had a full social calendar. We learned how to thrive socially without burning out. Thank God, I wasn't an introvert. After General Corns arrived, the hosting and entertainment dramatically slacked off. Up until that time, I was beginning to worry about "social burnout."

In early September, I received a telephone call from Carm Cavezza, my boss when I was with the 7th ID. "I'm going to Fort Lewis to command 1st Corps. If they don't give you a division, I'd like you to be my deputy commander." Any infantry general worth his salt should want to command an infantry division. I certainly did.

And when I didn't get that opportunity, I joined General Cavezza at 1st Corps. Serving with Carmen Cavezza again was a pleasure and getting back with troops was what I wanted most, and 1st Corps was the Army's warfighting corps for the Pacific.

On November 8, General and Mrs. Corns hosted a combination promotion and farewell ceremony for us on the Palm Circle parade field. It was a beautiful ceremony and concluded with the 25th ID Band playing *"Aloha Oe"* (*Farewell to Thee*) and the *Army Song*. I remember well standing on the field listening to that emotive Hawaiian song, "Here, I depart, until we meet again, sweet memories come back to me." Laden with beautiful Hawaiian floral leis, Linda and I departed Honolulu the next morning for Seattle.

Serving With America's Corps

Located in the Pacific Northwest in Washington State, Fort Lewis was 42 miles south of Seattle and 12 miles from Tacoma. It sat astride Interstate 5, allowing easy access to Seattle-Tacoma (Sea-Tac) International Airport. Contiguous to Fort Lewis was McChord AFB. Together, they formed the primary Army-Air Force projection platform west of the Rockies. In 2010, Lewis and McChord consolidated operations and the installation was later called Joint Base Lewis-McChord. Its 87,000 acres contained abundant training areas including live-fire ranges. Additional training space is available at the 327,000-acre Yakima Training Center, a sub-post of Fort Lewis located 165 miles east.

As deputy commanding general (DCG), I assisted the commanding general, Lieutenant General Carmen Cavezza, in command and control of I Corps, Fort Lewis, and its five sub-posts. I supervised 10 non-divisional major subordinate commands including the 9th Infantry Regiment, 199th Infantry Brigade (MTZ) and the 210th Field Artillery Brigade. I also exercised control over installation management, including the significant activation, deactivation, and restationing of units necessitated by the end of the Cold War.

When Cavezza and I arrived within days of each other in November 1991, the corps had been designated an "early deploying" corps for the Pacific and was converting to a permanently structured, no-mobilization warfighting contingency corps. With the end of the Cold War in 1991, the U.S. government was reassessing national priorities and Fort Lewis, ideally located for power projection in the Pacific region, was one of few military bases that didn't downsize. While most of the Army was downsizing, Fort Lewis and I Corps began to grow, receiving new units coming out of Europe. I was busy immediately welcoming and housing the new units while Linda was unpacking and setting up our new quarters in Fort Lewis. The corps and Fort Lewis would see a reshuffling of units in and out of the area for the next several years.

I Corps, nicknamed "America's Corps," because its units were widely spread across 46 states, was a unique command composed of active-duty, National Guard and reserve units. I recognized one of our challenges was to improve the readiness of I Corps reserve component (RC) commands. Cavezza made me chief liaison between I Corps headquarters and our wartime aligned Capstone reserve units.[3] The Army had just implemented a new program called Bold Shift to improve the training of high-priority RC units.

The innovative approach marked a bold shift for the Army. It included changes in training strategy, resource levels, and active-reserve relationships. "Get out and visit these units and get to know their commanders," Cavezza told me. I did, and we developed a trusting relationship with our National Guard and reserve commanders and units. They were spread throughout the USA from California to Pennsylvania, Alabama to Missouri to Hawaii and in between! I Corps staff was also busy developing new contingency plans should conflict emerge in the Pacific region. Writing our RC units into the wartime plans became a major undertaking.

The year 1992 was busy for our family. We flew to Philadelphia when our daughter, Becky, graduated from college on May 17. On October 16, Linda and I flew to Albany, New York, for our son Bill's wedding. We hosted a rehearsal dinner that night for our family and wedding party, and the next day Bill and Helen were married at Saint Margaret Mary's Church. Both occasions allowed us to return to the east coast and visit with family and friends we had not seen for four years.

After a successful year of triathlon competition in Hawaii, our daughter Heather won a spot on the United States Age Group Triathlon Team while racing in the U.S. Triathlon National Championships in Cleveland, Ohio. The team competed in the Triathlon World Championships in Ontario, Canada, in September. As busy as we were with our responsibilities to soldiers and their families, Linda always made sure that we were available for the major events in our children's lives.

I sometimes lost sight of what was important in life. I often got so wrapped up in my work that I let everything else suffer. Linda taught me to temporarily "mute the script" that wants to always put the Army first and pay more attention to the family. We did our best to support our kids and are proud of our three Army brats.

On January 13, 1992, I flew to the Enlisted Records and Evaluation Center, Fort Benjamin Harrison, Indiana, to be president of the CY 92 CSM/SGM/SMC Selection Board, which was a Department of the Army board convened to select NCOs in the grade of E8 for promotion to command sergeant major/sergeant major, or E9. Concurrently, I was also president of CY92 CSM/SGM Selective Early Retirement Board (SERB), the purpose of which was to select E9s for early retirement. SERBS were necessary to correct officer and NCO imbalances and overages and were an integral part of meeting congressionally mandated strength reductions.

Chairing the boards was an eye-opening experience for me! It wasn't the last Army promotion selection board I would sit on. I was president of the Board in 1995 that met to select captains to major, also a valuable experience. Other than fighting and training soldiers, there is no more important duty than selecting NCOs and officers for promotion.

Meanwhile, the shuffling and transfer of units mandated by law as part of the 1991 Defense Base Realignment and Closure (BRAC) Act kept us busy. The 199th Infantry Brigade (MTZ) relocated to Fort Polk, Louisiana, and was replaced by the reactivated 2nd ACR Armored Cavalry Regiment (Light) which later relocated to Fort Polk in 1993. BRAC also mandated the transfer of the 7th ID from Fort Ord to Fort Lewis. On January 19, 1993, I welcomed the first element of the 7th ID at a welcome briefing at Evergreen Theater.

Only one brigade ever transferred to Fort Lewis before the Army made the decision to inactivate the 7th ID in 1994. The restationing plan included moving 3rd Brigade, 1st Armored Division from Germany to Fort Lewis. The move to Fort Lewis was completed in September 1994, reflagging it as the 3rd Brigade, 2nd ID, posing a challenge, because the 4,000-soldier, 1,500-vehicle brigade was separated from its new parent division which was in Korea.

As a result, we got creative and reorganized the heavy unit as a split-based brigade combat team (BCT) with mechanized infantry, armor, artillery, and forward support battalions. The brigade didn't have the normal support of a division support command. Drawing from other corps units, we improvised and established an organization that met their echelon-above-brigade support requirements. Colonel Ed Dyer, the brigade commander, and I were happy when on March 29, 1995, the brigade was officially reactivated as part of I Corps. In 2000, the brigade was chosen to be the nation's first Stryker-equipped combat team.

In addition to the division/brigade/regiment restationing changes, an additional 12 separate company- and battalion-size units were transferred from Europe to Fort Lewis during 1992–93. At times, the extreme turbulence caused by the arrival, departure, activation and inactivation of units was overwhelming! The arrival of thousands of new soldiers, families, and pets placed a strain on the installation infrastructure, particularly the on-post housing. We couldn't have accomplished that significant expansion in the short timeframe without the superb leadership of Colonel Al Isaac, our garrison commander. His performance in the toughest job at Fort Lewis, in particular his balancing the many changing and diverse stationing requirements, was magnificent. I slept well at night knowing that calm, unflappable Al was on the job!

Not for a minute did the continuous, burgeoning restationing activity interfere with the corps' warfighting training, which was the priority. With the corps' many new and dispersed RC units, there was a need to improve command and control functioning at corps level. We did this by participating in the Army's Battle Command Training Program (BCTP).

BCTP was a program for division and corps commanders and primary staff. It included a week-long Battle Seminar and a computer-driven Warfighter Exercise.

The seminar and exercise scenarios portrayed real-world contingencies. The training featured extensive observation and critique in the form of after-action reviews. Retired senior generals acted as controllers/mentors. General Richard Cavazos, a larger-than-life Texan, Hispanic icon, and a very emotional man who genuinely loved soldiers, and Lieutenant General David E. Grange, a combat veteran of three wars, were the senior controllers for our BCTP exercises. As the senior mentors in the BCTP, they had a profound influence on me.

As the DCG, I was responsible for rear area operations including the rear battle. Sharing with me their personal Korean War experiences of how North Korean and Chinese forces infiltrated and disrupted U.S. units' rear areas helped immensely in our planning and executing rear operations during the corps' exercises. Listening to the stories by Cavazos and Grange about rear area operations during Korea was enlightening. We developed and integrated into the corps battle plan a new rear area battle concept that prompted the Forces Command CG, General Edwin H. Burba, to comment in April 1993, "I Corps executed the best rear area operations I've seen in a training exercise."

I am a strong proponent of BCTP as a training program for warfighting commanders and their battle staffs. A chief of staff and Fort Leavenworth initiative in 1987, it was renamed the Mission Command Training Program (MCTP) and remains the Army's premier program for training division and corps battle staffs. It was a naïve commander who didn't believe he was being graded while his unit and staff were undergoing a BCTP exercise. A not widely publicized benefit of the program was that it gave major command (MACOM) commanders and the Army chief of staff another means of assessing the potential of commanders and senior staff officers for future higher-level positions in the Army.

We had recently reactivated the 2nd ACR (Light) and needed to develop and validate an operational concept for employment of light cavalry in support of I Corps operations. The 2nd ACR had been a heavy mechanized/tank regiment in Germany and its ground squadrons, which were now equipped with Humvees, needed to adapt to Pacific Theater contingency-type operations. A challenge indeed!

Colonel Tom Molino was the ACR commander. I ensured he had priority for maneuver space and firing ranges, and he and the regiment spent six weeks training at Yakima under arduous winter conditions in 1993, developing new light armored cavalry tactics. At a memorable ceremony in May, I was inducted as honorary major general in the Honorary Squadron of 2nd Dragoons, 2nd ACR. The 67th colonel of the regiment, Tom Molino, presided. "We don't often induct someone into the Honorary Squadron. It's a rare honor," Molino said. Having been mostly in light infantry and airborne units, being inducted into the Honorary Squadron of the 2nd ACR and signing my name in the Honorary Squadron book was, indeed, an honor for me!

As when commanding the 7th ID, Cavezza saw to it that the soldiers of I Corps and Fort Lewis maintained a high level of physical fitness. The Corps Command

Sergeant Major, Rocky D. Houser, an Airborne Ranger and veteran of two combat tours in Vietnam, ensured that the NCOs fully participated in their unit programs. He had no sympathy for those who didn't. Companies and battalions conducted unit runs four or five days a week. With 10 brigades consisting of 29 battalions, there was always a unit to run with. We did a Corps Green Tab run each month with all units running the 4-mile course and hosted by a different major subordinate command. All company, battalion, and brigade commanders, along with their first sergeants and their command sergeants major, participated. "It gives the corps commander a chance to see the fitness of the subordinate commanders and their senior NCOs," said CSM Richard Novy, the command sergeant major of the 9th Infantry Regiment.

With increased emphasis on PT and running four or five mornings each week, and being in my early 50s, my polio leg was beginning to tire more quickly, just as my doctor had warned it would. I felt progressive weakness throughout my leg, and each stride was a little more painful than the last. I was becoming more prone to tripping, as a result of my weak drop foot, and there was increasing damage to the nerve cells causing muscle weakness and twitching. The thought of falling out of a unit run always gave me strength. Pain was my motivation! I continued running, but I became more selective in which units I ran with and was more careful not to join an aggressive macho unit. Some companies always wanted to show the general they could run faster and farther than he could.

Cavezza asked me to lead the I Corps delegation that would attend the Association United States Army (AUSA) annual meeting in October 1993. I was happy to do so as we had some I Corps soldiers that would be honored at the meeting. On Sunday, October 17, Linda ran in the 9th Army Ten Miler race across the Potomac River bridges, into Washington, DC and back to the Pentagon. I snapped a photo of her crossing the finish line wearing her Ten Miler entry #3764 sign pinned to her sweat-soaked shirt. She finished strong among the women runners and didn't seem to be the least bit tired. I was proud of her and wished I could have run with her.

In less than three months, we flew east again. Linda's father died, and we attended the funeral service and burial in Hamilton Square, New Jersey. My father-in-law, John Heal, was one of the finest men I ever knew. We miss him. The kids called him "Pop Pop." He always took a keen interest in everything they did and when I needed something fixed, his engineering skills paid off and what had been broken was fixed!

Included in my supervisor's responsibilities was the Fort Lewis hospital—Madigan Army Medical Center (MAMC). MAMC was a 414-bed tertiary-care teaching facility. It later became one of the Army's three remaining medical centers. When I learned I would be the supervisor/rater for the commander who was a medical doctor, I was somewhat taken aback. I wasn't in shock, but I was surprised to learn I was in this medical chain of command responsible for the efficient operation of a large teaching hospital that cared for 360,000 beneficiaries.

The senior rater was a medical doctor, the commander, Health Services Command, Washington, DC. I made it a priority to learn all I could about the medical center's mission and how Fort Lewis and I Corps could best support the hospital. I was relieved when one day the head nurse of one of the clinics kiddingly said, "Sir, you don't need to wear a stethoscope every time you visit." She was cleverly reminding me that I didn't need to be a medical doctor to carry out my responsibility to the hospital. She was right!

In the spring of 1994, Cavezza told me he would be retiring in the summer. Major General Glenn Marsh was nominated to replace him. On July 8, I Corps and Fort Lewis bid farewell to the Cavezzas at a relinquishment of command and retirement ceremony at Watkins Parade Field. General Marsh's nomination had not been confirmed yet by the U.S. Senate. The corps colors were passed to me, and I served as interim commander pending the arrival of General Marsh. General Dennis J. Reimer, commander of United States Army Forces Command, was the senior commander and reviewing officer for the ceremony. He later served as the 33rd chief of staff of the U.S. Army. "Carmen is a great soldier with 33 years of selfless service to his country," Reimer said.

Cavezza, after thanking everyone for their support, gave his final speech to I Corps and Fort Lewis. "When I assumed command, I stated that my vision for the corps was a go-to-war corps. Together, we have attained that." Carm was proud of the improved warfighting capability of the Corps.

Each commander took his place at the podium and farewelled the boss. It was fun "roasting" the CG. We also thanked Joyce Cavezza for her tireless support of soldiers and their families. Just as they had done in the 7th ID at Fort Ord in the 1980s, the Cavezzas opened their house and heart to all of us and the accolades from the commanders and their spouses that night reflected the love and deep respect we all had for Carm and Joyce.

Two days before the retirement ceremony, we had a farewell dinner to honor the Cavezzas. It was a particularly bittersweet, nostalgic time for me and Linda. We both felt that something happy in our lives was coming to an end. We had served with the Cavezzas at Fort Ord and then at Fort Lewis and had gotten to know them well. I would miss his raw wit, warm sense of humor and clever remarks. Most of all, I missed the trust he had in me to let me do what I knew needed to be done. I was never one to seek guidance or ask permission to do something. I don't hesitate to act. When something is broken, I fix it. He gave me full rein to do that. He trusted me, but I never took that trust for granted. We complemented each other well. We both understood that the mission always came first and we never deviated from that understanding. I missed that mutual trust and close working relationship.

The corps was extremely busy in the weeks immediately following Cavezza's departure. Our chief of staff, Brigadier General Robert R. Hicks, was unexpectedly reassigned in June, but his replacement didn't arrive until September, and our garrison commander, Al Isaac, retired on July 13. That left a significant void in senior leadership when there was much activity and turbulence requiring general officer participation and presence.

We were heavily engaged in joint planning with Eighth United States Army (EUSA) regarding North Korean "saber rattling," preparing to begin the ROTC Advanced Summer Camp which we hosted at Fort Lewis, all while we were receiving 400 soldiers/families per week from the 1st Armored Division in Europe which required general officer-led welcome ceremonies.

We were also heavily engaged in supporting and evaluating annual training for RC units. There were three scheduled brigade changes of command in July and August for which I was the reviewing officer. FORSCOM had also directed that we provide general officer oversight for the final closing and turnover of Fort Ord to Fort Lewis.

What weighed most on me as interim commander was the scheduled week-long BCTP Warfighter Seminar in mid-July. That was the most important training event a corps and division commander had. Our plate overflowed and would remain so throughout the summer. It was a busy time and with the loss of two of the three I Corps general officers and our garrison commander, it became a rather "sporty" time. I was feeling the pressure as the only general officer present. There were some days I felt all alone.

Each day I met with key commanders and staff at 0700 to review the requirements for the day to ensure nothing fell through the cracks. Our Warfighter exercise with the senior BCTP mentors went well, and all other events, ceremonies, etc., were executed without any significant loss of momentum. The staff and the battalion- and brigade-level commanders closed ranks, and their tireless and enthusiastic dedication made it happen. We got through the extremely busy period with no loss to mission productivity. I felt good about what we had achieved.

On August 20, 1994, our daughter Heather got married at the Main Post Chapel, Fort Dix, New Jersey. Her husband, Peter Jorris, a former Navy pilot, was living in Guam where he was a pilot with Continental Airlines. Heather lived in Hawaii where she was working as a fitness trainer at the Honolulu Athletic Club. While living at Fort Lewis, Linda and Heather planned the wedding to be held at Fort Dix so our many friends and relatives back east could attend. Linda arranged a reception after the ceremony at Club Dix. The fun lasted through the weekend. The life of the party was Lara Spencer, Heather's Penn State teammate and roommate. Lara would later be co-anchor and host of ABC's *Good Morning America* show. Due to the distances involved, the planning and execution of the wedding ceremony and the reception was tricky, but the Matz girls pulled it off in fine fashion! I was proud of them!

After a lengthy wait of several months, the United States Senate confirmed General Marsh and at a change of command ceremony on Watkins Parade Field, I passed the colors to our new I Corps and Fort Lewis commander. General Reimer was the senior commander and reviewing officer for the ceremony. Marsh was no stranger to Fort Lewis, having served two previous tours with the 9th ID, then a third as chief of staff I Corps. I was happy to finally have him aboard and on behalf of the corps expressed my feeling saying, "We held the position until reinforcements arrived."

I was fortunate the entire time at Fort Lewis to have a very dedicated and capable personal staff. My administrative assistant, Alice Cline, was meticulous at maintaining my schedule and arranging the numerous travel and temporary duty visits to our RC commands nationwide. She was particularly good at training my two aides and always checking to ensure they never forgot anything before we left for a trip. Captain Patrick Lee Fetterman and Captain Kent D. Savre were outstanding aides who were always there when I needed them most. Their wives, Sabina and Mary Beth, were also very capable and helpful, and Linda appreciated their gracious attitude and friendship. They were my personal team that assisted me during the busy years at Fort Lewis. I am proud of them.

Leaving the Army

After almost 33 years as a soldier, culminating with our time at Fort Lewis, in early 1995 Linda and I decided it was time for me to retire. General Officer Management Office (GOMO) notified me months earlier of a possible move to the headquarters staff of U.S. Army Europe (USAREUR). A senior general officer on the USAREUR staff called me twice wanting to know "how I felt about a possible assignment to Europe," strongly encouraging me and Linda to come to USAREUR headquarters in Heidelberg. He wanted badly for us to come. He was calling for his commander. I had never served in Europe and wasn't really comfortable or enthusiastic about going to Germany and serving with the heavy forces that late in my Army career. But Linda was and was ready to go.

About the same time, my good friend from our time together in the 82nd Airborne, Brigadier General George Landis, who had recently retired and was working for Raytheon Company, called about an upcoming opportunity with Raytheon. After much thoughtful contemplation, and still sensing some reluctance from Linda to leave the Army, I accepted an offer from Raytheon in Bedford, Massachusetts, to head the logistics program for a newly won billion-dollar contract to establish a complex surveillance system throughout Brazil's vast Amazon region. A program of that magnitude had never been attempted. It sounded challenging!

Leaving the Army after many years of fun-filled, challenging, and rewarding service to the nation wasn't an easy decision. My 33 years were coming to a happy close, and it was time to pass the mantle to others. I completed my required retirement physical examination at Madigan Army Hospital on May 17. Lieutenant Colonel D. E. Casey Jones, MD, chief Orthopedic Surgery, diagnosed me as having post-polio syndrome (PPS). At the time, I had heard of PPS but didn't know anything about it. PPS is a group of potentially disabling signs and symptoms that appear decades after the initial polio illness.

He wrote on the examination consult sheet, "It is highly probable that General Matz's career as an infantryman has caused far greater damage to his remaining muscular system than would've been the case with a less rigorous, less demanding occupation. Due to his military career as an infantryman, General Matz is at an even greater risk for progressive muscle weakness and degenerative joint disease, particularly in his lower extremities." It was the most comprehensive and thorough examination of my post-polio condition, leg and foot that I ever had.

I didn't want to hear it, but I wasn't surprised. Doctors had been warning me to slow down and ease up and conserve my energy. I had been battling the degenerative weakness for years, knowing the disease-generated loss of anterior horn cells would leave far less reserve in my polio leg than would otherwise be the case. Nevertheless, like the persistent and strong-willed German I am, and with the support of my family, I persevered in my military assignments, despite the physical difficulties! My toughest battle, coping with my post-polio symptoms and conditions while serving in the Army, would continue even as I pursued much less rigorous civilian employment.

We attended several farewell events during the week of May 21, including a fun-filled brigade and battalion commander's "roast" at the Fort Lewis Officer's Club. Three days later, Lieutenant General Marsh presided over a farewell and retirement ceremony on Watkins Parade Field at which time I retired from the Army. There was no better way to leave the Army than to be on the field with the soldiers and the colors of America's Corps and its major subordinate units flying at my final formation.

As I write this, I still see the colors gently waving against the majestic backdrop of Washington's snow-covered Mount Rainier in the distance. What began for me three decades earlier at Fort Benning, Georgia, came to a close that beautiful sunlit morning at Fort Lewis. It was a moment to cherish, and one I think of often.

Linda and I were glad that our children, cousins Nancy Hill and Kay and Clifford Reed, and friends, including two college fraternity brothers and their wives, Bill and Nancy Hemsing and Barry and Vonnie Kain, were able to travel from Pennsylvania and New Jersey to join us. Bill had been my roommate my last two years in college and was best man at our wedding. We remained close through the years. His and Barry's presence meant a lot to me. Linda, with the able help of my administrative assistant, Alice, secretly arranged for the attendance of our cousins and my fraternity brothers. Their presence was a complete surprise to me! They attended the farewell party given by our good friends the Campbells, Reeds, Haydens, and Wintrichs where they presented me with various bikini bathing suits hoping I would have time for a swim on Copacabana Beach in Brazil.

Raytheon wanted me in Massachusetts in early June so Linda and I departed Fort Lewis with a heavy heart on Sunday, May 28, arriving in Bedford, Massachusetts, seven days later.

PART III

Post Military Service

CHAPTER TEN

Defense Industry—Al-Qaeda Attack in Saudi Arabia

I never looked for a fight! My challenges, struggles, and battles came to me: overcoming polio paralysis, coping with boyhood bullying, completing Ranger School, and fighting and surviving infantry combat in Vietnam. Even when I didn't see the looming shadow of battle or the challenges it posed, when it came, I felt I was ready and would prevail, no matter the odds. Sometimes a fight is lurking, and the first punch hits you from an unexpected quarter. You don't know you are in a fight until it hurts you and you face it and fight back!

In 2001, I accepted the position as program general manager (PGM) with Vinnell-Arabia, the defense contractor assisting the Saudi Arabian National Guard (SANG) in their modernization program. For almost six years, I had been working for Raytheon, first as logistics manager on the SIVAM (Sistema de Vigilancia da Amazonia) contract to develop and install a surveillance system in Brazil's Amazon River basin, and later as vice president for Army programs in Washington, DC, and I was looking for the next challenge. During those years with Raytheon, the first three of our seven grandsons were born in 1998: twins Will and Ben Matz and Max Jorris. Indeed, we were blessed!

In 2001, Frank Carlucci was chairman of the Carlyle Group, an American multinational private investment and financial management organization. After I had served as his executive secretary when he was SECDEF in 1988, we kept in touch. The Carlyle Group had acquired the Vinnell Company, and while at lunch in DC, he asked me if I would be interested in managing the Vinnell program in Saudi Arabia. "See what Linda thinks," he said. In his typical low tone and persuasive manner, it was clear, he wanted me to manage the SANG program.

I met with the president of Vinnell, Tom Fintel, a retired U.S. Army colonel, and veteran of military assignments in the Middle East, and conducted an in-depth analysis of the program. I discussed it with Linda and agreed to be the PGM. Other than participating in airborne and training exercises in Egypt while with the 82nd Airborne, I had never been to the Middle East. The fact that 15 of the 19 Al-Qaeda terrorists that attacked America a few weeks earlier on 9/11 were Saudis didn't bother me. Linda could accompany me to Riyadh. That was an interesting and challenging opportunity I couldn't pass up.

The U.S. government had and would continue to have a contract managed by the United States Army to assist the SANG. The contractor hired by the U.S. Government was Vinnell-Arabia, a joint U.S.-Saudi company, that had been assisting the SANG since 1975.

The contract called for Vinnell-Arabia, supervised by U.S. Army Program Management organization OPM-SANG (Office of the Program Manager-Saudi Arabian National Guard) to assist the SANG with development of capacity and capability. That would encompass the manning, training and equipping of the SANG.

OPM-SANG, led by a U.S. Army brigadier general, was engaged by the Saudi government through a Saudi-funded U.S. foreign military sales case to function as their agent. That was to ensure that the program they purchased from the U.S. government provided the best value for their investment, and guaranteed a well-trained and well-equipped military force.

The SANG was a standalone military force, forged out of Bedouin tribal elements loyal to the House of Saud under the command of the crown prince. In some ways, the SANG could be compared to the U.S. Army National Guard. The personnel, leaders, and SANG organizations were connected through their tribes and regional allegiances to the royal family through shared cultural, historical, and economic interests.

The development and modernization of the SANG not only provided a robust defense capability but also served the purposes of sustaining internal alliances through distribution of resources, jobs, money, and favor amongst those who demonstrated their loyalty to the royal family by serving in and supporting the SANG. It has been described as a "praetorian guard" to defend, protect, and ensure the continuity of the royal family.

I became PGM, and after arriving in January 2002, assumed command of Vinnell-Arabia in March, which had a solid reputation with both U.S. and Saudi governments for exceptional support and successful performance. I "assumed command" as Vinnell-Arabia was a quasi-military organization, manned and organized similarly to a U.S. Army organization. All the senior leaders were former Army officers. We wore distinguishing work uniforms, which were basic desert tan fatigues. We gave the outward appearance of a military organization without being one. We were recognizable as Vinnell-Arabia, not to be confused with regular U.S. military personnel. In April, our fourth grandson, Luke Jorris, was born. I would not see him until I returned to the states for a short visit later that year.

When I became PGM, the foreign military sales contract that Vinnell-Arabia supported was, inter alia, to assist the SANG in the reception, de-processing, integration, training, and organization of over 700 LAV (light assault vehicle) 8x8 wheeled armored vehicles into nine maneuver battalions. That was a huge undertaking and important to both countries.

In 2002, Vinnell-Arabia was a component of Vinnell Company, TRW Mission Systems. In December 2002, Northrop-Grumman Company completed the acquisition of TRW, and Vinnell became a Northrop-Grumman Mission Systems company.

On contract there were approximately 350 U.S. citizens, 300 hired for their experience, expertise and demonstrated success when they were serving in the military, predominantly the Army. Of the 350 U.S. citizens on the contract, approximately 80 worked in the administrative and services divisions supporting the contract.

Additionally, there were approximately 130 TCNs (third country nationals) serving on the contract, mostly Filipino, Bangladeshi and Sri Lankan. TCN was the term for contracted employees who were not U.S. citizens, or citizens of the country in which they were employed. The TCNs were hired to fill requirements for basic labor, support, maintenance, and other tasks for the SANG. Vinnell-Arabia TCNs accomplished those tasks in parallel with the development of a SANG capacity. Some TCNs also worked on Camp Vinnell in general support of the contract.

U.S. and TCN personnel assigned to the contract were housed in a SANG-owned and provided facility: Camp Vinnell, located in Riyadh. The capital was the administrative hub for the government of Saudi Arabia and home to the international diplomatic community, numerous international business headquarters, the Saudi Ministry of Defense (MOD) and OPM.

Camp Vinnell was an austere place to live when compared to the resort-like facilities provided to other contractors in the kingdom. The camp was rectangular in shape, two 400-yard-long sides and two 465-yard-long sides, basically ¼ of a mile on each side and enclosed by an 8-foot masonry wall. It had four-storey high-rise apartment buildings, stand-alone single-storey houses, barracks-type buildings, classrooms, maintenance and dining facilities, a restaurant, a "7/11" type convenience store, a gym, and a pool. Basic, but adequate. It was both our home and our place of work.

Vinnell-Arabia was responsible for management and security inside the camp, and the SANG was responsible for physical security outside of the camp. Per SANG policy, no Vinnell employee was allowed a weapon; even the Vinnell security personnel inside the compound were not armed. Only the SANG soldiers providing security outside the camp were armed. Most personnel on the contract were unaccompanied, although the contract supported a number of accompanied family members of the senior leadership and staff. All told, there were approximately 500 personnel living on Camp Vinnell.

The population expanded during workdays as Saudi nationals, living in Riyadh and the surrounding area, were also on the contract. The contract ensured a certain number of Saudis (approximately 400) would be hired. Saudi employees worked in administrative and other support type positions.

Many U.S. personnel worked on Camp Vinnell, but the majority left each day for various training and support tasks for the SANG at their installations.

Although much of the life and work was routine, life was especially difficult for the employee spouses and children. Saudi rules for conduct and deportment by women could be irritating and frustrating to westerners used to conducting themselves in ways we considered normal in American or western society.

The spouses participated in organized shopping and cultural excursions and were required to wear a black *abaya*. The *abaya* was a garment that covered the body from neck to toe. Women couldn't travel alone outside the camp unless accompanied by a male relative, spouse, or a child. The Saudis gave special permission to allow my Yemeni driver, Saeed, to drive my wife and other wives on occasion.

Women usually carried scarves to cover their head. Some spouses had the unpleasant experience of being screamed at by a Mutawa, a member of the Saudi religious police, to "cover your hair." Western women were not required to cover their hair, but if a Mutawa demanded she do it then the best option was to comply. The Mutawa were usually accompanied by a Saudi police officer, and they were authorized to detain persons who refused to comply. It was easier to carry a scarf and cover hair when a Mutawa was in the area rather than become involved in a scene. There were stories of Mutawa striking a woman or a man with a stick if they failed to comply with their demands. The Mutawa might also ask for proof that a man accompanying a woman was a husband or family relation.

I was shopping with Linda one day when a Mutawa carrying a long, flexible, stick gently switched her because her *abaya* wasn't completely covering her bare ankles. He had a Saudi police officer with him. We smiled and left the store. Noncompliance would have been foolish. Those rules made dating a challenge. Single men and women, mostly expatriate nurses working at various hospitals, had to be creative to not violate the law. During my time as Vinnell-Arabia PGM we bailed out of Saudi police custody employees who were not sufficiently creative. Flagrant violators were removed from the contract and sent home.

Our senior staff worked continuously to provide a comfortable, normal, American life for all employees and their families. I found that to be a challenge in itself, over and above the daily work challenges of meeting all contract requirements. Each week, there was an opportunity to "escape" Camp Vinnell and walk or run in the desert. Many of the employees and their families participated in the "Hash House Harriers" or "H3." It was a traditional outdoor running or hiking activity called "hashing" that originated in 1938 in the expatriate communities of mostly British colonies.

Each week volunteers would reconnoiter and mark a trail through an interesting part of the local desert and hills. On Friday, a non-workday in Saudi Arabia, anyone who was interested could participate in the hash. It was the proverbial breath of fresh, though very hot, air. The hash was an opportunity to meet other expats and for them and their families to see some of the rugged but enchanting countryside that they might not have experienced otherwise. Saudis didn't participate. The desert and the hills around Riyadh reminded me of Arizona and New Mexico.

For me, that was my saving grace! All week, I longed for that day in the desert sun when I could divorce myself from being the PGM and leave behind Camp Vinnell, the *thobes*, the basic Saudi male garment, and everything else Saudi, and enjoy a western-type outing with other Americans and expats. That day was like a new lease on life!

"Chad, what kind of course do you and Ron have planned for this week?" I asked. Chad, a retired Army infantry colonel, was my director of unit training. He and Ron Hindmand, a retired Army intelligence officer, were among the leaders of the Riyadh House Hash Harriers. Both were "take charge" guys and in excellent physical condition. The walking and running courses they set up were the best—creative and challenging! I always wanted to know the length and degree of difficulty of the trail they were charting.

Was it over rough terrain? How steep were the hills, desert wadis, and ravines we would have to climb? I wanted to reassure myself that my polio leg could handle it. Fighting the effects of polio continued to be an ongoing battle, and my once-paralyzed leg had not let me down yet! Each hash course was a challenge, but I managed to complete them all without too much difficulty. I was only in my sixth decade of life, and I needed the continued support of that leg before I would finally call it quits.

The hash could include barbecues, camping, and games. The military and U.S. government employees and their families were always welcomed at the hash. They might even bring a cooler full of beer and enjoy the so-called "tea ration" available to official U.S. government employees. They had access to beer, wine, and spirits, which we didn't, and sometimes they would share. International residents were known to make their own wine, beer, and even distilled spirits. The Aussies and Brits led that category and were the most daring. All of it was strictly illegal and cause for immediate expulsion from the kingdom. Saudi airport customs was always on the alert for the smuggling of yeast, a key ingredient for the brewing of beer or wine, into the kingdom.

The air was hot, and the constant heat was oppressive and contributed to the challenge of living and working in that environment. There was a round thermometer, about 24 inches in diameter, hanging under the eaves of the camp gym. On most days, the thermometer reached 120 degrees Fahrenheit, its maximum temperature, and remained there continuously, including throughout the night.

I experienced the natural tension that comes with executing a contract to high standards and having three bosses to answer to. At the same time was something new to me: I had a corporate boss, Tom Fintel, President of Vinnell Company, in northern Virginia to whom I was responsible for executing the contract to U.S. government and SANG standards, upon which the continuity of the contract and corporate compensation depended. And profit was no small matter with both Vinnell and their senior management, Northrup Grumman.

Tom Fintel's Vinnell-Arabia Arab counterpart was Salah M. Fustok, in Riyadh. He represented the "Arabia" side of Vinnell-Arabia. Salah, a Lebanese Sunni, was an astute businessman and every bit a gentleman, someone I respected and became very fond of. His sister was married to the King of Saudi Arabia, Abdullah bin Abdulaziz Al Saud. He ran interference for me with the Saudi hierarchy when necessary and was indispensable to the program's success.

Then, I had a U.S. Army boss, Brigadier General Martin Dempsey, program manager, OPM-SANG, to whom I was responsible for executing the tasks of the contract to the satisfaction of all parties. Dempsey would later become the 37th chief of staff of the U.S. Army, and subsequently, the 18th chairman of the JCS. There were times when I thought his staff overplayed their role in our relationship. Vinnell contract employees were highly experienced former U.S. military, subject matter experts, in many cases more experienced and more qualified than their U.S. Army contract supervisors.

For example, the Vinnell senior trainers/advisors were all former U.S. Army colonels with highly successful 30-year careers in their fields of expertise, including 11 war college graduates. Among our former enlisted men were 14 sergeants major. Most were Vietnam veterans. Additionally, there were other civilian employees who were not former military. Most of these employees worked for Larry Wright, our very capable business manager, who handled all matters pertaining to contracting, finance, personnel, etc. Indeed, it was a talented and experienced work force. General Dempsey would be replaced by Brigadier General Clinton T. Anderson in the summer of 2003.

My third boss was the SANG whose leadership served under and was responsible to the crown prince. My SANG contact, whom I met with regularly, was Major General Bandar bin Omar Al Nahil, chief of SANG Operations (G3). Rarely did I meet with the commander, Prince Mutaib bin Abdullah,[1] third son of King Abdullah,[2] who at that time was deputy commander of the National Guard.

OPM Sang didn't want me, the contractor, meeting directly with the customer, so those meetings, often called at the request of General Bandar, had to be carefully orchestrated so as not to ruffle feathers. I made it a point to always back brief my American boss at OPM. There were times when it got quite tricky trying to keep all three bosses informed all the time. I became very nuanced at telling each what I thought they needed to know. They all seemed happy, so I guess I did okay.

I assumed management of a high-performing contract and determined to sustain and improve that performance. In addition to the compensation in accordance with the terms of the basic contract, there was a monthly opportunity for Vinnell-Arabia to receive additional performance compensation based upon the evaluations of the OPM U.S. military personnel who oversaw the contract and graded our performance. That was the area where I thought we could do better.

We received a monthly grade in the form of a percentage of satisfactory performance. When I assumed the position, Vinnell usually received a grade in the mid-to-upper 80s. My goal, which I set, was to raise that grade consistently into the 90s, and I saw no obstacles preventing that. After several frustrating months, where we continued to receive grades in the high 80s, I managed to convince OPM-SANG that our performance was, in fact, superior, and we began receiving grades in the low 90s.

I had a couple of very contentious meetings with Dempsey's civilian deputy over performance grading. He oversaw the compilation of the several evaluator grades and determined our final performance grade each month. When we didn't agree, I went to Dempsey ... and the matter was sometimes resolved in Vinnell-Arabia's favor. He was always receptive to meeting with me. OPM guarded very closely their prerogatives when it came to evaluating and grading our performance.

As PGM, I had to fight daily to ensure my workforce, their families, and our customers were happy and satisfied. At that point in time, in the spring of 2003, I didn't realize another completely different type of battle was looming on the horizon—a challenge that would require all my cumulative experience, flexibility, ingenuity, and tenacity to fight.

Saudi Arabia, at times, has been a dangerous place for expatriates, ebbing and flowing over time. The security situation I found in 2002 wasn't overly dangerous or concerning, but I knew from experience that the situation could change quickly. On November 14, 1995, seven persons, including five U.S. citizens, were killed in a car bombing of the OPM-SANG headquarters building in central Riyadh; in 1996 a truck bombing of Al Khobar Towers in Dhahran killed 19 U.S. Air Force personnel,[3] and in 2002 two separate car bombs killed a British and a German citizen. I knew that the Islamist organization, Al-Qaeda, or their proxies, thought to have originated in Saudi Arabia, responsible for the 9/11 attack, had targeted Saudi Arabia.

In 2002, when I arrived in the kingdom, Al-Qaeda had been under continual pressure from the anti-terrorism campaign led by the United States, but it was still present and dangerous in multiple locations throughout the region. For all those reasons, I knew that our presence was necessary and desired by the leaders, and tolerated by most of the population, but there were enough who resented our presence to the point of killing us if we didn't leave. I considered that our personnel could potentially be a target.

For example, in early 2003, because of a perceived threat to non-Vinnell-sponsored employees and their families living in Riyadh, I directed that they move on to Camp Vinnell. We modified the facilities and made them as comfortable as possible, believing they were safer there.

I also participated in an unofficial security-information sharing forum with the American Businessmen's Group Riyadh (ABGR). Its members included the major expat businesses in Riyadh, such as British Aerospace, Boeing, Lockheed-Martin, and Raytheon. The consensus was that the main threat was the so-called "lone wolf," defined as a disenfranchised, radicalized, or self-motivated individual who would take it upon himself to attack an individual target of opportunity. From 2001 to early 2003, expats were shot at, with two killed, attacked by suicide bombers, and threatened individually. The official and unofficial expatriate communities believed that the main threat was against individuals or small groups by individual terrorists or small terrorist cells, not from larger, organized, and better-resourced terrorist units.

Vinnell-Arabia, among the other contractors, was in the unique position of directly supporting the SANG and residing in Camp Vinnell, a SANG-secured facility. But that supposed tight security wasn't failsafe. From the day of my arrival to the day we were attacked, the staff worked to improve our physical security. We could only physically control what happened inside the walls. I had no authority and was told every time I asked for more and better external security that the SANG was responsible and it wasn't my concern.

I first raised the question of security for the camp and whether we had weapons on camp for protection, during my initial pre-hire meetings with Vinnell leadership in Virginia in November 2001. I was told that "no Americans could carry arms and that the SANG was responsible for our camp's security."

I never felt comfortable with that condition. What could happen because of inadequate external security could have a catastrophic impact on Camp Vinnell. My instincts directed me to ask the SANG to allow Vinnell-Arabia U.S. citizen employees to be armed with the SANG providing the weapons and ammunition. My requests and concerns were summarily dismissed. We also requested external security enhancements from OPM-SANG and asked that a formal security assessment be done. In late 2002, a physical security team from United States Central Command (USCENTCOM) conducted a review and evaluation of Camp Vinnell security.

The purpose was to determine what additional physical security measures would be required and certified by the USCENTCOM evaluation team to enhance our security. As a result of this review and evaluation, OPM-SANG agreed to fund improvements to the camp walls, including the installation of cameras to monitor the area along the outer walls.

The work to improve and raise the walls had begun and the contract for cameras was prepared to be issued on May 12, 2003. Doubtful that the proposed changes would be enough, I personally requested that the SANG increase their security posture and readiness to support us. As a result, they established a reaction company sized unit in the adjacent SANG compound.

We were not able to learn its composition, training, or rules for activation or employment. All areas where we needed insight and could have influenced and enhanced increased security were denied access. The other area where we engaged the SANG on multiple occasions was security at the main gate. The presence of a SANG vehicle with a manned machine gun appeared impressive and provided a level of assurance, if not deterrence, but physical presence is not a substitute for personal alertness, readiness, and training.

We offered to provide training to the SANG guards and to inspect and evaluate their weapons. We were told by SANG that was their business and not our concern. Had we been allowed to work with the SANG guards, we would have learned the status of their training, their level of readiness, and their standing orders. In sum, had we had insight into those areas, we could have improved their capability to provide physical, armed security for our main gate.

The SANG strongly discounted the fact that terrorists would attack them directly, at their own facility. The SANG was a physical representation of the royal family, and no one would dare to attack the royal family. I was told by a senior SANG officer that they were the elite force and not to worry about terrorists attacking a SANG facility, including Camp Vinnell. "That will never happen."

As a career military man, I knew that the SANG security wasn't adequate to my standards. The challenge for me was to balance the security of Vinnell-Arabia with the execution of a contract in a business environment. I couldn't influence the SANG security as hard as I tried, so my only option was to accept the judgment of all my bosses, Vinnell, OPM-SANG and the SANG—"It wasn't my responsibility and not my concern." I never accepted that and worked continuously to improve those areas of physical security within my control.

Al-Qaeda Attack

By January 2003, the perceived threat level in the kingdom had calmed to the point that I directed those employees and families that had been moved on to Camp Vinnell be allowed to return to their homes in the Riyadh community. Once again, we settled into our normal routine, but I remained uneasy given the volatility of the area and my former reluctance to accept the security arrangements provided for us. Unfortunately, my instincts were proven correct. That myth of tight security was destroyed on May 12, 2003.

Without warning the first punch hit hard and with surprise. Camp Vinnell was attacked by terrorists at approximately 11:20 p.m. resulting in the death of 10 employees and injuries to over 70, both minor and life-threatening. The fight had begun! I was evaluating a three-day joint Vinnell-Arabia-SANG brigade-size exercise in the desert along with our operations and maneuver advisors/trainers when I received the first indication that there was a problem at Camp Vinnell.

I received a phone call at around 11:30 p.m. from Brigadier General Dempsey, the program manager, OPM-SANG at Eskan Village. "Bill, Marty here, are you hearing any explosions out your way?" Eskan Village and Camp Vinnell were 25 miles apart. I told Dempsey I was in the desert 90 kilometers west of the camp evaluating a SANG field training exercise, but that I would check immediately. "OK. Thanks. I'm going to walk over to my operations center and see what is going on," he answered.

No sooner had I hung up, than the phone rang again. It was Dan Brownlee, my program coordinator, apprising me of the situation at Camp Vinnell. The program coordinator was similar to a chief of staff in a large military headquarters. Dan joined Vinnell shortly after I arrived. A former infantryman, he had recently retired from the Army as a Special Forces colonel and had the added advantage of being a Middle East Foreign Area Officer (FAO). He had served and travelled extensively throughout the Arab countries and was steeped in knowledge of their customs and military organizations.

A bright officer with good instincts, he was a rare find and critical to our mission's success. He was a can-do, proactive officer with good judgment and common sense. He reported, "Sir, we were just attacked by an unknown number of terrorists using small arms and a VBIED (Vehicular Borne Improvised Explosive Device), and we have reports of continued small-arms fire. We have casualties and are assessing the situation and have begun search and rescue efforts," he said in a calm but clear, firm voice.

It was at that moment that I learned and tried to comprehend that there had been a terrorist attack on Camp Vinnell involving a vehicle bomb and that there were casualties. I was very concerned that Dan's initial report indicated that the attack might not be over, and the security threat situation was still evolving. In an immediate subsequent call, he informed me that the residents had consolidated at the community pool and efforts were underway to account for employees and residents. He also said that search and rescue operations were underway at the damaged buildings despite not being sure if the attackers were still in the compound. I can't describe what was going through my mind other than a complete feeling of helplessness as I was 60 miles away in the desert!

I had to return as quickly as possible to Camp Vinnell. The SANG brigade commander in the field, Major General Mashouj bin Sayyah Al-Anazi, offered to drive me back in his Pinzgauer high-mobility, all-terrain command vehicle. Ron Hindmand, one of the Vinnell senior evaluator/observers for the exercise, returned with me. Major General Mashouj knew the terrain and correctly determined that the best and safest way was to drive cross-desert avoiding the major roads and highways as we learned they would be secured with Saudi police checkpoints as a reaction to the attacks.

He was correct, as we observed many checkpoints and ominous flashing lights in the distance as we drove to Camp Vinnell. Had he not driven me, I would have been delayed or even prevented from getting there. As it was, it took almost five hours of cross-desert driving with many deep wadi obstacles and detours on that dark night before we arrived. I was anxious and nervous, not knowing what I would find.

"Drive faster!" I kept yelling. "Drive faster!"

My anxiety was causing me to catastrophize and assume the worst. I remember my worst thoughts turning to possibly dead employees and family members. We had wives and children in the camp, including my own wife, prompting me to jump to the worst possible conclusion. Thoughts of 9/11 and the burning towers in NYC crossed my mind. The general was speaking in Arabic, and I was yelling in English as we continued to Camp Vinnell. By that time, we both knew something terrible was happening in Riyadh as his vehicle's tactical radios were on, and we were receiving reports of repeated attacks and explosions. There was even a report of a gun battle at a Saudi highway checkpoint. It was all in Arabic, so I couldn't understand the transmissions. General Mashouj, while driving, interpreted when he could.

En route, I was advised to first stop by SANG's King Fahd Hospital where our casualties were being taken. Upon arrival as I recall, we were met by Dr. Bandar Al Kanawy, chief executive medical officer. He and his staff were busy receiving, triaging, and treating the many casualties still arriving from Camp Vinnell. He gave us a quick report on their status. We then drove the short distance to Camp Vinnell.

We arrived around 5:30 a.m. I saw a vehicle pressed against the left-hand door of the main entry gate which wasn't normal. It looked like one of the silver-colored Crown Victoria sedans used by Vinnell-Arabia. I was puzzled by its presence and condition. As we entered the gate, I didn't initially see any major damage. At the first intersection inside the camp, I had a choice to turn right and go to the apartment building which housed the employees with families, or I could turn left and travel in the direction of the employee apartment buildings. As sunrise approached, I saw that the damage was extensive. I remember the pungent stench from the explosives that hung in the still morning air. A quiet pervaded the camp.

We were directed left to the scene of the explosion. I would later understand that the terrorist truck bombers had been faced with the same decision and had turned left and proceeded in the direction of the employees' apartment buildings instead of turning right and approaching the family building. The decision by the terrorists to turn left and detonate the truck bomb next to the employee buildings rather than right and detonating the bomb at the family building would be a subject of much discussion and consideration, as that chance decision caused the death and injuries to the employees instead of the families.

We arrived where the bomb had detonated. The entire nearside wall of the four-storey apartment building was blown down. You could see into the many rooms which housed the employees. There was a large crater at the foot of the building where the bomb detonated. The last of the casualties had just been removed from the building. I was met by Dan Brownlee, Glenn Edmonson, the Vinnell-Arabia assistant security manager, and other Vinnell-Arabia employees still actively conducting search and rescue operations. Our medical advisors and staff were triaging, stabilizing, and treating the injured and directing their movement by any available vehicle to the SANG hospital, as no ambulance support had yet arrived.

The triage, organization, and movement of the injured was organized and led by Russ Cotton, the Vinnell medical operations and logistics advisor. His leadership ensured the injured were treated and moved to the hospital expeditiously, and his work probably saved lives. I also saw the bodies of some of our employees who had been killed. They had been covered by blankets and treated with dignity and respect, even as the staff worked to conduct initial identification. "How many were killed?" I asked.

I can't overemphasize the heroic efforts of the Vinnell-Arabia employees in response to the attack. Ed Nicks, the Vinnell logistics advisor, personally organized and led the search and rescue effort. Fellow employees selflessly entered a seriously damaged, still burning four-storey building, the collapse of which might have been imminent, to search for and recover both the dead and injured and bring them to safety. Among the search team was Glenn Edmonson, who made several trips through the bombed and burning building to either rescue wounded employees or to carry out and identify the dead.

The only entrance into the most seriously damaged building was the rear fire escape, which was no longer securely attached to the building, the interior staircase having been destroyed. Nicks, Edmonson, and the others carried the dead and the injured down that fire escape with little thought for their own safety. I know from experience that it is times like that that reveal a man's true character. This was a courageous, passionate effort on their part.

As the sun rose on May 13, 2003, I could more clearly see the actual damage to the infrastructure. The windows of most buildings and vehicles were blown out; the entire facades of two employee four-storey apartment buildings were missing. Most single-storey buildings were damaged, and many had collapsed. There were multiple destroyed vehicles, burned out, still smoking, tossed around like toys.

I went to the community pool to see the employees and families gathered there. I stopped for a quick moment to speak with my wife and some of the families and employees. She and a few of the other wives had worked through the night to ensure all families and children were accounted for and comforted. I was impressed at how calm and positive in spirit they all were, knowing what they and their friends had just experienced. One employee passed around his cell phone among the residents sheltered in the pool area that night so they could call their loved ones to say, "I'm all right, don't worry," realizing the attack would be international news.

Linda never saw me look so sad and in disbelief. I was forlorn and devoid of hope. I was ashen in color but burning with rage inside. I will never forget that brief moment when her eyes met mine. She felt so badly for what had happened. Words were few, but we were feeling each other's pain, like only a husband and wife could. And for a moment, as I stood among my employees and their families gathered at the pool, a sense of guilt came over me, regretful that I was safely in the desert and not with them during the attack. My inner critic was telling me that I should have been there!

I remembered the work to raise the walls and to place barbed wire on top had begun, and, ironically, the contract for cameras had been prepared to be issued on

May 12, 2003. While each of those physical security enhancements would have improved Camp Vinnell's security, their completion would not have prevented the attack.

Later that morning, I met with Dempsey in my damaged headquarters office. The roof had collapsed onto my desk and floor. We stood while talking. In a generous and magnanimous act, he agreed to host our employees and families in vacant quarters and in the homes of his employees on Eskan Village, the OPM-SANG compound. Standing next to him, he called and told his staff, "These people have no place to sleep or eat, make preparations now to receive and house them." They did!

That act was a welcome gesture by fellow Americans and solved a serious immediate problem of where to house the employees and their families in a safe and secure location as the bombing made our buildings unsafe for occupation. The camp was in ruins. It reminded me of the devastation to the city of My Tho during the Tet Offensive years earlier.

That morning, I hosted the U.S. ambassador, the Honorable Robert Jordan, and Secretary of State Colin Powell, who had just arrived in the kingdom for an official visit as part of his swing through the Middle East. I first met Powell when I was executive secretary to SECDEF Weinberger in 1987 and again, several times during Operation *Just Cause* in 1989, when Powell was FORSCOM commander. We also attended the same church service as the Powells at Saint John's Episcopal Church in McLean, Virginia. We sat a few pews apart every Sunday and would sometimes gather in the church social hall after the service for refreshments. He knew me, but not well.

I had just finished touring the camp when his entourage with Saudi escort arrived. Throwing his arms around me and pulling me close, he said, "Bill, how are you holding up, my God, what did these guys do to you?" Since I wasn't present during the attack, I had Dan Brownlee join us for a quick walking tour briefing him and the ambassador on what happened.

The attack had occurred only nine hours earlier, and the buildings were still smoldering. Only a short time before, the last body had been removed. Powell, standing amidst the rubble, with the bombed-out High Rise 1 building in the background, gave an on-site press interview with the many reporters and cameramen travelling with him. The ambassador and I were standing right next to him. He was very composed, but angry as hell, and like me, in disbelief. His eyes were laser focused on the still smoldering buildings.

The May 14, 2003, editions of the *New York Times* and the *International Herald Tribune* described Powell as being "shaken as he toured an eerie scene of carnage." He came "to promote peace in the region and instead found himself inspecting a scene of carnage," which included eight dead, seven of them Americans. These preliminary casualty figures were later revised to 10 dead, eight Americans and two Filipinos. Powell called the attack "a despicable act" and said it had "the fingerprints of Al-Qaeda." Hours later, President George W. Bush denounced the attack as "ruthless murder." "We will find the killers, and they will learn the meaning of American justice," he said.[4]

Powell turned to me and the ambassador and talked briefly about the nature of the enemy we were up against and was beside himself that there were people so hateful that they would "kill innocent people in their sleep, even those trying to help others." A serious thinker and strong Christian, Powell was trying to fathom the mind and soul of someone who would do such a dastardly act. I read his mind and I have never forgotten his words. And I needed his presence, his words, and strong embrace that day!

Some months later, a longtime close friend of ours who had seen the live newscast on TV of the bombing and the Powell visit that day, said to Linda, "I knew that was Bill walking with General Powell, I could tell by his limp." She knew about my polio and its effects, but I was surprised that she noticed the limp. Also visiting that day was Crown Prince Abdullah, commander of the SANG, who became King Abdullah in 2005. He was accompanied by Prince Sultan, who became the crown prince in 2005. They graciously expressed their condolences on behalf of themselves and the people of Saudi Arabia.

I won't forget Abdullah's expressions of concern, as we visited the pool area where the dependents were still gathered, particularly for the children, and his relief that none had been casualties. He was visibly upset. Several times, he asked about the children. We told him none had been hurt. Some children were killed in the earlier attacks on the other two compounds. The crown prince's interpreter grabbed my arm and kept saying to me, "We will find them, we will find them!" Later, the crown prince denounced the attackers as "monsters."

A consequence of the coordinated attacks on May 12, 2003, was that in the words of the U.S. government the Saudis finally acknowledged that they had a serious internal terrorism problem that they needed to address, and our assistance could be useful. It was reported that at Powell's behest President Bush called King Fahd and offered our assistance, and the president requested that the FBI be allowed into the kingdom and assist in the investigation. King Fahd agreed and within 24 hours the FBI was in the air en route to Riyadh.[5]

In days, FBI agents arrived to participate in a joint U.S.-Saudi investigation. The cooperation between the FBI and Saudi authorities enabled the identification of the perpetrators and their methods and means of attack. An official Saudi government statement on June 7 identified, by name, 12 Saudi men as the attackers. According to that statement, the identification was based on DNA found at the scene of the three attack locations. Two others were killed in a gun fight with Saudi security forces following the attack.[6]

Over time, more information arrived. Several days prior to the attack, the local news announced that Saudi authorities had found and arrested a criminal cell with a truck bomb and numerous weapons. The significance of that arrest would soon be known.

On May 12, three separate compounds were attacked by three separate terrorist elements. In each case, the attack consisted of a truck bomb, a sedan with weapons, ammunition, grenades, and multiple attackers. The joint investigation revealed that the "cell" the Saudis arrested was a fourth attack element. When that cell was arrested, the other three cells decided to execute their attacks quickly rather than risk discovery by Saudi authorities.

The two other compounds attacked were Al Hamra Oasis Village, approximately 10 miles from Camp Vinnell, and Dorrat Al Jadawel compound, approximately 3 miles away, housing mostly non-U.S. expats. Camp Vinnell was the third target and contained the Americans. The method employed was similar in all three attacks, although each unfolded differently.

The attack cell was able to penetrate the gate of the first compound, Al Hamra. The sedan and the truck bomb proceeded to the housing units. The attackers moved out and began to execute any residents they found in or outside their residences, shooting some at point-blank range, including women and children. They then detonated the truck bomb, killing themselves and over 20 compound residents.

Later interviews revealed that Camp Vinnell residents had heard a "noise." That noise was the Al Hamra explosion. Although it seldom rains in Riyadh, it is not uncommon to have heat lightning and thunder. Within minutes, a second loud noise was heard closer than the first. That was the explosion at the Jadawel compound. It was clear to those on Camp Vinnell that it was an explosion.

At Jadawel compound, the attackers detonated a truck bomb killing two Saudi security guards. Considering the possibility that a bomb had been detonated in the vicinity of Camp Vinnell, our program coordinator climbed the steps from his first-floor apartment to the fourth-floor rooftop to survey the area.

Upon reaching the top of the building, he asked the guard if that was an explosion, and could he point to the location? As he asked the question, firing was initiated at the Camp Vinnell gate. Multiple rounds of small-arms ammunition, including tracers, were fired. As the entrance was shielded by trees, it wasn't possible to see what was happening. The next thing he saw from his vantage point on the roof was what appeared to be a pick-up truck with a cloth cover over its rear bed proceeding down the street from the gate at high speed with several armed men chasing it.

Additional firing was heard, and eyewitness statements revealed that the men chasing the truck down the street fired their weapons randomly and threw hand grenades. The FBI report stated, "several witnesses heard male voices shouting 'Allahu Akbar!' 'Allahu Akbar!' immediately prior to the main explosion in front of High Rise 1."

The truck turned left at the intersection, and the next thing the PC saw was the explosion at HR1 as the truck bomb was detonated. What he observed from his rooftop location was the detonation of what the FBI determined to be an approximately 400 lb IED containing the highly explosive organic compounds of nitroglycerine, PETN, and RDX, as detected by the Saudi laboratory chemists. He was located approximately 600 yards from the explosion, and his main recollection was that unbelievably there was no sound, and he didn't feel the concussion as the blast radiated outward reaping destruction in its path. The rising cloud of dark black smoke was filled with flickering embers. All the windows in the building on which he was standing had been blown inward by the concussion.

One of the most poignant (and I should say fortunate) occurrences was that after the Jadawel explosion the residents of our family apartment building began to flow into the hallways. My wife, Linda, Dan's wife, Colleen, and another employee's wife, Connie, gathered in the first-floor hallway outside of my apartment. Connie began experiencing an anxiety attack, so Linda invited her into our apartment to give her a drink of water. Colleen returned to her apartment to look after her son.

They had just cleared the hallway when the Camp Vinnell blast occurred. The concussion blew in the large double entrance doors and smashed them on the other side of the lobby directly through the space and location where the three women were standing moments earlier, a near miss if there ever was one! The sole family member casualty was a wife on the third floor who looked out her balcony doors just as the bomb detonated. The concussion dislodged the sliding door, and she was struck in the nose, receiving a minor injury.

Approximately three hours after the attack, the security reaction force from the SANG unit compound, adjacent to Camp Vinnell, was activated, and moved into Camp Vinnell. Their presence "to restore order and ensure our safety" was too little and way too late. Their presence caused friction between the residents and the SANG as the SANG military attempted to enforce control and constrict movement. We asked the SANG unit to leave Camp Vinnell and after some initial resistance and hesitation they departed.

It soon became clear to me what role the sedan played that I saw at the gate when I arrived the day of the attack. Investigation and interviews revealed that the sedan had approached the inspection station as usual for an arriving vehicle, and closely resembled one of the vehicles used by employees. As our inspection crew exited the pedestrian door of the gatehouse and approached the vehicle, it sped forward toward the gate. The Delta barrier (a high security iron barrier system that prevents vehicles from passing) was in the up position, and the sedan became entangled with the barrier, becoming stuck hard against the gate door.

Four of the five terrorists in the sedan immediately exited and opened fire on the SANG guards. Normally there were two to three SANG personnel manning a Pinzgauer, parked beside the gate with a mounted .50 caliber machine gun. The attack was so fast and unexpected that the SANG guards didn't have time to return fire. One of the SANG guards was killed instantly and the other two wounded.

The sedan occupants forced their way through the personnel door and moved to the gate house. Pushing aside the guard, they lowered the Delta barrier and opened the right-side door of the entry gate. Investigation and interviews revealed that the attackers were familiar with the gatehouse and knew exactly where the controls were located and how to operate them to lower the barrier. Almost immediately, a pickup truck zoomed out of the dark and drove through the open right-hand side of the gate. The sedan occupants were unable to get their car through the gate so they ran down the road following the truck indiscriminately firing their automatic weapons, throwing grenades and shouting "Allahu Akbar, Allahu Akbar." The FBI Riyadh Bombing Report described the attackers as wearing either the "traditional Saudi thobe (white robe) or dark-colored sweat suits."

The presumed plan based on the other attacks was to get both vehicles through the gate and wreak carnage among the residents with small arms and grenades prior to detonating the bomb. Upon examination of the sedan, we found weapons, ammunition, grenades, ammunition vests and ammunition pouches taped back-to-back in a quick-reload fashion in the car. Their failure to get the sedan through the gate prevented more severe damage, injuries, and loss of life. It is my belief that the sedan was to drive to HR3, the building where our families were living and attack it. There were 16 family apartments in the building, including mine and the PC's, with many women and children. The attackers were bent on killing Americans.

As to whether all the attackers perished in the attack, the investigation by my staff and the FBI indicated that possibly one or more escaped over the rear wall after the explosion. The FBI report in coordination with the Saudi investigative authorities determined through DNA analysis that six of the possible nine Camp Vinnell attackers perished in the explosion. That supported our belief that not all the terrorists died in the explosion.

With resolution of the immediate issues my attention turned to the injured. The SANG medical system and SANG hospital had a good reputation for quality care, capable, mostly expat, personnel, and good facilities. The care and compassion they provided over the next several months was exceptional and alleviated much of the stress and anxiety that we all felt.

The injuries from the blast included concussion injuries, stone and glass shrapnel, burns, and smoke inhalation. Two employees were so seriously injured that we couldn't positively identify them until they had been treated for several days. Command Sergeant Major (Ret.) Albert Mallet lost sight in one eye. Another employee died a few days later while in intensive care. Chad Chadwick and I had visited him and held his hand just moments before he died.

Among the TCNs who were injured was Egyptian Mohamed El Hefny, who had been on the contract for over 11 years and oversaw English language instruction for SANG members. Mohamed was a retired Egyptian Army brigadier general. As a tank commander, he had been wounded and captured in the 1973 Arab–Israeli War.

Another fortuitous circumstance, which resulted in fewer employees killed and injured, was that many of the American employees were in the desert with me conducting the SANG training exercises. Had they been in their apartments during the attack, our casualties would undoubtedly have been far greater as most resided in HR1, where the bomb was detonated.

I visited the hospital almost every day until the last employee was discharged. In all, over 30 Vinnell-Arabia employees were admitted. Their injuries ranged from non-critical to extremely critical. Once, while driving to the hospital, we stopped at a red light when something suddenly hit my front passenger side window with a loud bang and blew it out completely, scaring the hell out of me. "What in the hell was that?" I screamed at Saeed, my driver, who was also aghast with shock.

It smashed the safety glass window of our GMC Yukon SUV. Small pieces of glass were on the dashboard, on my lap and all over the floor. Miraculously, neither of us were hurt, thanks to the reinforced tempered glass windows which ensured there were no large, jagged pieces. Saeed put his foot to the pedal, made a quick left turn at the light, and sped away. We never found out what hit the window. He thought someone in another vehicle following us came alongside quickly and either shot or threw something that blasted the window. I was clearly the target. It happened so suddenly, we didn't see or notice any vehicle or person who might have caused the blast. It remains a mystery. Luck and Providence were with me that day!

In the early days after the attack and during the investigation, a line of inquiry was whether the attack had been facilitated or supported by "insiders." As I stated earlier, there were Saudis who were employees of Vinnell-Arabia, and other Saudis who would have access to the camp for business or training. Even Powell hinted during his visit that morning that "the facility had been cased, as were the others."[7] Powell never stopped believing that there was insider help. The May 16, 2003 edition of the *Daily Telegraph*, the British daily newspaper, reported American intelligence sources believed the "bombers depended on a significant level of 'insider' knowledge of the compounds."[8]

The FBI and the Saudi investigation didn't conclude whether there was inside help. Some of my employees believed that the Saudi terrorists that attacked the compound that night had some insider assistance. I agree.

What happened to Vinnell-Arabia was indeed the result of a planned and organized attack, which we never saw coming. Some were saying Vinnell-Arabia would never recover from the dastardly act. The naysayers and those few who lacked the courage to remain left. I wouldn't allow those timid, pessimistic souls to stand in the way.

I knew that if someone is not on board with the plans, their bad attitude can spread like cancer, and it was best that they leave. A few were removed from the contract involuntarily. Yes, I was shocked and angry. But never was there a moment when I didn't think we would rebuild and recover. I also believed that we owed it to those dead and injured employees and their families to "rise from the ashes" and carry on. To emerge renewed, revitalized, would stand as a monument to their strength and dedication. Their deaths gave us a renewed sense of purpose. I quickly got over any initial feeling of helplessness I might have had and directed my energy toward rebuilding.

I knew what needed to be done, and the onus was on me, the on-site PGM, to rally the troops. It is not my nature to hesitate when faced with a challenge. I had faced adversity and challenges earlier in life, particularly fighting polio, and combat in Vietnam, and that cumulative experience prepared me to fight and succeed. I had a strong team led by Brownlee, Chadwick, Hindmand and Wright—all positive-thinking, can-do men. I knew we would recover.

Key to a timely and complete recovery were the employees and their families. The employees had serious questions to be addressed: How can I be assured of my safety and security after I was attacked in my own home? Where will I live? Do I stay or do I leave? Security, of course, was an immediate concern. How can we restore confidence? Can we establish effective physical security? Can Vinnell-Arabia be armed? We certainly didn't want a repeat of the terror we experienced. We had to determine what could be salvaged. Where and when could we establish new living and working facilities?

It was a complicated mosaic where all the pieces had to fit in order for us to systematically continue. My senior staff and I wrestled with the matters, and we spent every waking minute addressing those questions and concerns as they developed. Family members were directed to depart Saudi Arabia immediately. Employees were authorized to depart. Initially, there was hope that the situation would calm down, and family members would be able to return. It soon became clear that recovery would take more time than initially thought and that it was unlikely that corporate headquarters would allow the return of dependents.

A suicide bomb detonated outside the Al-Mohaya housing compound just outside Riyadh five months later, on November 8, 2003, killing 17 people and injuring 122, among them women and children, sealed any chance of our family members returning. Terrorists were bent on killing westerners. It was indeed a tense time, fraught with danger.

To evaluate the Vinnell-Arabia camp facilities' habitability again, we went to work assessing damage and organizing clean-up, identification of salvageable facilities, and all activities required to assess what needed to be done. I was briefed continuously, gave guidance, made required decisions, and issued instructions for continued work.

To honor and remember our fellow employees, fellow Americans and Filipinos, we conducted two memorial services. The first service was to honor and pay tribute to the nine American and Filipino employees who had lost their lives on May 12. The Filipino presence in Saudi Arabia was so large that they had a diplomatic-level representative in their embassy solely to support their countrymen. I met the representative several times to arrange the transportation for the two deceased Filipino employees back to the Philippines. We conducted a second separate memorial service to honor our employee who had died later in the hospital. Some employees delivered beautiful eulogies on behalf of their co-workers. Those memorial services were organized and led by another of our senior advisors, Freemon "Don" Donley, a retired U.S. Army colonel and lay preacher. His heartfelt compassion provided comfort and solace to all who attended.

Several times, I have been asked as a combat veteran of Vietnam to compare that experience to my return to Camp Vinnell the morning of May 13, 2003. Without hesitation, my answer is always, "There is no comparison! In Vietnam, we were fighting an enemy and I expected to get shot at and to sustain casualties. In friendly Saudi Arabia, I didn't. In Vietnam, I had my own weapons and ammunition and could fight back. In Saudi Arabia we were unarmed and at the mercy of our hosts for protection!" We depended fully on them for timely intelligence and for our physical security.

Vinnell-Arabia wasted no time in getting back to work. In business it is referred to as "continuity of operations." While still in the critical "assessment and evaluation" phase, we again became productive. My administrative and training staffs organized their remaining assets and set out to meet the contractual requirements. No one thought we could go back to work so soon, if at all, but we did.

Simultaneously we were repatriating the families to their location of choice. The families were traumatized and faced with a host of questions and options, and we supported them and helped them work through them. Some of the wives, children and employees who experienced the attack that night still suffered with post-traumatic stress later, including my wife, who occasionally experienced flashbacks of the event like it was happening all over again. All families returned home by late June 2003.

I wanted as many employees as possible who were willing to remain to stay. A few decided to leave for their own reasons. Vinnell Company was supportive, even though employees, especially qualified experts, were hard to recruit even under normal conditions. Vinnell Company decided to offer "continuity pay" through the end of 2003 and that had a positive effect on near-term retention.

The issue of a permanent residence for the employees to live and the facilities required to support all the training and services provided before the attack turned out to be most challenging. My highest priority was to find secure accommodation for over 500 employees and get them off Eskan Village. I wanted to end the growing friction and irritation between the employees, Vinnell-Arabia, and OPM-SANG. I knew full recovery wasn't possible until we were back in our own facilities.

We started searching for an adequate compound. The SANG provided, in effect, a Saudi real estate agent who showed us what was available. His idea of available and ours were radically different. He showed us the structural skeleton of what appeared to be an unfinished apartment complex. It could be ready in less than five years, but even the most vivid imagination failed to see how an adequate facility could be constructed. I emphasized that we needed an adequate and secure compound now, not in years. I was reasonable, but I would only accept an adequate facility.

Finally, I was informed of a vacant compound that was available. It had been the Boeing Company compound for their employees during the Saudi Air Force support program, which concluded years earlier. It was an opportunity to take over an entire compound. Upon inspection, it met our requirements.

At the same time, I was working on the idea with the SANG to allow our U.S. employee security and reaction force in the new compound to be armed, a totally new concept for any expat contractor residing in the kingdom. At first blush, the Saudis were not receptive. An armed U.S. contractor in Riyadh was unheard of!

That was a non-negotiable stipulation. Never again would my men not be able to defend themselves against an attack. After weeks of steady, unrelenting pressure, reluctantly, the Saudis finally relented. The SANG issued us six German-manufactured G3 7.62mm rifles, with one magazine each, ammunition, and cleaning kit. They intended to issue one load of 20 rounds for each of six magazines, 120 rounds total. We convinced them to issue three magazines per weapon and 2 x 200-round cans of ammunition. The only SANG stipulation was that men with weapons not be visible from outside the compound. The Saudi government didn't want Saudi citizens to know the Americans were armed inside the compound.

We agreed to that stipulation, which later caused a minor incident. A former Marine, Tom Williams, had joined the contract as the U.S. citizen employee who would be responsible for managing and securing the weapons, and organizing and leading the armed security element. A dynamic and energetic Marine, he was actively patrolling the compound when he climbed up on a ramp overlooking the gate entry point. He was observed by a Saudi to be carrying a rifle and reported to the authorities. We quickly addressed the violation and there were no further incidents.

The staff, led by Brownlee and Chadwick, developed a plan to first secure the compound, and then determine requirements and allocate residences based upon employee population and other requirements. A major improvement in the amenities to those of Camp Vinnell was that all the new residences were detached houses, which could house three to four employees.

For the first time, we were responsible for our own physical security, and we had weapons and ammunition to do it. We could tell future employees with certainty that Vinnell-Arabia managed their own security, not the SANG. For the first time in over six months, I was confident we were on the road to our complete recovery. There was finally light at the end of the long tunnel.

We set about developing a physical site security plan. I wanted to integrate best practices and apply the lessons learned from the many attacks in the kingdom. We didn't want the compound to resemble a prison or a fort, but we were determined to make it completely impenetrable. We installed Delta barriers which could only be raised or lowered from a location removed from the outside gatehouse. We installed a vehicular entry point that required a vehicle to enter the inspection point where a gate was closed behind the vehicle. The gate beyond the Delta barrier could only be opened by a U.S. citizen security employee who was located away from the gate.

We began to occupy the new compound in February 2004. By the end of March, all employees were off Eskan Village and in the new compound. I was proud of our team. There had been highs, lows, and many frustrating moments when it seemed as if we would never make progress. They persevered through the difficulties with both the Saudis and OPM-SANG and responded as true professionals. They never got discouraged, and they never quit! Finally, after a year, Vinnell-Arabia had fully recovered from the May 2003 bombing. Despite our recovery, Al-Qaeda, or groups inspired by it, continued their attacks vowing to drive foreigners from the kingdom.

The *New York Times* reported on June 9, 2004, that since May 1, 28 foreigners, most of them contractors with western companies employed in Saudi Arabia, had been killed by militants. Among them was Robert Jacobs, a seven-year Vinnell employee shot in Riyadh on June 8, and the week before a gunman had fired on two Vinnell employees, wounding one of them. Riyadh remained dangerous for Americans.

After negotiating the new contract, I left Riyadh in mid-June of 2004. My team had done everything required and more to restore Vinnell-Arabia to the dynamic, successful contractor support organization that it was when I arrived in 2002. Today, it remains a strong program where the U.S. continues to assist its strategic partner of 80 years, Saudi Arabia, in their modernization of the SANG. It was an emotional moment for me when I bid farewell to my co-workers and Saudi friends at the farewell dinner in my honor hosted by Brigadier General Anderson. Call it what you want. For me, as the PGM, getting through the aftermath of the bombing and building a new secure compound ended a long, arduous 13-month battle—that we won.

CHAPTER ELEVEN

Fighting for Our Veterans

My American and Saudi bosses at Vinnell–Arabia and Northrup Grumman were most appreciative of our rebuilding efforts after the bombing of the compound in 2003; therefore, they included as part of my end-of-contract bonus, a generous 10-day vacation for Linda and me. I was elated and called Linda. We met in Zurich on June 11 and toured Switzerland, Austria and Germany.

The opportunity that I had but didn't take advantage of nine years earlier of going to Europe with the Army seemed to be partially fulfilled by the wonderful European vacation. While there, we followed the funeral events of President Reagan, our Cold War crusader, who died at age 93. Days after returning home, the kids and I hosted a belated surprise 60th birthday celebration for Linda on June 26 at our daughter's house in Yardley, Pennsylvania. After almost three challenging years in Saudi Arabia, and aging into my mid-sixties, I looked forward to reuniting with family and having time to enjoy my six young grandsons, two of whom, Alexander Wall and Josh Matz, I hadn't seen yet.

The precious time I had hoped for didn't last long. On October 13, I accepted the job as president of the National Association for Uniformed Services (NAUS). My good friend from the 82nd Airborne, George Landis, told me NAUS was looking for a new president and thought that I would be a good fit. He gave them my name. I wasn't interested in taking on another job so soon, but after meeting with the chairman of the NAUS Presidential Search Committee, Lieutenant General (Ret.) Robert G. Yerks, USA, former personnel chief for the Army, I accepted.

I never served with General Yerks in the Army and didn't know him. He was an infantry combat veteran of Korea and Vietnam and as deputy chief of staff for personnel, he launched the Army's very successful "Be All You Can Be" recruiting campaign in 1981. At my interview, I never forgot what Yerks said: "You have done a lot in your lifetime, but you aren't 'all you can be' until you step up and lead NAUS." I liked Yerks the moment I met him. I sensed that he and the Board Executive Committee felt that I was the right person among the final four candidates in consideration. Later, at my request, Yerks became chairman of the NAUS Board of Directors. What a pleasure it was working with such a great leader who always put the soldiers first!

NAUS was a non-profit, non-partisan military-affiliated association founded in 1968 and chartered under Internal Revenue Code Section 501(c)19. What attracted me most was its uniqueness compared to other military and veterans' organizations. It was the only military/veteran affiliated association that represented the entire military/veteran family, including, at the time, all seven branches of the federal uniformed services: Army, Marine Corps, Navy, Air Force, Coast Guard, the commissioned corps of the United States Public Health Service (USPHS)[1] and the National Oceanic and Atmospheric Administration (NOAA).[2] Its membership included all grades and ranks—officer and enlisted—and all components: active duty, reserve, National Guard, and all military veterans and their family members. Other military and veterans' organizations typically had a narrower focus, even though several boasted larger membership numbers than NAUS.

All these organizations worked to advance the cause of their respective members, but as president of NAUS, I could always say that our "all services, all grades and ranks" representative membership gave me a little broader perspective when talking about issues on Capitol Hill. An affiliate organization of NAUS was the Society of Military Widows (SMW), a separately chartered organization that served the interests of women whose husbands died on active military duty or during retirement from the armed forces.

I not only served as chief executive officer of NAUS, but also as administrator of SMW. NAUS had a staff of 12, which included a legislative team of four, so when adding myself, we had the highest "tooth to tail" ratio of staff dedicated to working on legislative issues than any of the other veterans' associations.

The mission of NAUS was twofold: to support a strong national defense and promote and protect benefits earned by members of the uniformed services for themselves, their families, and their survivors. We did this through extensive advocacy and lobbying activities to ensure that members' concerns were heard and acted upon. NAUS had its own modest Political Action Committee, which allowed us easier access to senators and representatives for advocating for our members' interests.

I liked the fact that when we dealt with Congress, the White House, the Pentagon, and the Department of Veterans Affairs, we spoke for all members of the uniformed services community—not just one segment of it. We were the "service member's voice in government." We represented and spoke for the Total Force. That is what General Bruce C. Clarke, U.S. Army, the first chairman of the Board of Directors and other founding officers and NCOs had in mind when they organized NAUS. I took the job knowing that I was in a position with an organization where I had some clout and could fight for all service members and their families.

I took over as president in January 2005, which coincided with the convening of the first session of the newly elected 109th Congress on January 3. Our work was cut out for us, especially as with each new Congress convened during my time at NAUS, we saw the percentage of veterans among those elected decline. We had to start educating the newly elected members, particularly those who never served in the military, about the unique requirements of life in the military and why it is critical that our government keeps its promises to those who serve in uniform.

The NAUS legislative team and I began visiting the key members of the Armed Services, Veterans Affairs, and Appropriations Committees in both the Senate and the House, briefing them on NAUS legislative goals. That included office meetings with Representative Steve Buyer from Indiana, the new chairman of the House Veterans Affairs Committee, and with Representative Duncan Hunter from California, the chairman of the House Armed Services Committee.

I also met with Senator Daniel Inouye from Hawaii, who talked about the importance of improving veterans' health care. Inouye said to me, "a promise is a promise" and told me that the promise made to our veterans must be kept. He remembered me from my time serving in Hawaii, specifically, his visit to the Pohakuloa Training Area in 1991, when I escorted him through the training ranges. He said he was glad to see me in Washington fighting for veterans. We had a photo taken together in his office standing in front of the Nisei 442nd Regimental Combat Team colors, his unit in World War II.

On February 16, I testified before the House Veterans Affairs Committee. It was an early opportunity, and I used it to push NAUS priorities that fell under their oversight, especially the importance of full funding for VA medical care, challenging recent member proposals to initiate a TRICARE enrollment fee and raise copays which service members pay. "This is absurd," I told the committee in open session. After the hearing, our legislative director, Master Gunnery Sergeant (Ret.) Ben Butler, USMC said, "Sir, you did well, but telling them their proposal is 'absurd' may piss them off."

I also met with the crusty Senator Ted Stevens from Alaska, president pro tempore of the Senate and a strong supporter of the military and veterans. "Nothing gets approved for the Pacific that doesn't come through me and Senator Inouye," he reminded me. He had an appetite for risk and a passion for service. He was a scrapper and fought hard for veterans. I admired that and enjoyed the times I was in his presence. He was a strong supporter of NAUS.

Another visit that impacted me personally was with the commandant of the Marine Corps, General Michael W. Hagee, where we discussed the importance of taking care of the most "catastrophically disabled" Marines as they leave active-duty medical care. That transition wasn't going well. NAUS got on it immediately and worked with Representative Markey of Massachusetts and others on drafting bills ensuring that all wounded service members would be fully cared for, not just immediately, but for the rest of their lives.

These meetings and hearings gave me an opportunity to introduce myself and learn what was important to Congress and the veteran community. We had also laid the groundwork for a positive working relationship between NAUS and Congress that would be writing legislation to support our troops and veterans. I also visited NAUS chapters in Merced, California, and Tampa-St Petersburg, Florida. I felt good about these initial visits and what we accomplished during those first several weeks.

I was surprised at how easy it was to get an appointment with representatives and senators and how receptive they were to military and veterans' issues. They listened and wanted to know what NAUS members were thinking. They liked the fact that our association represented all service persons. Representative Chet Edwards, after my initial visit with him, gave me his Inaugural Committee invitation tickets to the 55th presidential inaugural ceremony for George W. Bush and Richard B. Cheney on January 20, 2005. He wasn't able to attend. Linda and I had never attended a presidential inauguration and enjoyed the ceremony and inaugural ball that marked the beginning of President Bush's second term.

NAUS association membership and "grassroots" chapter development were always a challenge. There are dozens of military and veterans' organizations all working hard for their members. We all worked together in advocating our members' needs and concerns, which often coincided. At the same time, we were in competition with each other for membership among veterans, active-duty and reserve component communities. NAUS had been working hard to increase its membership.

That became a major priority for me, and in October we hired Lieutenant Commander (Ret.) Steve Hein, United States Coast Guard, to help boost our marketing efforts. He became the association's first marketing director. Steve was a certified marketing executive and had 16 years' experience in marketing and advertising management. His most recent position was marketing director for the *Army Times* Publishing Company.

His experience was a great match for NAUS as we were striving to grow our membership continuing to be "the service member's voice in government." He also brought the "voice" of the Coast Guard to our staff. His responsibilities at NAUS included membership recruitment, coordinating fundraising activities, and managing our member benefit programs. We also used his vast and diverse talent to assist in chapter management and publishing the bimonthly NAUS magazine, the *Uniformed Services Journal (USJ)*.

A month earlier, we hired Richard "Rick" Jones, an Army veteran, as our new director of legislation. Rick had broad experience in congressional legislation activity and had been the National Legislative Director for the American Veterans Association (AMVETS). Prior to his time with AMVETS, he had 20 years as a legislative staff aide to key senators and House representatives. He also had an exceptional network of trusted relations across Congress.

"Rick Jones is the most influential veterans' advocate on Capitol Hill," said Chuck Partridge, NAUS acting legislative director. Those two leaders brought with them a wealth of professional experience in their respective fields. Both Steve and Rick possessed a drive to overcome challenges with the ability to clearly and persuasively articulate our goals. More important was their personal commitment to the values that generations of Americans have defended in military service to our country. They served NAUS well and the team was becoming stronger and more capable.

One area we made notable progress in during my tenure was our bimonthly magazine, the *USJ*. Almost every organization has such a publication, and ours was pretty standard when I came aboard as president. Good basic information was presented in magazine format with a few photos. Once Steve came aboard, his military recruitment advertising background, and recent position as marketing director of the *Army Times* Publishing Company, was noteworthy.

His experience included the *Army, Navy, Air Force* and *Marine Corps Times* newsweeklies, as well as the bimonthly magazine the *Armed Forces Journal*, and he brought some fresh ideas and leadership to the table for our editor. Our *USJ* immediately got better. The most important, anticipated, and widely read section was the "Legislative Update" written by Rick Jones and his team. Steve hounded me as each deadline approached for my "president's message," and usually gave me a few ideas to write about or include.

Another big change happened when we had to replace our editor, and we hired Tommy Campbell—another *Army Times* alum. Tommy had served as creative director at *Army Times* for several years and brought his unique military-flavored design eye and creative experience to our magazine. Of interest, the *Army Times* leadership was beginning to get upset with NAUS taking some of their key personnel.

The *USJ* won a MARCOM Gold Award one year for best small nonprofit magazine following those changes. We began mailing a copy to each senator and representative. It was indeed a best seller! Members and friends looked forward to receiving it every other month.

On September 27, 2005, the White House announced my appointment as a commissioner to the Veterans Disability Benefits Commission (VDBC). Some weeks before, Chairman Buyer of the House Veterans Affairs Committee asked me if I would serve on this newly formed commission. As U.S. casualties began returning from combat in Iraq and Afghanistan, Congress debated longstanding issues regarding the most effective ways to deliver benefits and care to our wounded veterans.

To help resolve those concerns about veterans' benefits, President Bush and Congress created the independent commission under the National Defense Authorization Act of 2004. It was the first study on veterans' disability benefits since the General Omar Bradley Commission in the Eisenhower administration 50 years earlier. I told Chairman Buyer that the study was "long overdue," and speaking for the Vietnam veterans, I said it was needed and would be greatly appreciated by veterans of all wars.

The commission was composed of 13 former officers and enlisted members, representing all services, including nine combat veterans from the Vietnam War. Two, First Sergeant Nick D. Bacon, USA (Ret.) and Major General James E. Livingston, USMC (Ret.), were both recipients of the Medal of Honor. A friend and former infantryman, Lieutenant General James Terry Scott, USA (Ret.), was the chairman. You couldn't have had better, brighter, more qualified people. The purpose of the commission was to "carry out a study of the benefits under the laws of the United States that are provided to compensate and assist veterans and their survivors for disabilities and deaths attributable to military service, and to produce a report on the study."

Whatever I was able to accomplish in life until this point would pale in comparison to the opportunity I had as president of a national veterans' association and an appointed commissioner of a federally established Veterans Benefits Commission, to help our service-disabled veterans and their families as they rehabilitate and reintegrate into civilian life.

Many of the NAUS legislative goals for the 109th Congress addressed the same issues and benefits the VDBC would analyze. Both gave special attention to the care of severely wounded veterans, treatment and compensation for PTSD, and the timeliness of claims processing. The VA couldn't process and adjudicate claims in a timely and accurate manner. I personally devoted a significant amount of time and study to claims processing, as many NAUS members were complaining about the long wait times. As of 2006, the VA had over 500,000 compensation and pension claims pending decision. In addition, nearly 25 percent of those had been in the VA system for more than 180 days. The processing system for disabled veterans was broken! The government bureaucracy burdened the system, and our veterans faced long, intolerable, and unnecessary delays.

The commission heard extensive testimony from veterans, advocates, and family members regarding the current disability system. We met and deliberated as American service members were being wounded and killed in Iraq and Afghanistan while fighting the global War on Terror. Not a week went by that an aeromedical evacuation flight didn't arrive at Andrews AFB with severely wounded warriors who were transported to Walter Reed Army Medical Center in DC or to the Naval Hospital at Bethesda, MD. During Vietnam, it typically took more than a month to move wounded troops from Vietnam to advanced treatment facilities in the United States. Now it was down to three days.

It was with the on-going global reality that the commission conducted its work. Many of the commissioners had their own combat experiences from Vietnam to draw from as we watched a new generation of troops engaged in war in the Middle East. The IED was the signature weapon used by the enemy. It was killing and maiming our troops by the thousands. Each day I went to work, I was reminded of that and driven by our responsibility to ensure that the care and services available to our wounded troops were the right ones and that they were applied in a timely manner. As Vietnam veterans, we understood what these younger combat veterans struggling with physical and psychological injuries needed. Having stayed in touch with some of my badly wounded soldiers from Vietnam, I knew what they and their families were coping with in their everyday lives.

Two and a half years after it began, we delivered our final report to Congress, the president, and the American people on October 3, 2007. It was a comprehensive 544-page document. We spent the morning briefing the Departments of Defense and Veterans Affairs and the staffs of the House and Senate Armed Forces and Veterans Affairs Committees.

At 3:00 p.m., Chairman Scott, I and several other commissioners held a press conference in the Cannon Office Building. When asked by a rather obnoxious, flippant reporter if the Commission was aware of the costs required to implement the recommendations, I stated, "There is a cost to providing these benefits, but if we get our priorities straight, they are not over-whelming. Stop federal spending on free healthcare for illegal aliens, for Groundhog Day, for the Rock and Roll Hall of Fame, and for a vast number of similar, non-federal pet projects." Additionally, I said it wasn't our intent to produce a report to be shelved. "We have the chance to do right for all generations of veterans. It's the right thing to do, so let's get on with it!" I can still see that reporter sitting there dumbfoundedly absorbing my answer.

At a hearing on October 10 the Commission briefed Chairman Bob Filner and the House Committee on Veterans Affairs on our findings and recommendations. The 21 members present were effusive with praise and appreciation for the VDBC's work, and Chairman Filner thanked us for answering "this call to duty" and for the two years of dedicated work "that honors the sacrifice that our men and women in uniform have made."

In making the 113 recommendations contained in the report, we prioritized those we thought needed to be implemented soonest; specifically, developing and implementing new criteria specific to PTSD and mental health in the VA Schedule for Rating Disabilities was of utmost importance. Chairman Scott emphasized the priority recommendation throughout his testimony, and it clearly got the attention of the committee members.

As is evident by the wide range of priority recommendations, the commission was very thorough and clearly had the best interests of veterans, their families, and survivors in mind. I have never been part of a more serious, productive study group than the VDBC. NAUS, as well as the other DC-based veterans' and military organizations, worked every way possible to assist Congress in developing new laws and regulations needed to ensure that the well-thought-out and documented recommendations were acted upon. As I said to that reporter, "Let's get on with it." NAUS did!

Most of the recommendations were implemented either in their entirety or in some fashion. It was, indeed, an honor and a privilege to serve our veterans and their families through that effort. As a Vietnam veteran, I felt good about being able to assist our younger-generation Iraq/Afghanistan veterans. They and their families sure deserved it!

An aging membership base, the continuing evolution of communications via email and smartphones, and the growth of special interest lobbying efforts on the Hill all contributed to a persistent question about the continued relevancy and effectiveness of such organizations in our day and age. We all faced that challenge in our membership recruiting efforts, and we banded together in The Military Coalition (a sort of "association of associations") to help maintain our relevancy and present a united front.

At the NAUS end-of-year annual meeting, I was pleased to report to our members that we were beginning to see an increase in membership and in "grassroots" chapters. Our association was financially sound. Our investment portfolio was well diversified, and our income sources provided the capital to operate efficiently and effectively. Our members and supporters continued their generosity that enabled NAUS to work toward our legislative goals. One senator noted "NAUS's reputation as a leader in the fight to protect the service members' hard-earned benefits gets better all the time."

During my years with NAUS, the staff and I logged tens of thousands of airline miles traveling around the country and overseas (we had chapters in Hawaii, Alaska, Guam, Puerto Rico, and England) to attend and speak at chapter meetings and military Retiree Appreciation Days (RADs). RADs were almost always held on Saturdays at military posts and bases throughout the country and were always well attended by military retirees and their families from the local area. Because retiree benefits and their potential erosion were always top of mind, local Retiree Activities officers welcomed our offer to come and speak about legislative issues affecting retirees, so it was easy to keep my calendar full attending these RADs.[3]

We gave the keynote legislative update and talked to the hundreds of retirees and their family members about their concerns and what legislation was important to them. The NAUS legislative update was a "best seller." Attendance was often standing room only. Veterans knew NAUS fought for them on Capitol Hill, and they wanted to hear directly from a NAUS spokesman.

Of great assistance on these visits was my wife, Linda, who attended the NAUS information table with local members, greeting folks and handing out literature and membership information. She signed on more new members than anyone else at the NAUS table. We often returned with as many as 30–40 new memberships, which I provided to our very diligent treasurer, Tamie Boone, at our Monday morning staff meetings. Linda and I traveled an average of 19 weekends a year for NAUS, leaving on Friday and returning Sunday. For six years, our bags remained packed and ready for the next trip. We were, in fact, living out of our suitcase!

Visits to overseas RADs and chapters took 4–5 days, although the 8,000-mile trip to Guam was longer. I felt it important to get out of my comfortable "foxhole" in Virginia and meet our veterans at the grassroots level and learn directly from them what their concerns were. They appreciated that. On several occasions, soldiers who had served with me in Vietnam or elsewhere came to our information table to say hello … and I wouldn't let them leave until they completed an application and joined NAUS.

Based on the early success we had recruiting members at RADs, we decided to try exhibiting at some other events. Steve Hein suggested we try the annual Public Health Service (PHS) Symposium—an event that would typically draw close to 1,000 USPHS officers, usually including the U.S. surgeon general. Steve came back from the first such NAUS-attended PHS Symposium with 36 new, active-duty NAUS members, which not only paid for his trip, but with so few active-duty members in NAUS overall, was a big success. It became an annual event for us, and the Commissioned Corps of the PHS.

Even though NAUS members nationwide reacted very positively to the challenge to help grow our membership and add new local chapters, membership growth, especially among younger retirees, veterans, and active-duty members, remained our biggest challenge. We added four new chapters in 2005, with many other prospects on the horizon. By the end of my time with NAUS in 2011, we had reactivated or added several new chapters.

One very active chapter was our Groton/New London, CT, chapter, presided over by retired Navy Master Chief Petty Officer Paul Dillon, who also served for a time as the co-chair of the NAUS Board. My marketing director, Steve Hein, was a Coast Guard Academy graduate and happened to have a classmate serving as the assistant superintendent of the Coast Guard Academy in New London. Steve worked with him and Paul to establish an annual award sponsored by NAUS for a graduating cadet presented at the annual awards ceremony during graduation week.

Another real success story in chapter growth was the addition of Albany, New York's, Rainbow Chapter. Tom Quinlan, whom I served with in the 9th ID in Vietnam during the Tet Offensive in 1968, called and volunteered to start a local chapter in the capital district of New York. Tom's leadership, abundant energy, and many contacts brought about the establishment of the Rainbow Chapter and their adoption of the New York Army National Guard's 42nd Infantry (Rainbow) Division. At the time, the division was deployed on Operation *Iraqi Freedom*. The chapter adopted the division for wartime support. Tom seized the opportunity to provide the Rainbow soldiers at home and in Iraq their own veterans' organization chapter. That relationship provided NAUS the opportunity to reach out to a younger generation of American fighting men and women. As he said, "It all fit quite naturally," and NAUS gained a chapter and many New York guardsmen as members.

We made it a priority to visit new members assigned to the Veterans Affairs and Armed Services Committees in the House and Senate. On January 3, 2007, Virginia Senator James Webb was sworn into the 110th U.S. Senate and was assigned to both the Veterans Affairs and Armed Services Committees. On his first day in the Senate, he introduced the Post-9/11 Veterans Educational Assistance Act. The legislation expanded some key provisions of the Montgomery GI Bill for Iraq/ Afghanistan veterans.

Both bills provided educational assistance benefits. A NAUS legislative priority for the 110th Congress was to support the upgrade to the Montgomery GI Bill. Immediately, we met with Senator Webb to learn about his bill and to brief him on NAUS's priority to expand educational assistance benefits for veterans. Webb explained his bill was modeled on the World War II GI Bill and designed for the Iraq and Afghanistan war veterans—exactly what NAUS was fighting for. We fully got behind his bill and had our members support this legislation at the grassroots level.

Webb's bill passed the Senate on September 6 and was signed into law by President Bush on June 30, 2008. The bill paid educational tuition costs and provided a monthly living stipend to service members with dependents. NAUS members were proud of their strong support of these educational benefits that helped our newest veterans fighting the War on Terror like the way the GI Bill helped the Greatest Generation in shaping America after World War II. Webb's office reached out and thanked the association for its leadership and dogged support of his new bill. Today's servicemen and women are now enjoying the benefits.

Topping our list of legislative priorities was always the preservation of military and veterans' healthcare benefits. At the end of 2008, NAUS and our partners in The Military Coalition celebrated a great legislative victory—blocking TRICARE fee increases again for the third year in a row. Together, we defeated the administration's attempt to drastically raise TRICARE fees, copays, and prescription drug costs. In my end-of-the-year president's message, I congratulated our legislative team and our members, telling them, "You worked this issue hard on Capitol Hill and with your letters, calls, emails and our postcard campaign over the summer—we won!"

Again, with our Military Coalition partners, we also blocked a drastic cut in the Medicare physicians' reimbursement rate. The rate for physicians who treat veterans under Medicare was to be reduced by more than 10 percent, even as costs continued to rise. That would have led to many physicians refusing to accept veteran Medicare patients. As a soldier and American citizen, I couldn't believe the constant attacks on American veterans' benefits by our own government! What were they thinking?

The start of the 111th Congress in January 2009 also saw the end of the Bush administration and the inauguration of President Barack Obama. It also closed our association's 40th anniversary year—2008. On May 20, 2009, I and several other veterans' organizations' presidents were invited to meet with First Lady Michelle Obama, at the U.S. Chamber of Commerce building in Washington DC. She addressed several issues that day and emphasized that the president and she were very interested in military veterans' issues and that she would be working on programs to assist military families.

The Obama administration was quick in wanting to learn about and support key veterans' legislation. I received a phone call shortly after the Obama administration took over; a person from the White House Chief of Staff Office called me at home one evening wanting to know what I thought the key veterans' issues were. I thought it strange that he would call me at home, but after I was convinced that it was a real inquiry and not some prank call, I was happy to let him know NAUS's priority issues.

After taking office, the Obama administration worked hard to ensure that America fulfilled its obligations to veterans and their families. General (Ret.) Eric K. "Ric" Shinseki, U.S. Army, had just been confirmed as Obama's secretary of Veterans Affairs. He and I had worked together on the Army staff in the Pentagon in 1986–87. I was quick to visit him and let him know what our priorities were for veterans.

In a few months we saw a dramatic increase in funding for veterans' healthcare, specifically care for women veterans and for our wounded warriors from Iraq and Afghanistan suffering from PTSD and traumatic brain injuries (TBI). It was the largest percentage increase in the VA budget in more than 30 years and included funding for the new post-9/11 GI Bill, ensuring our younger veterans could pursue their education and live out their dreams. Despite that, a serious budget problem continued to plague timely funding for veterans' health care. Over the past two decades, the VA budget had been months late nearly every year due to budget battles. It had to be fixed and was high on NAUS's list of priority legislation.

Finally, in the East Room of the White House on October 22, 2009, the late budget plague ended. Obama signed the Veterans Health Care Budget Reform and Transparency Act. The White House press secretary's release that day reiterated what President Obama said earlier: "This advance funding bill ensures that veterans' health care will no longer be held hostage to the annual budget battles in Washington."

That was a victory for all veterans' organizations. At the invitation of the White House, I stood on the podium with other veterans' organization representatives at the signing. When placing us at the podium, NAUS was honored when the White House representative in charge of the ceremony told me to stand next to the president in the line of veterans' organization representatives.

"We want NAUS right next to the president," she said. Just prior to the signing ceremony, I met with Obama where he thanked NAUS for our strong support of the advance funding bill and for "NAUS's leadership in fighting for veterans." I recall his words to me were, "Keeping faith with our veterans and promises made is never truly done … thank you for being with us today." Secretary Shinseki was present at the signing.

Several days later, I received an invitation for Linda and me to attend the 2009 Veterans Day Breakfast at the White House hosted by the president and first lady. A contingent from the Japanese American Veterans Association (JAVA) were invited as special guests. Some were NAUS members. Linda and I were seated at the table with their executive director, Terry Shima, veteran of the famed World War II Nisei 442nd RCT (Regimental Combat Team), and President Robert Nakamoto, who, as a 10-year-old child, was interned with his family in Camp Topaz in Utah during World War II.

The Obamas visited our table group to welcome all of us and for a brief chat. What a pleasure it was to meet those proud Japanese American soldiers who fought so courageously in both Europe and the Pacific during World War II. After that, we worked closely with JAVA on legislation and invited their members to the NAUS annual meetings.

Throughout each legislative cycle, NAUS worked closely with Arizona Senator John McCain, an ardent supporter of the military and veterans. We met with him several times to discuss NAUS issues with the annual National Defense Authorization Act (NDAA). McCain had just lost the 2008 national election to President Obama, and he largely opposed actions of the Obama administration. He felt the war in Afghanistan was winnable and criticized Obama for dragging his feet in deciding whether to send additional troops. We didn't know at that point to what extent he would support Obama administration-sponsored veterans' legislation.

McCain was the ranking member of the Senate Armed Services Committee in 2010 and 2011 and was always interested in knowing our issues and why we felt the way we did. In two meetings on August 4 and October 4, 2010, we met with the senator and reviewed the issues. He was a strong proponent of the TRICARE program for veterans, and NAUS credits him with the much-appreciated provisions in the post-9/11 GI Education Bill that allowed transferability of earned benefits to spouses and dependents—a top NAUS goal.

"When you think I need to know something, don't be afraid to call—you know the number," he said. He gave me his private number and insisted I call him directly—a fast, easy conduit to the ranking member! I did on several occasions, and he always answered. He was always willing to listen to NAUS. I found him to be stubborn at times, reminding me of my service with Weinberger—he never gave in when he wanted something and knew it was the right thing to do. I appreciated his tenacious spirit, and our staffs worked closely.

In late 2009, I was shocked to learn of a DOD plan to raise by 20 percent hospital inpatient copayments for retired service members and their families covered by TRICARE, which ran contrary to assurances given by Obama that TRICARE fees would not be increased in fiscal year 2010. In an October 1 letter to SECDEF Robert Gates, I wrote, "This type of behavior breaks a sacred trust with those who have worn the country's uniform, and it should not be allowed to stand." The NAUS effort, combined with letters from our coalition partners, produced the desired effect. Within the week, Congress nixed the ill-thought-out DOD plan.

Until the time I left NAUS in June 2011, we were constantly battling attempts by our own government to raise TRICARE fees and reduce payments to doctors who accepted TRICARE and Medicare patients. That recurring problem posed a serious threat to the access of earned military healthcare benefits and was addressed daily. In a meeting with Chairman Ike Skelton of Missouri, we discussed ways to stop DOD from increasing any TRICARE fees or copayments. He was adamant about it and didn't have kind words for DOD.

At that meeting in his office, he and I talked about our respective bouts with polio. He was a polio survivor of the 1940s, also, and he shared with me his teenage bout with the disease and how it shaped his life. It left his arms and shoulders largely disabled. "I was perfectly healthy, then one day in 1947, I fell at home and couldn't move," he said. I told him about my falling at my grandmother's house in 1944 and not being able to move my leg and being diagnosed with poliomyelitis that day. He asked, "Were you treated at Roosevelt's Warm Springs polio clinic?" He was. I wasn't. He, like me, tried to hide his defects and never talked about it.

I left that meeting thinking about the similarities of our polio onset and how we coped with its effects through life. NAUS was pleased when Skelton inserted a provision in the fiscal year 2011 National Defense Authorization bill to prohibit DOD from ever increasing fees and copayments. You could never let your guard down. The bureaucrats in the Pentagon never rested. They always thought of ways to harvest dollars each budget cycle by taking away earned benefits from veterans. Political promises sometimes fell to political expediency, and it was important that NAUS remained ever vigilant in this constant battle!

In the president's fiscal year 2012 budget, the DOD again proposed increases in TRICARE, wanting to increase TRICARE Prime enrollment fees by 13 percent. That would affect all military retirees between the age of 38 and 64. In addition, DOD wanted to boost copays of prescription drugs. When polled, 93 percent of NAUS members vehemently opposed that, saying, "Keeping costs as they are is a way for the government to honor its promises of lifetime health care, particularly when the country is at war." We were constantly having to fight our own government on behalf of our wartime veterans!

We carried this message to the Hill and stepped-up one-on-one meetings with key members of Congress. The meetings were frank and candid. One supporter, Representative Joe Wilson from South Carolina, chairman of the House Armed Services Personal Subcommittee, relayed his own concerns at a February 15 meeting with me and invited NAUS to testify at a March special meeting on the matter. Representative Wilson, like Ike Skelton, was losing patience with DOD. Wilson, a NAUS member, was an ardent supporter of our military and veterans and pulled no punches when telling the DOD bureaucrats how wrong they were.

Another strong supporter was Senator Webb who said, "lifetime healthcare for career military personnel is part of a moral contract between our government and those who have stepped forward to serve." NAUS legislative director Rick Jones presented NAUS testimony to Representative Wilson's House Personnel Subcommittee at their special hearing on military healthcare on March 16. Rick's testimony was powerful. It got the attention of every member on the panel. Jones urged members to "hold the line against the Pentagon plan to increase TRICARE Prime rates."

He continued, "A modest increase now is 'a nose under the tent' designed to divide the voice of retirees and members of Congress to start the roll-out for substantial increases in TRICARE fees and copays." He informed the panel, "One Pentagon undersecretary said that the costs of earned benefits have gotten to the point where they are hurtful and are taking away from the nation's ability to defend itself."

This raised eyebrows, causing some Congressmen to react with surprise or mild disapproval. What a callous statement for a top government official to make—that benefits earned by American soldiers in honorable military service threaten our national security! I was livid and burning with rage that the Pentagon bureaucrat made such an insidious, disingenuous remark. I went after that bureaucrat who never served a day of his life in uniform. I made it a NAUS war cry in our fight to prevent the raising of TRICARE fees!

The NAUS testimony struck a chord with DOD, and within hours of the March 16 hearing, a Pentagon spokeswoman released a statement blaming rising healthcare costs on the introduction of the TRICARE For Life benefit for over-age-65 retirees. We saw the statement as a clear signal that DOD had more in mind beyond its so-called "modest" increase in TRICARE Prime. Attacks on veterans' healthcare benefits would continue.

In fighting them, on March 30 we visited with newly elected Representative Allen West from Florida's 22nd District. West was a retired Army officer and veteran of the Persian Gulf and Iraq Wars. A "firebrand" conservative, he was a high-profile member of the Tea Party movement and strong supporter of the recently passed Patriot Act. He became a strong supporter of the Military Healthcare Protection Act. As an aside, my seven-year-old grandson, Josh Matz, had just read the story of Flat Stanley[4] in his second grade reading class and as part of the program asked Grandpop to get a photo of a congressman with a paper cut-out of Flat Stanley.

I had the cut-out Josh had sent and asked Representative West if he would mind having a photo taken with this very popular figure among children. He agreed. "Please don't show this to the press; they will accuse me of playing with paper dolls," he said as we snapped the photo. He enjoyed the moment and told us he was familiar with the Flat Stanley children's books as his wife was a schoolteacher. "This is about character development for our young. I was raised very conservatively. I wish the best for your grandson." Representative West showed his human side and how in touch he was with the important things in life. The photo remains part of our family archives. The press never got the photo!

NAUS took on another battle which other veterans' organizations refused to tackle. Following World War II, the men of the U.S. Merchant Marine received none of the benefits offered to others that served. The denied benefits included education and home loans under the World War II GI Bill of Rights. Those mariners, who transported men and equipment through enemy submarine-infested waters to every corner of the world during the war, experienced the highest casualty rate—1 in 26 mariners died in the line of duty—of any branch of the armed services.

Merchant mariners are civilians except in time of war when, under a 1936 law, they are considered military personnel. Legislation in 1986 officially made all merchant mariners who served in wartime military veterans. Upwards of 9,300 of the 240,000 merchant mariners who served in World War II died during the conflict. Congress didn't even thank them or recognize their service for 40 years. NAUS wanted to heal the wound of being overlooked for such a long period of time. We were in the vanguard of joining and supporting Chairman Bob Filner's House Committee bill entitled "Belated Thank You to the Merchant Mariners of World War II."

Despite broad bipartisan support for the bill in the House, the Senate and some other veterans' organizations were not behind it, and Congress failed to pass the bill in 2009. Some thought it a "special privilege not available to other veterans." One of our veterans' organization coalition partners said, "They're not veterans … we have real heartburn with the idea we're going to give them a monthly gratuity; it's not equitable."

I was bothered by such troubling and specious arguments. And NAUS put on a full-court press in the uphill battle to support both the House and the Senate Belated Thank You to Merchant Mariner bills. We instituted a nation-wide letter campaign to rectify this egregious error when our government failed to make mariners eligible for the GI Bill. Generals Eisenhower and MacArthur praised the Merchant Marine. They said, "The war would've extended in time or been lost without the service of the Merchant Marine."

I was the keynote speaker at the 24th Annual American Merchant Marine Veterans Association Convention on May 12–13, 2010 in Las Vegas, Nevada. The mariners were in their 80s and 90s, accompanied by their families. I don't remember ever being among a group whose appreciation for what we were doing for them was so enthusiastically shown.

During the six and a half years I was president of NAUS busily fighting for military and veterans' benefits, Linda ensured that family obligations and responsibilities were also met. Our seventh and last grandson, Ian Wall, was born in June 2005. In 2009, we took the entire family of 15 on a cruise to the Bahamas. I had Agent Orange-related prostate cancer surgery on January 26, 2010, at Walter Reed Army Medical Center, and three weeks later we joined some friends, Chad and Darleen Chadwick, for a six-day visit to southern Germany.

We also managed to get away twice for mini-Gettysburg College reunions at Sally Foreman Reed's lovely home in Jamaica with Pat and Bob Smith. We returned to

Gettysburg for my 45th and 50th college reunions in 2006 and 2011, respectively, where I was part of the reunion planning committee. Despite my years in the Army, I always tried to never let anything divert me from maintaining close ties with my Gettysburg College classmates or from providing valuable service to the college.

I remained an active member of the Eisenhower Institute National Advisory Council, a committee chairman for the college's Commission on the Future, Army ROTC Commissioning speaker, and participated in leadership and Vietnam War panel discussions. Linda and I hosted both Gettysburg College and University of San Diego summer send-offs for college freshmen from the Washington, DC, area, hosted several college fraternity reunions in our home, and co-hosted with our friends, the Smiths, annual Gettysburg College "seashore reunions" in Ocean City, New Jersey, to name a few events. I mention these certainly not to talk or boast about myself, but rather to emphasize that while my job was always paramount, I found time for family, friends, and community … and to enjoy the pleasures they brought!

In 2007, we joined Linda's family, including her brother Tom and his wife Madelyn, and her many cousins in Hamilton Township, New Jersey, for the 100th anniversary of the Tindall family Christmas Day gathering. It had been held for 100 consecutive years and continues today. The reunion took place in the original farmhouse known as the John Abbott House in recognition of its first owner. It was built around 1730 and was added to the National Register of Historic Places in 1976 for its significance in military history. During the Revolution, the colony's treasurer, fleeing the British advance on Trenton in December 1776, removed the money in his care to the home of John Abbott II. British troops searched the home but failed to find the monies hidden in the bottom of tubs in the cellar. Linda's family, the Tindalls, purchased the home in the 1800s. Forty-one attended the reunion. The historic home remains open to the public for visits.[5]

Sandwiched in between family events and brief vacations were the many NAUS-related events, including 120 mostly-weekend visits to RADS and NAUS chapters during the six and half years of my presidency. I never realized the number of events we participated in until I reviewed my seven NAUS desk calendar diaries and several NAUS journals that I had saved. I would not trade a single moment as each visit brought me in touch with the grassroots of America and the thousands of veterans and their families whose earned benefits we were fighting for.[6]

In recalling those many RAD/chapter visits, I remember fondly working with the many NAUS chapter and board leaders who volunteered their time and energy. I also had the pleasure of working with four outstanding NAUS Board of Directors senior chairmen: Major General Bill Gourley, USA; Lieutenant General Marc Reynolds, USAF; Lieutenant General Robert Yerks, USA and Vice Admiral Jim Zimble, USN.

As administrator of the SMW, I worked closely with widows who had lost their veteran husbands. Linda and I attended their annual conventions. NAUS staff members, Gene Willis and Hal Grant worked with them as they developed their legislative goals and planned their convention programs. Linda chaired the planning committee for the SMW Convention held in Washington, DC, in October 2007. The convention included a tour of the White House. The Executive Committee made Linda an honorary member of SMW at their banquet that night. I was glad it was only honorary, since members were women whose husbands were dead!

Another legislative battle waged was the repeal of the Survivor Benefit Plan/Dependency and Indemnity Compensation (SBP/DIC) offset, known as the "widow's tax," which reduced the paid-for SBP annuity payment benefit received by widows who also received DIC payments. We didn't win that battle despite our good fight every year I was at NAUS, though Congress finally saw the injustice, and approved its elimination in 2021. I am relieved that others continued the battle to right the wrong and accomplish President Lincoln's reminder that it is our duty to "care for the widows."

My initial employment contract with NAUS was for four years ending December 31, 2008. The Board of Directors asked me to stay for another four years. Although I couldn't have been happier and more satisfied with what I was doing—helping veterans and their families—I wasn't ready to commit for another four years. Instead, I agreed to two years.

The significant amount of time, including the many weekends both Linda and I spent away from home on NAUS business, compelled the decision. Although our kids have always been very understanding of the importance of Dad's service and of supporting veterans, they and our seven grandsons and their many activities were begging for our time and attention (their soccer, basketball, and lacrosse games were on weekends) ... and rightfully so!

Knowing I would be leaving, I asked Secretary Ric Shinseki to be our keynote speaker at the NAUS 42nd Annual Meeting in November 2010. He agreed. Our members wanted to hear from the secretary, and I asked Ric to emphasize the range of issues the VA was addressing, specifically those dealing with PTSD. I asked him also to address veterans' access to benefits and what he was doing to eliminate the large claims backlog.

Wounded warriors recuperating from their wartime wounds in local hospitals and their families attended. Shinseki delivered an outstanding address before a record attendance and followed it with a very informative question and answer period. His presentation struck a wonderful chord and was well received.

Approaching my final mile with NAUS, I was feeling good about our advocacy efforts on key quality of life issues, including health care and veteran and survivor benefits. Defeating major TRICARE fee hikes, saving military retirees thousands of dollars, rewarded the staff's persistence and good work.

At a farewell dinner at the Fort Belvoir Officers Club on June 18, 2011, NAUS staff, board members, and friends gathered to offer their farewells. Our children and their spouses attended. Board Chairman Admiral Jim Zimble started the program with his farewell address. His speech was heartfelt and humorous! He praised Linda for her participation and recruiting efforts throughout the country during our tenure with NAUS. After presentations by others, Admiral Zimble yielded the floor to me.

I told them, "We won many battles in our mission of protecting and enhancing earned benefits of uniformed services. And while we haven't won them all—we know we have always been on the right side of the service member and their families and can take the fight to the next hill when the time comes."

It was a fun-filled evening, and I can't remember a time when I left an organization feeling as good as I did as when I left NAUS. I remember General Yerks telling me that "I wouldn't be all I could be" if I hadn't accepted the offer to lead NAUS a few years before. He was right! I was confident that I had done all I could fighting the good fight for our nation's men and women in uniform … and feeling satisfaction that we had helped generations of military veterans and their families live a better life! I was thinking retirement could really stick this time.

Fred Fielding, my college classmate and fraternity brother of over 50 years, and White House counsel to two presidents, and his wife Maria were among the attendees. He came up to me afterward and said, "This is your third retirement since retiring from the Army. What's next?"

That question would be answered six years later!

Until then, both Linda and I were looking forward to being able to devote more time to our three kids and our seven grandsons living in New York, Washington, DC and Texas by attending their activities. A wise septuagenarian told me, "Happiness now spikes in your 70s!" We were gaining in years and wanted also to enjoy some European river cruises and visit long-time friends.

In 2010, we purchased a condominium on the beach in New Jersey and Linda wanted to decorate it. We had a busy social life with our close friends at the beach; among them were long-time high school friends, Warren Eberlein and Bill Cowperthwait, and their wives, Jan and Cathy, and our many Gardens Plaza condominium neighbors, including Rita and Gabe, Mary Ellen and Garrett, Pam and Scott and Bernie, who we got to know well. There was more than enough to keep us busy. We would not be idle!

On July 2, 2016, we celebrated our 50th wedding anniversary with a dinner party at the Washington Golf and Country Club in Arlington, VA. My college fraternity brother, Bud Dougherty, was a member and sponsored us for the event. Eighty guests attended. It was an opportunity to honor our family and friends who have been an important part of our lives. Linda produced a video featuring our times with them. It was a memorable occasion as family, friends, and neighbors through the years joined us.

CHAPTER TWELVE

Guardian of America's War Dead

On a cool day in late November 2016, I was sitting comfortably at home in Great Falls, Virginia, reading the newspaper when the phone rang. "Hello sir, Mike Meese," the voice said. Mike was chief operating officer of the American Armed Forces Mutual Aid Association (AAFMAA) and became the association's president in 2020.

AAFMAA is a member-owned association that provides diversified financial services and life insurance to the U.S. Armed Forces communities. I was on the AAFMAA Board of Directors. However, his call had nothing to do with that.

He was calling as a member of the presidential transition team. Both presidential candidates, Donald Trump and Hillary Clinton, had transition teams, which started in August 2016. Donald Trump had just defeated former secretary of state Hillary Clinton on November 8 in what was considered one of the greatest upsets in American history. Trump had received 304 electoral votes to Clinton's 227.

Mike was chief of the Department of Veterans Affairs 2016–17 transition team. He knew me well, and recalling an earlier conversation about serving in the new administration, Mike asked, "Would you be interested in serving in the Trump administration?" The phone was on speaker, and Linda overheard his question, and unlike her, in full-throated voice, she answered, "Of course, he would!" I had been retired for some time, and she wanted me "out of her space"!

I missed the competitive fray and challenge of the workplace and told Mike, "Yes, I would be interested, but it would depend upon the job and location." He proceeded to explain his understanding of his responsibilities and that he was still in the process of identifying the appointed positions to be filled and learning the qualifications required of candidates to fill those positions. His team was responsible for suggesting candidates for filling 12 positions in the VA that required presidential appointment and Senate confirmation.

In a subsequent call, he talked to me about the senior positions in the VA, namely the undersecretary for memorial affairs, thinking I would be a good candidate. He also mentioned the VA's deputy secretary job. The VA headquarters is in downtown DC, and I wasn't particularly enthralled about having to drive into DC every day, so I demurred whenever a VA position arose.

Mike's team worked very closely with the DOD transition team, and I learned that my friend, Lieutenant General Keith Kellogg, U.S. Army, retired, was chairing this team and provided oversight for Mike's VA team. Keith and I had served together in the 82nd Airborne and in the 7th ID during Operation *Just Cause*, and before long, I was engaged with him regarding candidacy for a job in DOD.

After he discussed several Pentagon positions with me, I asked about the secretary of the American Battle Monuments Commission (ABMC) position headquartered in Arlington, Virginia. That position had not come up on his radar yet, but he agreed I would be a "very good fit," because, among other things, the White House wanted an infantry combat veteran to fill that position. I knew some officers who had served in ABMC, and I always thought what an honorable duty that would be as "guardian of America's war dead." We both agreed on my candidacy for the position and in March 2017 Kellogg set in motion my nomination with the presidential appointee vetting process. So began an amazing three years of my serendipitous life!

There were roughly 4,000 political appointments that any new administration must fill. Each candidate had to be thoroughly vetted before proceeding with a formal nomination or appointment. Those responsibilities were handed off to the White House Counsel's Office and the White House Presidential Personnel Office (PPO). PPO's mission is to provide the president with the best applicants possible for presidentially appointed positions.

As part of that vetting, I was called to the White House on June 21 for an interview with the director of PPO and another White House official. Both Meese and Kellogg were following my nomination progress and would check with me occasionally. Finally, on December 1, PPO informed me that I had been approved by the president to be the secretary of ABMC. I didn't know President Trump, nor did he know me. I was pleased and ready to serve. I was sworn in as ABMC's new secretary by the agency's deputy secretary, Robert "Rob" Dalessandro, on January 9, 2018.

Following the armistice ending World War I, the War Department established the Battle Monuments Board in 1921 to mark U.S. battlefields in Europe. As an outgrowth of that board, the ABMC was established by Congress in 1923. An independent agency of the U.S. government, it is charged with commemorating and honoring the service, achievements, and sacrifice of the U.S. Armed Forces where they have served overseas and in the U.S., when authorized by Congress.

Today, the agency administers 26 overseas cemeteries and 32 battle monuments, memorials, and markers located in 17 foreign countries, the U.S. Commonwealth of the Northern Mariana Islands, and the British Dependency of Gibraltar. Four of the memorials are in the U.S. Indeed, it has a global presence.[1] The cemeteries and memorials are among the most beautiful, reverent, and best maintained shrines in the world. There are 218,000 U.S. war dead interred or memorialized on the Walls of the Missing and more than 15,000 additional members of the U.S. Armed Forces, veterans, and others interred in the sites. Indeed, ABMC is the U.S. government's premier agency for commemorating America's war dead, fulfilling the promise of General of the Armies John J. Pershing, the agency's founder and first chairman, when he first said in the 1920s, "Time won't dim the glory of their deeds."

ABMC has a board of commissioners also appointed by the president for an indefinite term and who serve without compensation. The commissioners establish the strategic direction for the agency. Daily operations, to include management of the agency staff, are carried out by the secretary, the commission's CEO. ABMC's headquarters office is in Arlington, Virginia, and an overseas operations office is in Paris, France. It is a small agency in terms of staffing and budget, but our mission—"guardian of America's war dead"—was an enduring and sacred one transcending all else. The character of a nation is defined by the way it honors its war dead and is a measure of its heart and soul. None do it better than America!

I wanted to meet the Arlington office staff as soon as possible, so we arranged an "all hands" meeting prior to the scheduled staff call on Tuesday, January 16. At the first meeting, I told the staff how honored I was to accept the president's appointment as secretary of ABMC and looked forward to working with them in advancing our sacred mission. Referring now to my exact notes I used that day, I told them, "It eclipsed any other mission I've ever been associated with." It is a mission that resonates deeply with me as an infantryman of 33 years.

I have an up-close, strong, and personal feeling for what these men and women may have endured as they fought our nation's wars. And now to be personally involved in commemorating their deeds as they rest peacefully in our cemeteries is an "awesome and humbling responsibility that I welcome." I was full of strong emotion and told them, "This is not a job, rather, this is a duty, a sacred calling ... and I am very happy to join you."

We then proceeded with my first staff meeting. Each staff principal gave an update. I wasn't bashful and had many questions. I was particularly pleased to see that the agency had a strategic plan—a working document with focus areas that laid out best business practices and served as a blueprint for the future. The meeting also gave me a first glimpse into how the staff was organized, what the functional areas were, the importance of close interface between our Paris and Arlington offices, etc. I left the meeting with an appreciation for the issues confronting us.

With the initial briefings out of the way, my first priority was to get out of my office and start visiting our overseas cemetery and battle monument sites and meet the frontline "troops" who toil day-in and day-out managing and maintaining the sacred sites. They are the heart of ABMC and where the most meaningful work is done. These sites are where the mission is executed!

My predecessor, Max Cleland, a former U.S. Senator and Vietnam veteran, was only able to make a few overseas visits during his seven-and-a-half-year tour owing to his severe physical incapacitation from an accident and subsequent long, hard illness. Another year lapsed after Max left when the deputy secretary served as acting secretary. As the new secretary, it was vital that I get overseas and "show the flag." John Wessels, our overseas chief operations officer, told me that "some of the cemetery superintendents, as well as some of the Paris office staff, had never seen the secretary and were looking forward to the visit." I departed Dulles airport for Paris on United Airlines Flight 915 at 1740 hours Saturday, January 27, with our deputy, for a busy 10-day itinerary.

My visit began with an "all hands" meeting with our Paris staff. As an ardent admirer of our oldest ally, France, I enjoyed my meeting with this enthusiastic, culturally diverse group of mostly French and Americans. I sensed their deep appreciation for our mission and thanked them for their good work. In the following days, I visited multiple ABMC sites including Suresnes American Cemetery, the Lafayette Escadrille Memorial Cemetery, Normandy American Cemetery, including the near-by Pointe du Hoc Ranger Monument, Meuse-Argonne American Cemetery, Chateau-Thierry and Montfaucon Monuments, as well as our Aisne-Marne and Oise-Aisne American Cemeteries. I enjoyed most meeting the "troops," the superintendents, associates, the cemetery foremen, and their green space staffs (grounds keepers), at each site. These are the people who perform the mission every day. I thanked them on behalf of the American people and told them they are the indispensable "tip of the spear."

Since the establishment of our cemeteries, particularly those in France, Belgium, and the Netherlands, generation after generation of local families have been employed by ABMC to maintain and care for these sacred sites. "Sir, my grandfather worked here after the war, and my father after him, and now I am helping," said one of the grounds workers as I walked down the line shaking the hand of each worker at one of our cemeteries in France. The pride they had in their work was evident. Many see their work as a way to continue to show their gratitude for the sacrifice and courage of American soldiers in liberating their countries during the World Wars. Our cemeteries are staffed by local citizens except for the superintendents and assistant superintendents, who are Americans fluent in the host country's language.

While at Meuse-Argonne Cemetery, the assistant superintendent, Jim Bertelson, a former Navy SEAL, gave us a walking tour of some of the World War I battle sites in the vicinity of the cemetery. We had to climb a long, steep hill on a cold, rainy, February day. The ground was icy and slippery, and we had a difficult time reaching the top, particularly me. I wasn't wearing my polio leg brace and slipped several times. Each time, a helping hand quickly reached out for support.

The descent back down the hill was equally sporty and upon reaching the base I breathed a huge sigh of relief. I didn't want the humiliation of falling on my butt in front of our younger staff and the Navy SEAL due to the strain on my polio leg. Visiting the nine sites was, indeed, inspirational and gave me a firsthand appreciation for the world-class architecture, artwork and landscaped grounds; most importantly,

I experienced the humbling beauty of these sites as well as the pride and dedication of the men and women who maintain them.

The itinerary also included a courtesy meeting with the newly appointed American ambassador to France and Monaco, Jamie D. McCourt. She had just assumed her new duty and was most gracious. She invited us to attend an embassy award ceremony that afternoon. I learned early the importance of a close U.S. embassy–ABMC relationship for mission accomplishment. I also visited a British Commonwealth War Graves Commission military cemetery in Bayeux, France, run by our sister organization for the United Kingdom.

It was a wonderful and informative 10-day visit. I didn't want to leave! I would make many similar visits to our sites across Europe, North Africa, the Mediterranean, the Americas, and the Pacific. An added dimension to the visits was the side trips to the World War I and World War II European battlefields and historic towns that had played a role in the history of ABMC. Our deputy, where appropriate, integrated these trips into each itinerary.

Rob's knowledge as a former U.S. Army chief of military history was most helpful! Earlier, I mentioned the agency's Board of Commissioners. The authorizing legislation for ABMC specifies "the president will appoint not more than eleven commissioners." Seven were appointed with me in January. I had the pleasure of swearing them in. Appointments for three others would follow for a commission of 10. In May, the president appointed our eighth commissioner, David Urban, whom I also swore in. Urban at the time was president of the American Continental Group, a premier bipartisan government affairs and strategy consulting firm.

Commissioner Urban's appointment was followed by a July 10, 2018, letter from the White House director of presidential personnel, stating, "The president would like to designate David Urban to be the Chair of ABMC." ABMC had our chairman. The commissioners were a diverse and geographically dispersed group of patriotic Americans who enthusiastically embraced their responsibilities as stewards of our noble mission. Like me, they "served at the pleasure of the president" for an indefinite term. They clearly had a vested interest in how our nation honors its fallen, and I was glad to finally have them on board. My only disappointment was that none had been an enlisted person in the armed forces, which would've further enhanced the commission's role and reputation.

A West Point graduate, David brought three decades of experience at the highest levels in public policy, business, military, and public service sectors, which served ABMC well. He clearly understood the higher strategic role of the commissioners and my role as CEO to lead and manage the agency's daily operations. They were separate and distinct.

That was most comforting and allowed for a non-contentious, seamless operation as we stayed in our lanes, working harmoniously to achieve ABMC's goals. There are stories of previous commissions where that wasn't always the case and a contentious, sometimes hostile, relationship developed between the chairman, commissioners, secretary and staff. When that happens, nothing is accomplished. David and I worked diligently to ensure that didn't happen. We scrupulously followed the agency's bylaws, carefully drafted by our counsel, Edwin Fountain. Our focus was always on the mission and what was best for ABMC. We were a strong, cohesive team that accomplished much during our tenure.

I mentioned earlier the importance of working closely with the American embassies in the countries where our sites were located. John Wessels and our director of cemetery operations, Christine "Tina" Young, did a good job maintaining a strong and trusting relationship with their embassy counterparts. Tina, a retired Navy captain, ensured that was a priority with our 26 cemetery superintendents and that they understood the absolute importance of a close working relationship.

The chief of mission (COM), with the title of "ambassador" or "charge d'affairs," headed the mission's "country team" of U.S. government personnel. ABMC was a member of the country team. Responsibilities of the COM included directing, supporting, and coordinating all executive branch offices and personnel in their respective countries. Our Paris office and all our overseas personnel greatly depended on embassy support. I always made it a point to include a personal visit with the host ambassador or the *chargé d'affaires* when visiting our sites.

Since it was a new administration, all the American ambassadors had also just been appointed. We were all new at our jobs and meeting each other for the first time. The Trump- appointed ambassadors were extremely supportive of ABMC's mission. They could always be counted on to attend and support the various ceremonies and events hosted at our cemeteries and monuments. I formed some lifetime friendships among them.

One overzealous ambassador would frequent one of our sites often, bringing visitors or wanting to hold events of his own. Our young superintendent was becoming intimidated by his somewhat aggressive nature, sometimes giving the appearance of taking control of the cemetery, and I had to step in and run interference. He had good intentions, but we couldn't open the cemetery for visits during the lockdown quarantine periods for the COVID-19 pandemic to the extent that he wanted.

He reminded me in a letter that as COM, he had "full responsibility for supervision of all executive branch personnel and operations" and further, that "ABMC, established by Congress in 1923, is an agency of the Executive Branch." I didn't need that reminder and suggested politely he call our chairman if he felt that strongly. He didn't call! We worked it out, arriving at an amicable solution, although not to his complete satisfaction, and he was able to conduct his visit. I cite this to emphasize again the importance of a close professional relationship with our supporting embassies. It was an honor to work with our U.S. ambassadors and their staffs as we daily represented America abroad.

My visits overseas would continue, but before the next trip, I wanted to visit our National World War I and World War II museums to further explore the possibility of a more formal working arrangement with them and discuss ways to enhance their exhibits with texts and images of our cemeteries and monuments. It was an excellent way to reach the public in keeping with the part of our mission to educate the American public and to provide teachers with content to use in the classrooms. The materials illuminated the service and experiences of those honored at our sites worldwide while helping explain the history of America's role in past conflicts.

At my first staff call, I was briefed on the status of our education programs. I had many questions and saw the potential to expand and do more. I discussed that with our deputy and our history folks, and we all felt visits to these national museums would be appropriate. On February 26 we visited the National World War II Museum in New Orleans, Louisiana, and on March 19 the National World War I Museum and Memorial in Kansas City, Missouri. I had been to neither. We were well received by their leadership, and the seeds were sown for a formal education program partnership between ABMC and the World War II Museum. We visited the Smithsonian American Art Museum later, which also resulted in an education program partnership.

Maintaining relevance was important. ABMC was established to honor the American Armed Forces where they fell in battle. Immediate next-of-kin visited their loved ones' grave sites to grieve and honor them and to say farewell. Although the primary mission would always be to "commemorate and honor our war dead," it was incumbent upon ABMC to seek new ways to further its relevance for future generations.

Our chief of staff, Mike Conley, said it best. "We can't just be content with making our memorials beautiful, we must keep that sacrifice in the public conscience." In December 2020, we launched the "Re-Discover ABMC" campaign, a new series of videos highlighting our commemorative sites. Educating the American public as well as peoples of other nations about the deeds and sacrifices of American service members was the path we took.

In 2007, ABMC began constructing visitor/education centers at some of its cemeteries and monuments. There were six when I came on board. As explained in the ABMC *Commemorative Sites* booklet, they were created "to add historical context to the commemorative landscapes of the cemeteries and battle monuments by explaining the events of the wars and role Americans played in them."

Each center focused on the battles represented by those Americans interred in its plots and memorialized on its Walls of the Missing. Personal stories, photos, and wartime letters of American men and women buried or memorialized at the cemeteries helped visitors foster close personal connections to those who fought there. The centers, like the hallowed ground they are on, are maintained beautifully and are well received by the public. We finished building and dedicating three more visitor centers before I left ABMC in 2021.

We had two cemeteries in Italy and the new American ambassador, Lewis Eisenberg, wanted to meet with me. The staff developed the itinerary and my deputy and I departed with our wives on April 14, arriving in Rome on Sunday, April 15. We had a tour of the Vatican and environs of Rome. During the visit with the ambassador, he talked about his relatives who fought in World War II. A successful financier and philanthropist, he became very supportive of the ABMC mission.

We next visited our Sicily-Rome American Cemetery near the town of Nettuno, about 3 miles south of Anzio and 38 miles south of Rome. There are 10,953 Americans interred there or listed as missing on the walls of the chapel. I toured their small visitor center that explained the significance of Allied operations in Italy during World War II.

The next day we went by train to Florence and met with the consular officer at the American consulate. We visited our Florence American Cemetery, just outside the city, where 5,802 American servicemen are interred or memorialized on the Walls of the Missing. Most of the dead fell breaching the Gothic Line, the last major German defensive position in Italy's Apennine mountains. Both cemeteries in Italy were undergoing major landscaping and other work.

It gave me a firsthand appreciation for the importance of horticulture in the overall care and maintenance of our cemeteries including landscaping, tree care, grass management and more. That gave me a better understanding of exactly what comprised daily cemetery operations. Again, I mostly enjoyed meeting the "troops," the superintendents and the green space staffs, and thanked them for their outstanding work in maintaining the sacred sites. We returned home on April 22.

We have two cemeteries in England, and I wanted to visit them before returning to France for a scheduled series of Memorial Day events in May. I left for London on April 28. Our two cemeteries are Brookwood American Cemetery located 30 miles from London, and Cambridge American Cemetery, 3 miles west of the city of Cambridge. Cambridge University donated the land for the cemetery.

Brookwood is one of our World War I cemeteries and contains the graves of 468 Americans who died throughout the United Kingdom and Ireland during World War I. On one of the walls of the chapel were inscribed the names of 564 missing Americans. Among them were the crew of the U.S. Coast Guard Cutter *Tampa,* sunk by a German torpedo in 1918, representing the largest single American naval loss of World War I.[2]

Our Cambridge Cemetery is a World War II cemetery and contains 8,938 American war dead or missing. Most of the individuals interred or memorialized there died in the battle of the Atlantic, the strategic air bombardment of Europe, and the Normandy invasion. Among those listed on the Tablets of the Missing are the crew of the USS *Reuben James,* the first American warship lost in the battle of the Atlantic.[3] Joseph Kennedy, Jr., the older brother of John F. Kennedy, and former bandleader Glenn Miller are also listed among the missing.

When I met the green space workers at both cemeteries, I was able for the first time to speak to a cemetery staff without using an interpreter, as they all spoke English. One of the groundskeepers reminded me that his English was "the King's English," the "proper English." I enjoyed the banter of the moment as we had a witty and enjoyable conversation. It reminded me of the time I was commanding 3/187 (Rakkasans) in the 101st Airborne at Fort Campbell, Kentucky, when we hosted the 600-man British 1st Parachute Battalion for a month-long training exercise. It was a "bloody good" four weeks, and before it was over, I was speaking like a "Cockney." You've gotta love the British.

ABMC Ceremonies and Events

Every May, over the course of three days, ABMC marshaled its forces and hosted Memorial Day ceremonies at each of its 26 cemeteries. It was a huge annual undertaking, led by the Paris office, which planned and executed the ceremonies. The local populace looked forward to the events which were eagerly supported by local and national government officials. U.S. military theater commanders supported us by providing troop units, bands, and color guards, as needed. It was an American effort supported by our Allied host nations to pay tribute to and remember the sacrifices of our fallen.

Memorial Day 2018 was like no other! Being the World War I centennial year, it gave the day added significance. We hosted 29 ceremonies in 10 countries on four continents.

I participated in three of the ceremonies, giving welcoming remarks and introducing the dignitaries and keynote speakers. This proved to be a feat as all three ceremonies were at different locations on the same day, May 27.

The first ceremony at 9:45 a.m. was at the Aisne-Marne American Cemetery in Belleau, followed by the ceremony at 3:00 p.m. at the Oise-Aisne American Cemetery in Fère-en-Tardenois. The last ceremony was at 7:00 p.m. at our Château-Thierry Monument, 17 miles southwest of Oise-Aisne. That ceremony included the dedication of the new Château-Thierry Monument Visitor Center. Each ceremony was followed by a reception/Vin d'Honneur (Wine of Honor reception). Participants that day included the U.S. ambassador to France, Jamie McCourt; the commandant of the Marine Corps, General Robert B. Neller; and numerous French, German, and British military officials. I attended all receptions since ABMC was the host.

My driver that day did yeoman's work. I had no idea where I was as we traveled the narrow and crowded French country roads going from site to site. "Sir, they're putting you through the wringer today; this is worse than the Marine Corps marathon," said Bert Caloud, a retired Marine and superintendent of the Oise-Aisne

Cemetery. Shuttling from one site to the next gave me time to go over my speech and practice my French, which I would agonizingly stumble through in my welcome and introductory remarks. I had once enrolled in a basic French language course in 1977 and after two classroom sessions, my very patient professor discovered I would never be a francophone … not even close! I agreed and dropped the course. However, there were times like the present, that I wish I had toughed it out!

The weather was perfect, and thousands attended the Memorial Day ceremonies, with over 6,000 taking part in the Aisne-Marne proceedings. As John Wessels recounted in an interview published in the September 2018 edition of our magazine, *ABMC Voice*, "The ceremony was marked by an air of redemption. Former foes came together as genuine allies. A German officer gave a very moving speech. The anthems of the U.S., France and Germany were played with equal dignity."

Constant Lebastard, at the time a 30-year-old French employee of ABMC and of the generation of French people determined to keep the memory alive, said, "My family fought and lived through both world wars. They tell stories of the occupation and liberation. The emotions remain very strong." Indeed, emotions were high that day. The honors and spoken tributes to our fallen were even higher. The World War I Centennial Year Memorial Day ceremonies, attended by thousands of French citizens and overlooking the thousands of perfectly aligned white marble headstones, impacted me like no other had.

My visits continued through the summer and fall. I travelled to our World War II cemetery in Tunisia in late July to meet the staff and to host a retirement ceremony for a long-time ABMC employee, Foued Bouaziz, who held down the cemetery during the dangerous days of the Arab Spring. I then concentrated on visiting more World War I sites since it was the centennial of the Great War. I visited and participated in the 100th anniversary ceremonies at the St. Mihiel, Meuse-Argonne and Flanders Field American Cemeteries, as those were my priority. In early August, we visited our Somme American Cemetery in Bony, France, 100 miles north of Paris.

We also visited ABMC's Bellicourt Monument, 3 miles from the cemetery, commemorating the achievements of 90,000 American soldiers who served with the British forces in France during 1917–18. The monument contained a map and orientation table illustrating the American operations. When climbing the steps, I noticed significant cracking in the concrete, and weather damage and discoloration to many areas of the monument. I had noticed similar defects during my earlier visit to the Montfaucon Monument. It needed cleaning and wasn't to ABMC standards.

The weather in northern and western France could be severe, eroding even the hardest concrete and granite. Those monuments required constant vigilance and maintenance to prevent erosion, discoloration, and structural damage. I asked about a maintenance program. "Who maintains the monuments? Who inspects them? How often are they cleaned?" The answers from the staff were unsatisfactory. I requested that an inspection survey be done of all monuments and markers and a maintenance plan, including timelines for completion and cost estimations developed.

Our Paris office and engineers made maintenance a priority and developed a program for the repair and restoration of those damaged monuments. Emphasis was placed on daily care and inspection. Indeed, the commission would be restoring those historic military heritage sites to a condition commensurate with the service and sacrifice they honor. To further emphasize the importance, we had "Monument Maintenance" added as a separate focus area in our strategic plan. Chairwoman Debbie Wasserman Schultz was most supportive when we briefed her House Appropriations Sub-committee on Military Construction, Veterans Affairs and Related Agencies, requesting funding for our Monument Restoration Program.

Most of Europe was marking 100 years since the end of World War I. Ceremonies and parades were being planned. The key ceremony was Sunday, November 11, Armistice Day, at the Arc de Triomphe in Paris. ABMC was alerted in early October that President Trump (POTUS) and First Lady Melania (FLOTUS) would be travelling to France for the Paris ceremony. Their itinerary details were being discussed with our embassy in Paris.

The White House was looking for appropriate side visits for POTUS while in France. We, of course, suggested visits to our World War I cemeteries. The embassy and White House were in full agreement, and after many variations of which venues President Trump would visit, what type of ceremony there would be, the White House agreed to visits to the Aisne-Marne American Cemetery (adjacent to the Belleau Wood Battlefield) and Suresnes American Cemetery.

The White House Chief of Staff Office requested we brief them on what we were proposing for the president's visits. I briefed the White House personnel including the team that would be travelling with POTUS, while our overseas operations director briefed our embassy in Paris. It was agreed that the Aisne-Marne visit would be a very private wreath-laying ceremony by POTUS and FLOTUS on November 10, followed by a POTUS visit to Suresnes on November 11, which would include a ceremony for the public and remarks by the president.

During the weeks leading up to the events, we worked closely with the White House and Secret Service teams planning every aspect of the visits. It is a rare opportunity when the commander in chief and first lady visit our overseas cemeteries to honor our war dead. Our Paris office and our Aisne-Marne and Suresnes Cemetery staffs addressed every detail. They were excited, particularly the cemetery superintendents, to have the commander in chief visit their sites!

On the morning of November 10, I was at the Aisne-Marne Cemetery awaiting the POTUS/FLOTUS arrival. The cemetery superintendent and I would greet them. Our overseas chief operations director, John Wessels, was with me. The weather wasn't good. There was a drizzling rain and dark gray overcast sky, so we had our coats and umbrellas with us. The president would be arriving by Marine One. We were in communication with the president's travelling team and Secret Service team in Paris. An element of the Secret Service team was also with us at the cemetery. I looked at Wessels and said, "With this low ceiling, I doubt the president will make it." Within moments we got the call "POTUS visit cancelled, repeat cancelled, bad weather."

When the president suggested the option of coming by motorcade, the Secret Service said, "No, too far from the airport and big Paris shutdown" as reported in the November 13, 2018, edition of the *New York Times*. There was insufficient time to clear the highways, ramps, and overpasses and ensure a safe route for the entire 60-mile drive and return for a presidential motorcade. White House Chief of Staff John F. Kelly, and Chairman of the Joint Chiefs of Staff General Joseph F. Dunford, accompanied by their wives, arrived later that day and placed wreaths honoring our soldiers and Marines interred and on the Walls of the Missing at Aisne-Marne.

They also visited the Marine Monument at Belleau Wood adjoining the cemetery, which honors the 4th Marine Brigade, consisting of the 5th and 6th Regiments of the United States Marine Corps, part of the U.S. 2nd Division, that captured much of the ground in 1918.[4] Chairman Urban and our superintendent greeted them upon arrival. I was called to attend a late afternoon final rehearsal for the Suresnes ceremony with the president the following day, so Wessels and I departed. The weather was still miserable.

The following day, Armistice Day, the president spoke as scheduled at ABMC's World War I Centennial Ceremony at Suresnes American Cemetery just outside Paris. Among the many dignitaries attending were Ambassador McCourt, Secretary of State Mike Pompeo, General Dunford and 12 congressmen. Our chairman was also present. The president arrived in the chilly pouring rain that continued throughout the day. I met him and escorted him into the cemetery plot area.

We were both wearing overcoats, but the damp weather penetrated and chilled through to our bones. In the plot area, he was greeted by the superintendent, Keith Stadler, who spoke briefly about the cemetery and those honored there. As President Trump stood alone among the headstones staring down at one headstone in particular, he was visibly moved. It was the headstone of a soldier from New York City. When I finally was able to get him out of the plot area to escort him onto the dais, he had a tear in his eye. Those few moments standing among the headstones of our dead soldiers elicited emotion; I could see in his face he was moved by the moment.

While sitting on the dais he leaned over and asked me to show him where the World War II veterans were seated. I did. He looked up at them and gave them a nod and a smile. They returned the gesture with a wave. One saluted. There were six World War II veterans, all in their 90s, some donning their service caps, sitting together on a raised covered balcony. Two were in wheelchairs. My welcome remarks and introduction of the president were brief as the sound of the rain pinging on the podium microphone was amplified.

The president spoke for 10 minutes in the pouring rain, honoring all those who "gave the last full measure of devotion" and paid tribute to the brave Americans "who gave their last breath in that mighty struggle," citing the over 1,500 service members interred at Suresnes. He told heart-rendering stories of soldiers and Marines buried there. Descendent family members of World War I veterans, who were present, were singled out. He mentioned each World War II veteran by name, saying, "You look so comfortable under shelter as we're getting drenched … you are very smart people; you look great!" he said as he looked right at them. These old soldiers loved it, as did the audience. It was a light, heartwarming moment.

It was a personal, very moving speech, striking the right tone and emphasizing the gallant and courageous effort of American troops who fought in World War I. His mentioning the names of the World War II veterans present was well received. I spoke with them afterward, and they were effusive with kind remarks and appreciation for the commander in chief, who would take the time to recognize them individually.

During the ceremony, President Trump presented ABMC with an American flag that had recently been flown over the White House for display at Suresnes American Cemetery, marking the historic occasion. He ended his remarks by saying, "This is certainly the highlight of my two-day visit to France." He departed for the airport and Air Force One to return to Washington.

Meanwhile, the cancellation of his wreath-laying visit to Aisne-Marne Cemetery the day before prompted widespread criticism on social media and from some officials in Britain and the U.S. that Trump had "dishonored" U.S. servicemen. Outrageous, unfounded comments poured in, and in the ensuing months, the media, when convenient, would cite the cancellation of the wreath-laying ceremony at Aisne-Marne as an example of President Trump "not caring about veterans."

This came to a head in September 2020 when *The Atlantic* magazine published a story citing multiple anonymous sources, of how President Trump disparaged U.S. troops, veterans, and missing service members. It was followed by other stories by the Associated Press, *Washington Post* and the *New York Times*—all citing alleged incidents of Trump not respecting our military and veterans. *The Atlantic* story specifically mentioned the president's trip to France during the commemoration of the 100th anniversary of World War I ending and that he didn't go to Aisne-Marne that day "because he feared his hair would become disheveled in the rain and because he didn't believe it important to honor American war dead according to four people with firsthand knowledge of the discussion that day."[5]

The false reporting was explosive! What a disgraceful attempt to smear President Trump 60 days before the presidential election. White House Chief of Staff Mark Meadows said, "It's sad the depth that people will go to during a lead-up to a presidential campaign to try to smear somebody." He was looking for support to factually counter *The Atlantic* story about Trump not going to our cemetery on November 10.

I could no longer take the lies and insults being told each day about why Trump didn't go to our ABMC cemetery. In a telephone conversation with Mark Meadows, I gave him the "facts" as I lived them that day, being the person on the ground awaiting his arrival and receiving the status reports from his travel team in Paris. I joined 13 other officials who were on the trip to France with the president, refuting the anonymous sources and allegations in *The Atlantic* article. Among them were:

Jordan Karen, former personal aide to President Trump: "This is not even close to being factually accurate. Plain and simple, it just never happened. Again, this is 100 percent false. I was next to POTUS the whole day! The president was greatly disappointed when told we couldn't fly there. He was incredibly eager to honor our Fallen Heroes."

Tony Ornato, then head of the Secret Service's Presidential Protective Detail, said the story was "false."

Dan Walsh, former White House deputy chief of staff: "There was a bad weather call in France, and the helicopters were unable to safely make the flight."

Jamie McCourt, U.S. ambassador to France and Monaco: "He was devastated to not be able to go to the cemetery at Belleau Wood. The next day he spoke at the ceremony in Suresnes in pouring rain."

Stephen Miller, White House senior adviser: "The accusation is a despicable lie … the president deeply wanted to attend the memorial event in question and was deeply displeased by the bad weather call."

Following is my September 2020 on-the-record statement I provided the White House:

> I was the host for the event discussed by the false and despicable article published in *The Atlantic* magazine on 3 September. On 10 November 2018, I was at the Aisne-Marne American Cemetery awaiting the arrival of President Trump. As a former infantryman, who has flown on many helicopters, I knew that morning the weather was bad and the ceiling too low for a safe landing that day. When the president's visit was appropriately canceled due to weather, I received word also that he was upset he would not be able to make the wreath-laying visit and to pay his respect to 2,300 fallen soldiers and Marines interred there. Those who know President Trump know that those anonymous smears peddled by *The Atlantic* have no basis in fact or reality and do a terrible disservice to journalism and to our veterans, living and deceased.

I was infuriated!

Additionally, *The Atlantic*'s reporting was refuted by White House email and Navy documents that directly show a "bad weather call" was the reason for the canceled presidential trip to Aisne-Marne Cemetery in 2018.

As a republic, we are, indeed, in perilous times when anonymous sources are believed, and no one even questions them. This is not journalism and is a grave disservice to our country and our people. When a country can't trust its news sources to report the unbiased truth, the people are left lost and devoid of truth. We are in an age of fake, mistrustful, reporting and credible journalism is in a battle for survival.

To cap off the year, Linda and I joined our chairman and others in Philadelphia on December 8 for the annual Army-Navy football game. David was busy that day helping to escort President Trump as he made his many visits throughout the Army and Navy sections of Lincoln Financial Field Stadium. We were in a box suite overlooking the seating area. Among those 40 or so people in the suite was Secretary Pompeo and his wife. I asked Pompeo, who graduated first in his class, if he had played football while at West Point. "No, no, I was too slow," he answered.

I came to know the new secretary of state that day. A bright, engaging and very modest guy! President Trump spent considerable time with us, and, for the first time, I was able to chat with him in a relaxed, unofficial venue. We talked briefly about ABMC, our veterans, and the military. He enjoyed waving at the fans seated immediately below who turned to look up at the president waving in support and trying to get a picture of him.

The commander in chief was very relaxed that afternoon as he met many Army and Navy fans while cheering on both academy teams. Of course, everyone wanted their photo taken with the president, and he obliged. This was a fun event which I enjoyed after 12 months of steady overseas travel. The best part of the day was Army won the hard-fought battle 17–10 and the Commander in Chief's Trophy for 2018!

The year 2019 was another busy year for ABMC. Looming on the horizon was the 75th anniversary of the D-Day landings (Operation *Overlord*) in Normandy, France. Our Normandy American Cemetery sits on a cliff overlooking Omaha Beach and is one of five landing beaches used by Allied forces for the June 6, 1944, invasion. ABMC would be intimately involved in the planning and execution of the commemorative events, but before the real work started, I wanted to visit our sites in the Americas and the Pacific, as well as additional World War II sites in Europe. It had been years since the secretary had visited our sites in Panama, Mexico, and the Philippines. I needed to go!

I did make one trip to the Pacific immediately after our ceremony in France with the president. I went to our Honolulu Memorial on Oahu, Hawaii, for the rededication of the memorial and for the commemoration of the 75th anniversary of the battle of Tarawa. I had been there before when I was deputy commanding general of U.S. Army Pacific to attend events and burials.

Our memorial honors the achievements of American Armed Forces in the Pacific during World War II and the Korean War. It was dedicated in 1966 and grew in 1980 to include the missing of the Vietnam War. We were rededicating the World War II and Korean War Courts of the Missing that had been refurbished. The battle of Tarawa was among the fiercest the Marines and Navy fought in World War II, losing over 1,000 dead and nearly 500 missing in action. Joining us at the ceremony were several family members of those MIAs.

Two outstanding staff were Mike Conley, the agency's chief of staff and Pat Harris, our front office executive secretary. Mike, a retired Air Force colonel, had been with

ABMC for 20 years. He was our continuity and "go-to" guy. A writer and journalist of the first order, I relied on him to draft my speeches and remarks. Pat coordinated and made all the travel arrangements, including all travel for our 10 commissioners … not an easy job!

In March 2019, I visited our cemeteries in Panama and Mexico. I had last been to Panama during Operation *Just Cause* in 1989–90 when we removed Noriega. Our Corozal American Cemetery was 3 miles north of Panama City. When I arrived, I recalled moving with our infantry units sweeping the cemetery grounds, ferreting out the last of Noriega's PDF.

I said to John Wessels who was with me, "I remember this cemetery. I was here in 1990 with the 7th Infantry Division." ABMC assumed responsibility for the maintenance of the cemetery in 1982. Many of the 5,500 burials were workers who contributed to the building and operation of the Panama Canal in the early 1900s. Corozal is one of only two ABMC cemeteries that is still open to new burials of qualified veterans.

I next visited our Mexico City National Cemetery. Established in 1851, it contains the graves of Americans from the Mexican–American War, Civil War, and the Spanish–American War. Young Lieutenant Ulysses S. Grant fought with the U.S. 4th Infantry Regiment against Mexico in 1846 on the grounds immediately adjacent to the cemetery. Some of his soldiers were buried there. Located in the heart of Mexico City, it is our smallest cemetery. These visits helped me better understand some of the issues unique to our Central American sites; being far removed from mainstream ABMC events they needed that visit and a pat on the back.

It was important that I visit our sites in the Philippines. ABMC proudly administers three sites there: the Manila American Cemetery, the Cabanatuan Memorial, both World War II sites, and Clark Veterans Cemetery. America's relationship with our Philippine ally is a strong and storied one. Two former commissioners, when they called after my appointment to wish me well, urged me to visit those Philippine sites as soon as possible. They, like I, understood the importance of a secretary-level visit, and I wanted to go.

Manila was our largest cemetery, and had, in the late stage of construction, what would become our ninth visitor center and first in the Pacific. I wanted to check its progress, and the staff wanted the final stamp of approval for the interior layout of the exhibits. Our Philippine cemeteries were our most distant in terms of miles and travel time from both our Arlington and Paris offices. The Manila cemetery superintendent was a retired Navy chief who had been with ABMC for many years and was retiring, and I wanted to personally thank him before he left.

It was a long 8,600-mile flight. There were 17,000 war dead interred and 36,000 listed on the Walls of the Missing in the one cemetery. Those numbers reminded me of the terrible price we pay in war, the human costs being the highest. The mosaic maps recalled the achievements of American forces in the Pacific, China, India, and Burma.

The Cabanatuan Memorial marked the site of the Japanese prisoner of war (POW) camp where 20,000 Americans were held captive. It also honored Filipino servicemen who fought side-by-side with us. Co-located on the site was the West Point Monument, which paid homage to the 170 American and six Filipino graduates of the U.S. Military Academy who lost their lives fighting in the Philippines or while POWs at Cabanatuan.

Our last visit was to the Clark Veterans Cemetery, north of Manila, formerly Clark Air Base Cemetery. The eruption of Mount Pinatubo in 1991 led the Air Force to evacuate Clark AFB. The cemetery fell into disrepair and was turned over to ABMC in 2013. Besides Corozal Cemetery in Panama, mentioned previously, it is the only other ABMC cemetery that is still open to new burials.

I couldn't have been more pleased by the six-day trip to the Philippines. The Filipino site staffs were immensely proud of their work, and I detected that they saw it as a reverent duty to take care of our American service members who gave their lives in the defense of the Philippines. I gave each an ABMC coin!

I had been secretary for a little over a year and was bursting with enthusiasm and pride. Logging thousands of air miles, every visit to one of our sacred sites opened a new door. I remember saying to Linda how lucky I was to be part of the ABMC team as we went about caring for our war dead every day!

The commissioners met twice a year, usually at our offices in Arlington. Their counsel and guidance were excellent as we progressed. I wanted to get them more personally involved in our overseas events. The upcoming World War II 75th anniversary events and dedication ceremonies for our new visitor centers were perfect venues for their participation.

Following are some of the more memorable events in 2019 where the commissioners represented ABMC and gave the keynote address: Ben Cassidy in Rome at the 75th anniversary of the Anzio and Nettuno landings, John McGoff in Paris at the dedication of our Lafayette Escadrille Memorial Cemetery Visitor Center, Bob Wefald in Manila at the dedication of our Manila American Cemetery Visitor Center held on the 75th anniversary of MacArthur's return to the Philippines.

Commissioner Evans Spiceland hosted a commissioner's board meeting in New Orleans in 2020. Commissioner Bob Ord and other commissioners gave presentations to local groups and organizations telling the ABMC story. The staff prepared those presentations, tailored for each group. Commissioner Deecy Gray hosted a lovely dinner reception at her high-rise condominium in Arlington, Virginia, for the commissioners and staff. Our commissioners were fully involved, engaged and were great ambassadors for ABMC. Their hearts were in it, and I appreciated their enthusiasm and dedication.

As we did for the 100th anniversary of World War I, ABMC would join the Europeans and others in celebrating the 75th anniversary of World War II events in 2019. The key event would be the celebration of the D-Day landings at Normandy.

President Trump and the first lady would again be travelling to France to participate in the 75th anniversary of D-Day ceremony at our Normandy Cemetery. That ceremony would be jointly hosted by Presidents Trump and Emmanuel Macron of France. It was a major head of state event. Our Paris office once again went into high gear planning and preparing for the once-in-a-lifetime event. Thousands would be attending, including hundreds of American World War II veterans.

For me, it was poignant and possibly the last time many of the 90-year-old veterans would be with us—often in wheelchairs—representing the "Greatest Generation." I travelled to France with Linda, our deputy and his wife, sailing on Cunard's *Queen Mary 2*. We departed New York City on May 24 and arrived in Southampton, England, on May 31. To commemorate the 75th anniversary of D-Day, Cunard partnered with the Greatest Generations Foundation to pay tribute to the heroes of World War II on board their flagship. Twenty World War II combat veterans, ages 94–100, many veterans of the D-Day invasion, sailed with us. One celebrated his 100th birthday during the crossing. All would attend the ceremony at Normandy on June 6.

During the seven-night crossing, the veterans participated in many activities and programs to celebrate the momentous occasion. Guests aboard *Queen Mary 2* would have the opportunity to attend lectures and listen to personal wartime stories by those veterans, including Q&As and interaction with the heroes. I participated on the panel for the program "Remembering the Fallen."

Every session, held in the ship's theater, was packed. Many of the guests on the crossing attended the 75th anniversary events in France. Fox News and ABC News anchors joined the voyage and served as moderators for the panels and lectures. Through the veterans' vivid firsthand accounts of their wartime experiences, history came to life and all guests had a better understanding of what these men faced 75 years ago.

Those 90-year-old veterans didn't tire easily. They danced, some dressed in their World War II uniforms, into the wee hours of the morning. The ladies were standing in line waiting for the next dance. It was a celebratory voyage enjoyed by all; a crossing like no other, filled with dignity and respect for the American heroes on board.

Arriving in Southampton, we took the train through the Channel Tunnel from London to Paris. Our commissioners arrived on June 3. My deputy and fellow historian, Mike Knapp, conducted a bus tour of the Normandy battle areas for the commissioners. I was busy at our Normandy Cemetery in meetings with the White House advance team and security folks, doing interviews and attending rehearsals. John Wessels and Tina Young, our director of cemetery operations, were with me most of the time. David Smeigh, our D-Day coordinator, and Scott Desjardins, the Normandy American Cemetery Superintendent, often seemed overwhelmed, but never lost focus on what needed to be done during those final 48 hours before the ceremony began.

Public affairs, news media, parking, VIP and public seating, audio-visual, stage and ramp construction, security, press boxes, hanging of the national flags and banners and more, were all being addressed. We also had to contend with the strict rules demanded by the French government concerning registration and passes for all people attending the event. We were expecting as many as 12,000 participants.

Our ABMC security chief, U.S. Secret Service and French security officials disagreed at times over the security arrangements, all wanting the best security possible. Control measures were tight! My head was spinning as I kept remembering what the White House representative said to me during my initial briefing to them earlier in Washington. "Secretary Matz, you are the president's on-site host responsible for ensuring 'everything' in the Normandy cemetery is up and ready for the POTUS ceremony. Everything must be pristine and correct. The world will be watching."

It was late afternoon on the day before the ceremony when I was inspecting the stage, backdrop and ramps and the memorial platform that was just built where the two presidents would speak. Tina looked up and noticed that the large national flags hanging from the semicircular colonnade behind the memorial stage, where the veterans would be sitting, didn't look right. I looked up at the flags, also. Something was, indeed, wrong. The contractor who set up the stage and hung the flags had mistakenly hung Dutch flags instead of French flags alongside the American flags!

The rectangular flag of the Netherlands has the same red, white, and blue colors as the French flag. The only difference is that on the Dutch flag, the stripes are horizontal, while the French are vertical. These were huge 20-foot flags draping the height of the colonnade needing a crane to hang them. We immediately got the contractor and pointed this out. I went to bed very nervous that night, worrying like hell, not knowing if the flags had been exchanged. Thankfully, they had!

Had they not, I don't want to even imagine what the next day would have been like. It would have been a joint United States–Dutch not a United States–French ceremony, and I would have lost my job. The media would have latched onto that, embarrassing both countries and making worldwide news media headlines while finding a way to blame it on President Trump. I still get nervous thinking about the potential consequences. The contractor was a German company, and at a luncheon the day after the ceremony, our French host, well aware of the incident, remarked, "Those darn Germans are still after us."

The weather was beautiful the day of the ceremony, as 171 World War II veterans, many proudly displaying their medals and unit insignias, and wearing their unit caps, sat high on the memorial stage and looked out at the over 14,000 guests. Fifty-nine of the veterans on stage had landed on Omaha or Utah beaches or had parachuted onto the DZs 75 years earlier. Some were in wheelchairs, and some sat clutching their canes. For some, it was the first time they had returned to Normandy. They received a four-minute standing ovation from the crowd before the ceremony even began. It was awe inspiring and spectacular as every guest showed their appreciation for what these war heroes had done 75 years earlier, not just saving France, but saving the world.

While on stage awaiting the official arrival of the two presidents, I looked out at the audience and spotted Linda just as she tripped over a chair leg while making her way to her seat in the fourth row. Trump's two sons, Don, Jr., and Eric were standing close by in the next row and reached out quickly to help her up. She was wearing a bright yellow suit and easy to spot in the crowd. She wasn't hurt, but certainly embarrassed. She was an admirer of the president's children and finally met two of them. One of the TV crews panning the audience captured the entire incident. I worry about the embarrassment of falling or tripping during a public event because my polio leg sometimes suddenly, without warning, buckles beneath me. Yet my wife, who is very athletic and stable, is shown on TV falling in front of an audience of thousands!

After my brief welcome remarks, President Trump spoke: "You are among the greatest Americans who will ever live. You are the pride of the nation. You are the glory of our republic, and we thank you from the bottom of our hearts," he told the veterans gathered there. The president told the story of Russell Pickett, 94, who was with the first wave on Omaha Beach and was wounded multiple times during the war. Mr. Pickett was on stage in his wheelchair as both Trump and Macron walked over to him and helped him up. Trump hugged him tightly pressing his right temple to Mr. Pickett's and whispering to him. It was a touching, emotional moment for both the commander in chief and the D-Day veteran!

Trump and Macron helped the aging veteran back to his seat as the applause got louder. "Our cherished alliance was forged in the heat of battle, tested in the trials of war, and proven in the blessings of peace," Trump said. And with Macron sitting a few feet away, Trump said, "Our bond is unbreakable." And turning and facing Macron said, "We will forever be together."

President Macron followed with an equally gracious and heartfelt speech. He, too, shared stories of bravery and generosity of the American veterans on stage. In speaking about the importance of the trans-Atlantic alliance he said, "America is never greater than when it shows its loyalty to universal values that the Founding Fathers defended." That was a powerful statement drawing strong applause. As reported in the June 7 edition of the *Wall Street Journal*, "On Thursday, the two leaders greeted each other warmly and struck a conciliatory tone."

I had a difficult time getting Trump and Macron off the stage as they and the veterans were embracing each other with warm hugs of affection. The veterans didn't want either president to leave. Almost 900 credentialed members of the press and international media attended. The reporting and stories were upbeat and reflected the moment of the day, and the deeds and sacrifices of the soldiers that fought in Normandy 75 years before. Many U.S. cabinet members and senior military leaders, including 14 four-star generals and admirals, and the largest U.S. congressional delegation to ever visit France, were present. Hundreds of thousands, if not millions, watched the ceremony live streamed on television and other devices. It was the largest event ever hosted by ABMC!

There are moments and stories in our lives that seem to define us. Moments we return to. This is one of those moments for me. To be on that stage in our Normandy American Cemetery where 9,382 American war dead rest peacefully, with 171 veterans of America's Greatest Generation, is a seminal moment for me … and one I remember often and will never forget! It was a proud day for the ABMC and for the common values America shares with its oldest friend and ally, France.

The following day, we rededicated a completely refurbished Normandy Visitor Center which had just received a second-place award from the National Association of Interpretation. Chairman Urban opened the ceremony followed by Ken Gilman, a board member of our recently formed American Battle Monuments Foundation (ABMF). Ken's keynote remarks were most fitting for the official occasion marking the re-opening of ABMC's first visitor center.

In the fall of 2019, I had one of my most memorable, sobering meetings during my time as secretary. I met with Rabbi Jacob J. Schacter, Steve Lamar and Shalom Lamm who represented "Operation Benjamin". We met to discuss ABMC's role in their mission. Operation Benjamin is a program devoted entirely to preserving the memories of American-Jewish soldiers who made the ultimate sacrifice while defending the cause of freedom during World War II. Their aim is to locate Jewish soldiers at American cemeteries who, for various reasons, were buried under markers and headstones incorrectly representing their religion and heritage.

Their dignified mission is simply to correct those mistakes. Their program is named for Army Private Benjamin Gradetsky, a Jewish soldier mistakenly buried under a Latin cross at our Normandy Cemetery years earlier. Private Gradetsky's case was the first for which their team's efforts led to a successful marker change. His grave was finally marked with a Star of David at an emotional ceremony at Normandy on June 20, 2018. For a moment, we discussed the question "How could such mistakes have been made?" Some were clerical errors.

But it was also not uncommon that American-Jewish soldiers didn't want a Jewish identification on their dog-tags because of the terrible consequences they could suffer if captured. Rabbi Schacter so eloquently said, by placing a Jewish Star over the graves of these Jewish soldiers, "We welcome them back into the bosom of their ancestral faith." Beautiful words! After that meeting, we formalized our partnership with Operation Benjamin, arranging appropriate marker change ceremonies in the ABMC cemeteries. When I left ABMC in 2021, we had completed nine change of marker ceremonies and were working on several others. If ever there were two similar missions, they are ABMC's and Operation Benjamin's. Both are dedicated to honoring our fallen American service members.

There were three remaining World War II 75th anniversary events in which ABMC played a major role. The ceremony commemorating the 75th anniversary of the Southern France landings (Operation *Dragoon*) was on August 16. The ceremony was hosted in our Rhone American Cemetery in the town of Draguignan, 40 miles west of Cannes. It was the final resting place for 858 Americans.

In my brief remarks, I reminded all that, "The world knows well the names of 'Normandy' and 'Omaha Beach,' but today we remember beaches named Rosie, Camel, Delta, Alpha, Garbo, and Romeo, across which U.S. and Free French forces launched Operation *Dragoon*." Brave French Resistance forces played a major role in that operation. I recognized those Resistance forces veterans that were present.

At a luncheon at Restaurant L'Andruono that day, hosted by The First Alliance Foundation, a French-American partnership, I recognized our director of cemetery operations, Tina Young, for her herculean effort and leadership in support of the D-Day ceremony two months earlier at Normandy (especially for noticing the incorrect Dutch flags and getting the French flags displayed!). I will be forever indebted to her.

Four weeks later, I was in the Netherlands in the village of Margraten where we hosted the 75th anniversary of the liberation of the Netherlands ceremony. In our Netherlands Cemetery rest 8,291 Americans. Most were killed in the airborne Operation *Market Garden* and ground operations in eastern Netherlands. The Eindhoven-Nijmegen-Arnhem corridor is close by where the fiercest fighting took place. Eindhoven and Nijmegen were liberated on September 20, 1944, but Arnhem proved to be a "bridge too far" and wasn't liberated until April 1945.

In 2020, I returned for the 76th anniversary ceremony and to also review the final plans for the construction of the long-awaited Netherlands visitor center which opened in 2024. The Dutch will never forget what American and Allied soldiers did to drive the Germans out and finally liberate them. Dutch families are in line to formally adopt an American soldier's grave site as if it were their own.

Our final World War II anniversary event was the commemoration of the 75th anniversary of the Battle of the Bulge on December 16, 2019, at our Luxembourg American Cemetery. It culminated a four-day international commemoration of the largest and bloodiest single battle fought by U.S. forces in World War II. Three days earlier we hosted a beautiful luminary event honoring those memorialized in our cemetery.

The citizens of Luxembourg joined the cemetery staff in placing candles atop the 5,073 headstones, including the graves of General George S. Patton, Jr., and 22 sets of brothers buried side-by-side. Also interred are five soldiers of Easy Company, 506th Parachute Infantry Regiment, the "Band of Brothers." Each candle placed is a torch of freedom held high by the sacrifice of those Americans who lie beneath—each a solemn reminder of our own responsibility to keep that torch lit for generations to come.

Whenever I visit one of our cemeteries, I recall the words of Archibald MacLeish, an American poet who fought in World War I, from his poem, *The Young Dead Soldiers*: "They say we were young. We have died. Remember us ... They say we leave you our deaths: give them their meaning."

The American ambassador to Luxembourg, J. Randolph Evans and I placed a candle at General Patton's grave. Linda and I placed a candle at the grave of PFC Reed Davis, killed in the battle of the Bulge, with his son and grandson who were in attendance. The town of Bastogne anniversary events included walking in their annual parade, where the people lining the streets threw nuts, commemorating Brigadier General Anthony McAuliffe's famous response to the German request to surrender in 1944.

On the morning of December 16, military and civilian dignitaries from Europe and the U.S. were in Bastogne, Belgium, to mark the anniversary. SECDEF Mark Esper and the chairman of the JCS, General Mark Milley, attended. They also attended the ABMC ceremony later that day at our Luxembourg Cemetery. At that ceremony, culminating the four-day commemoration in Belgium and Luxembourg, it was a special honor to have with us the king and queen of Belgium, King Philippe and Queen Mathilde, and His Royal Highness Grand Duke Henri of Luxembourg.

Our Belgian embassy and ABMC co-hosted a luncheon in honor of our veterans and were joined by a congressional delegation headed by Speaker Pelosi, which kept us waiting as they had just arrived from the U.S. and hurriedly came directly to the luncheon by motorcade. The commander of the 101st Airborne Division, Major General Brian Winski, and some of his soldiers attended. They were a welcome sight wearing their battle dress uniforms and maroon berets. Seeing them brought back memories of my time with the "Screaming Eagles" 40 years earlier! The Grand Duke hosted a lovely reception at his palace that evening.

We had a flyover scheduled as part of our final ceremony in Luxembourg. Wouldn't you know it? The planes were late. We were in the middle of the ceremony seated with the king and queen of Belgium and the grand duke of Luxembourg, all patiently awaiting the flight of airplanes, as the seconds ticked away. Finally, we heard the faint roar of engines in the distance. We all looked at each other and politely smiled. "They're coming, the planes are coming!" Heads and eyes turned skyward. The noise got louder, and finally they passed overhead. There was a silent, yet audible, sigh of relief from the crowd, especially among the front row of dignitaries. Believe me, as the host, it wasn't a comfortable feeling as I felt the glances and polite stares of many as the seconds, maybe minutes, ticked away awaiting the flyover.

The Battle of the Bulge ceremonies were the culmination of a year of many beautiful events. It left me with a wonderful feeling about ABMC and the many ways the commission had honored the deeds and sacrifices of our war dead. During the year's commemorations, the inspiration and crucial lessons drawn from the events took center stage to ensure that people everywhere never forget the wisdom gained from such extraordinary moments in time.

In March 2020, the COVID-19 pandemic struck. We closed all ABMC cemeteries to the public, for the first time since World War II. It curtailed travel, but we quickly adapted to the restrictions and telework environment, continuing our mission through innovative virtual ceremonies and social media campaigns. Personally, I was upset as I had scheduled several return visits to our sites and had to cancel. I was opposed to closing our cemeteries but was overruled by the host governments who closed all their similar sites.

Sometime in 2019, Ambassador Gidwitz shared with me his vision of the Mardasson Memorial in Belgium becoming the 31st federal monument cared for by the ABMC. He had discussed the idea with the State Department. He wanted my reaction and thoughts on the matter. The Mardasson Memorial is a Belgian monument honoring the American soldiers wounded or killed during the battle of the Bulge in 1944–45.

Designed in the shape of a five-pointed star, it is located near Bastogne, the small crossroad town where Brigadier General Anthony "Tony" McAuliffe made his now-famous response to the German request to surrender the town of Bastogne: "Nuts." It was built by the people of Belgium and dedicated in 1950. It bears the names of the 50 U.S. states. Insignia of most participating U.S. battalions are shown on the walls representing 77,000 American soldiers killed and wounded during the battle. The Latin inscription on the memorial stone translates to "The Belgian people remember their American liberators—4th July 1946."

I responded to the ambassador's desire to make it happen with a resounding "YES, we would be interested. What red-blooded American wouldn't?" Both in size and scope, it is one of the largest battle monuments in all of Europe and dedicated to the American soldiers who fought what Churchill described as "undoubtedly, the greatest American battle" of World War II. To have it under the management and care of ABMC was most fitting. It needed repair and to be brought up to the standards of an ABMC site, all of which was doable.

There were some who were hesitant and didn't share my fervent and absolute enthusiasm. Our commissioners were ecstatic over the possibility, as were some members of Congress who were personally briefed by Ambassador Gidwitz, who started a personal campaign to raise private funds for the initial refurbishment costs. ABMC served as the procurement agency and oversaw the initial contractor work.

Our Paris office was motivated like I had never seen, and working with the embassy staff, organized a Battle of the Bulge Memorial Project Team and developed various site plans for presentation to the Belgian authorities. Some local Belgian officials had to be fully convinced that turning over their monument to America was the right thing. I understood their concerns and worked closely with them. I received a call from a U.S. representative who had a relative who fought in the battle, wanting to know when the historic monument would belong to ABMC.

After all issues necessary for the official transfer of the Mardasson Memorial to ABMC were resolved, a signing ceremony arranged by our embassy and the government of Belgium took place during the week of the 76th anniversary of the battle of the Bulge at the memorial on Mardasson Hill outside Bastogne. That was during the COVID-19 pandemic lockdown/no travel period, and our embassy and Belgian authorities had to approve my travel for entry into Belgium.

On December 12, 2020, I signed the formal document for the government of the United States of America accepting the monument as ABMC's 31st federal monument. We renamed it, and from here on it will be the "Battle of the Bulge Memorial." State Secretary Mathieu Michel signed for the government of Belgium. I said in my brief remarks, "How honored we are today to assume responsibility for preserving for generations to come, this heartfelt tribute from the people of Belgium to the American soldier." A dedication plaque was unveiled, and wreaths were laid.

Upon my return home, I had a telephone call from General Merrill A. McPeak, a former chairman of ABMC and 14th chief of staff of the Air Force. He was ecstatic that we were able to add the Battle of the Bulge Memorial to ABMC's American monuments, honoring those who fought in America's largest battle in World War II. "How in the hell were you able to do that?" he asked, "What a great accomplishment for our country!" I shared his call with the commissioners and staff.

The driving force was Ambassador Gidwitz whose vision it was two years earlier to bring this monument under U.S. control. His personal generosity, perseverance and unwavering patience made it happen, while ABMC was a steadfast partner. Our commissioners, when asked about their three-year service with ABMC, say "of the many achievements, this is our greatest achievement!"

At the First Colors ceremony on April 16, 2021, the new National World War I Memorial located in Washington, DC, was dedicated with the American flag being raised over the memorial for the first time. The flag had flown previously over our nation's capital and over all nine of our World War I cemeteries in Europe. The new national memorial was built within what was once the site of the American Expeditionary Forces (AEF) Memorial, dedicated in 1981 to honor American soldiers who served in Europe during the war and their commander, General of the Armies John J. Pershing.

The memorial honors not just the AEF, but all Americans who served in the Great War, including those who served in civilian capacities. The final element of the memorial, a 58-foot-long bronze sculpture titled *A Soldier's Journey*, was installed in September 2024. That new memorial became the 32nd memorial maintained by ABMC and is jointly managed with the National Park Service.

This chapter of my life began with, "You serve at the pleasure of the president." On January 20, 2021, we had a new president and administration, and I received a letter from the White House in March requesting I resign. As secretary, I visited all 26 of our cemeteries, some of them multiple times. I left ABMC on March 15, 2021.

It was one of the most unique, busiest, and enjoyable experiences of my life. It certainly was the most gratifying, and most respectful! I witnessed firsthand our nation's legacy of service and sacrifice. No one does a better job of honoring their war dead than the American Battle Monuments Commission, and it was an enormous honor to serve as the Commission's secretary in the Trump administration!

Epilogue

My story has no final ending, but in telling it I have had the opportunity to reflect upon what I feel has been a life well-lived. It is not a self-congratulatory tale, and hopefully, it does not come across as boastful. It is intended to extend hope and promise to those who read it who face their own setbacks. My challenges came to me early, starting with overcoming polio paralysis when I first gathered the fortitude to confront them. I hope that my ability to "overcome" will serve as an inspiration to others and give them hope that through perseverance, faith, and the strong desire to succeed, it is possible, and eventually, gratifying. Take that first step, then another, and another until it is done. Never stop trying. Never give up!

My never-ending battle with polio continues with the worsening effects of post-polio syndrome complications. It will never be over. Conserving energy and muscle strength is paramount in my treatment. I continue to wear a plastic ankle-foot orthosis (brace) and use a cane for support and as the good soldier I feel I have always been, I will "soldier on," continuing to persist just as I encouraged my soldiers to do over decades of military service. The secret of a good and happy life is to not take yourself too seriously! I never did. Nor did I ever feel I was special or important. I never lost my sense of humor or my capacity to laugh at myself, which are essential to my contentment and well-being.

Never did I envision I would write my life's story and that it would include 33 years' service in the United States Army, followed by years of service in both the private and public sectors. How this happened while overcoming childhood paralysis and battling the effects of polio is the quintessence of the story you have read.

The purpose of life is to live it and to enjoy it to the fullest without fear! I have done that, and I continue to look for newer and more challenging experiences. It is how you cope with adversity and its setbacks that determine who you are and that makes life what it is. Happiness is a choice. I never put the responsibility for my own happiness in someone else's hands.

From the day of my polio onset, I fully understood the strength my parents gave to me. My father telling me to throw my brace and crutches under the bed was a pivotal moment for me that would define who I was and determine the course of my life. His confidence in me led to my becoming an Army infantryman.

It was with perseverance and resilience that I completed nine weeks of Ranger School—the Army's toughest training. The words that followed me for the rest of my life came from that experience: "How badly do you want the Ranger tab?" That is what we must ask ourselves, "How badly do you want ...?"

Self-discipline comes from within and is determined by how badly you want something. I could have given up fighting to overcome polio, or in the Army, I could have pursued a less-arduous, less-dangerous path than an infantryman and paratrooper. I didn't. I wouldn't settle for anything less than normalcy and full recovery.

I dealt with the taunts and indiscretions that came my way and moved on. My character and spirit were cultivated by an exercise of mental and physical will. I have been tried and tested. You can do anything you put your mind to! This is what I want the reader to take from my life's story. Friends and family sometimes ask, "Where do you get your energy? You're getting old. When are you going to stop working and enjoy life?"

I never let anything impede me—certainly not age. Age is only a number. One of my favorite Hollywood actors and a great American patriot, 90-year-old Clint Eastwood, said, "I get up in the morning, and I don't let the old man in!" Said another way, "I get up in the morning and I don't let the young boy out!"

Now in my 80s, and still embracing those boyish, daredevil tendencies of honesty, innocence, taking a risk, and not worrying what others think, helped me to not become too serious and to enjoy life. I'm much wiser and more mature but not afraid to admit I still carry that boy inside me! I'll never let him go. My memories of lying in a hospital bed with paralysis served me well while participating firsthand in some of America's most historic and defining moments these past 60 years.

I was privileged to have led and fought alongside the best and toughest American soldiers that ever wore the uniform. Sometimes, I felt that I learned more from them than they learned from me. What could have been my negative, ended up being my strength!

Leaving the Army after 33 years wasn't an easy decision. Later, being appointed by the president to be secretary of ABMC responsible for guarding and commemorating America's war dead, was the fitting culmination of my decades of national service to our country.

I cherish the inalienable rights our Declaration of Independence gives—among these is the "pursuit of happiness." It doesn't guarantee happiness, only the right to pursue it. Achieving it is up to you! It's not easy. There are obstacles along life's journey, but a stout heart and mind will overcome them. I will continue to pursue to the fullest, living a happy and good life of service to my country and family, and not let the burdening effects of polio or other setbacks stop me.

Afterword

In an effort to create context for my story and for today's reader, I have attempted to provide a comparison of the America of my youth and of the six decades I was serving and working, to the country we live in today. There is no comparison!

Today, too many people believe that they are "victims" because of past transgressions and are "owed" something. The mere thought of Americans who never owned a slave, paying reparations for an injustice that occurred 160 years ago, to Americans who never were slaves, is unacceptable and contrary to the very principle of fairness. Victimhood and the belief that "I am owed" something robs one of their very dignity!

I grew up in an era of accountability. You didn't blame others for past transgressions or for your personal setbacks. You took responsibility for your circumstances and worked to overcome challenges and adversity. Today, there is a more somber, melancholy and less optimistic spirit in our country. Some Americans are losing all hope. But we must keep our hopes and spirits high.

Being able to laugh at oneself is good medicine! There is no comedy in today's world. We have forgotten how to tell a joke. A good joke helps relieve pressure and lightens your workday load. Today, people become too easily offended. If you even glance at someone the wrong way, or use the wrong word, you'll lose your job, or worse, you'll be physically attacked or slapped with a lawsuit.

Today, we are a society that walks on eggshells. Americans can't continue to live this way. We are not a fragile, cautious lot. Tiptoeing through life is not who we are. I played and worked in an arena of "give-and-take." There were no "safe spaces" where I could run when someone hurt my feelings. I dealt with it. It toughened my mind and spirit and prepared me for the bigger, more ominous, bumps in the road ahead and the cruelties of life. Today's humorless, overly sensitive, delicate souls who are so easily bothered would not have survived in my day! They weaken the traditional and well-grounded ethos of America!

My teachers encouraged, guided, and pushed me. They didn't coddle me or feel sorry for me. My father reminded me and my sister that "no one owes you a living." You must play the hand you're dealt. Don't fold! With encouragement and tough love, I stood my ground in the face of adversity, cultivating an inner strength that served me well when times were dark.

Unlike today, in grammar school, the day began with students standing for the recitation of the Lord's Prayer and the Pledge of Allegiance to the American flag. Before every school event, we stood for the playing of our national anthem. This spiritual and patriotic aspect of our education was fundamental to teaching the values America was founded on.

What is in vogue today that I never heard of while growing up, is "diversity, equity and inclusion" (DEI). Why is there such an emphasis on this concept in every aspect of our society? An entire industry of bureaucrats has emerged on every academic campus and school to ensure all Americans adhere to the tenets of DEI. The military services now have DEI staffs. I grew up with kids from all races, ethnic and religious backgrounds ... and served in an integrated Army.

We didn't call it equity; we called it friendship. I also had, among my high school classmates, exchange students from France and Germany. I couldn't have had a more diverse experience growing up and going to school. There were no victims or persecutors. And the best "privilege" I ever had was being an American! DEI needs to die a natural death!

Unlike today, "merit" was a dominant force in my growing up. No one received special treatment because he or she was from an "under-served" or "disenfranchised" community. Victimizing and lowering standards doesn't foster high self-esteem. Rather, it destroys ambition and productivity.

Self-esteem comes from hard work and the accomplishment that results from it. Today's bleeding heart is more supportive of the lazy lawbreaker than the righteous, industrious, person. We all had the same opportunity, and we competed to win, to be the best we could be! We were judged on merit and skills, and "earned" whatever we got. There were no handouts. The bar was never lowered.

Failure was the best teacher. I failed often at my first and second attempts to do something. But I kept trying. Knowledge gained from failure lasted longer than knowledge gained from successes.

I disagree with those who espouse "diversity" as being the strength of America. America's cultural greatness and strength is its "oneness," described in our de facto motto, "*E Pluribus Unum*," "Out of many, one!" Immigrant groups, all learning English and assimilating as one America was my experience guiding me throughout my life.

Some are attacking our past, our country's character and all the values upon which America was founded. The best-selling author and political commentator, Heather Lynn MacDonald, describes it well: "We are turning on our own legacy and declaring it evil, oppressive without any redeeming characteristics."

Our society has devolved from being one of the best educated, most patriotic, and hardest working in history to one that denigrates the country, alters its history, destroys its memorials and cities, and sees less than 1 percent of the population in military service.

As a soldier, what is most disheartening is the "woke" agenda's anti-Republic social justice movement permeating American society, including the military, the nation's last bastion of decency! Of all segments of society, where social engineering will guarantee defeat for the country, it is the military, whose mission is to fight the nation's wars.

The more I compare the America of my youth to the America of today, the more frightened I am for our future! America's Founders wrote, "Only a moral, righteous, and virtuous people can be free." I believe that. We have lost the spiritual and religious qualities enabling us to do what is good and best for the country.

Despite today's feeling of gloom and hopelessness, the heavens have not fallen. America has faced worse times and overcame them. Our government derives its just powers from the consent of the people. As Thomas Jefferson wrote, "When the government fails to abide by the Constitution, is destructive and corrupt, and/or fails to serve the people, it is our right, and our duty, to change or abolish it and establish a new government." I'm up to the task as are others. Our "toughest battle" is ahead. We should not despair!

Don't tell me we will never see "one" America again or that America's decline is permanent! It is not! Too many have given their all and have fought, bled and died for this Republic for that to happen. Indeed, my experiences of 80 years have more than prepared me for joining other Americans in the fight for the "heart and soul" of our nation—in whatever form it might take. It is a battle that we can't afford to lose.

Glossary

AAFMAA	American Armed Forces Mutual Aid Association
AAR	After-action report
ABGR	American Businessmen's Group Riyadh
ABMC	American Battle Monuments Commission
ACR	Armored Cavalry Regiment
ADC	Assistant division commander
AFB	Air Force Base
AIS	Army Intelligence and Security
AIT	Advanced individual training
AMVETS	American Veterans Association
AO	Area of operations *or* Action officer
AP	Anti-personnel
ARG	Amphibious Ready Group
ARG/SLF	Amphibious Ready Group/Special Landing Force
ASOP	Airborne Standard Operating Procedure
ATC	Armored troop carrier
AUSA	Association United States Army
AVM	Arteriovenous malformation
AWC	Army War College
BCSB	Bayonet Combat Support Brigade
BCT	Battalion Combat Training/Brigade Combat Team
BCTP	Battle Command Training Program
BJU	Beach jumper unit
BLT	Battalion Landing Team
BT	Basic training
C&C	Command and control
CENTCOM	Central Command
CGSC	Command and General Staff College
CINCPAC	Commander in Chief Pacific
CNO	Chief of Naval Operations
COM	Chief of mission
COMPHIBGRUEASTPAC	Commander Amphibious Group Eastern Pacific

COMUSMACV	Commander United States Military Assistance Command, Vietnam
CPX	Command Post Exercise
CSM	Command sergeant major
DACG	Departure Airfield Control Group
DCG	Deputy commanding general
DCSOPS	Deputy chief of staff for operations
DEFCON	Defense condition
DISCOM	Division Support Command
DMZ	Demilitarized Zone
DODDS	Department of Defense Dependent School
DRF	Division ready force
DS	Direct support
DSC	Distinguished Service Cross
DTAC	Division Tactical Command Post
DZ	Drop zone
EDRE	Emergency Deployment Readiness Exercise
EPA	Environmental Protection Agency
EUSA	Eighth U.S. Army
FAO	Foreign area officer
FLOTUS	First Lady of the United States
FO	Forward observer
FORSCOM	Forces Command
FRAGO	Fragmentary order
FTX	Field training exercise
GOMO	General Officer Management Office
HAZMAT	Hazardous materials
HEAT	High explosive anti-tank (ammunition)
IED	Improvised explosive device
IET	Initial Entry Training
IOBC	Infantry Officer Basic Course
JACADS	Johnston Atoll Chemical Agent Disposal System
JAVA	Japanese American Veterans Association
JCS	Joint Chiefs of Staff
JTFSO	Joint Task Force South
JRTC	Joint Readiness Training Center
JRTX	Joint Readiness Training Exercise
JTF	Joint Task Force
KATUSA	Korean Augmentation to the United States Army
KIA	Killed in action
LBE	Load-bearing equipment
LID	Light infantry division

LST	Landing ship, tank
LZ	Landing zone
MACOM	Major Command
MAC	Military Airlift Command
MACV	Military Assistance Command, Vietnam
MAMC	Madigan Army Medical Center
MAU	Marine Amphibious Unit
MAW	Marine Aircraft Wing
MCTP	Mission Command Training Program
MDL	Middle Demarcation Line
METL	Mission essential task list
MI	Military Intelligence
MOD	Ministry of Defense
MOS	Military Occupational Specialty
MRB	Mobile Riverine Base
MRF	Mobile Riverine Force
MSR	Main supply route
NAUS	National Association for Uniformed Services
NCO	Noncommissioned officer
NLF	National Liberation Front
NOAA	National Oceanic and Atmospheric Administration
NTC	National Training Center
OB	Order of battle
OCFHO	Oahu Consolidated Family Housing Office
ODCSOPS	Office of the Deputy Chief of Staff for Military Operations and Plans
OPM-SANG	Office of the Program Manager-Saudi Arabian National Guard
OPTEMPO	Operating tempo
OVT	Operational verification testing
PACOM	Pacific Command
PDF	Panamanian Defense Forces
PGM	Program General Manager
PMS	Professor of military science
POI	Program of instruction
POTUS	President of the United States
POW	Prisoner of war
PPO	Presidential Personnel Office
PPS	Post-polio syndrome
PT	Physical training
PTSD	Post-traumatic stress disorder
PUC	Presidential Unit Citation

PX	Post exchange
RAD	Retiree Appreciation Day
RADM	Rear admiral
RC	Reserve component
RCT	Regimental combat team
RDF	Rapid Deployment Force
RDJTF	Rapid Deployment Joint Task Force
REDCOM	Readiness Command
ROE	Rules of Engagement
ROKA	Republic of Korea Army
ROTC	Reserve Officer Training Corps
RPG	Rocket-propelled grenade
RTO	Radio telephone operator
S3	Operations officer
SANG	Saudi Arabian National Guard
SEAL	Sea, air and land
SECDEF	Secretary of Defense
SECNAV	Secretary of the Navy
SERB	Selective Early Retirement Board
SGS	Secretary of the general staff
SMW	Society of Military Widows
SOP	Standard operating procedure
SP	Start point
STRICOM	Strike Command
TB	Training brigade
TBI	Traumatic brain injury
TCN	Third country national
TRADOC	Training and Doctrine Command
UDT	Underwater Demolition Team
USAREUR	U.S. Army Europe
USARPAC	U.S. Army Pacific
USCENTCOM	United States Central Command
USFK	U.S. Forces Korea
USPHS	United States Public Health Service
VBIED	Vehicular borne improvised explosive device
VC	Viet Cong
VDBC	Veterans Disability Benefits Commission
WESTCOM	Western Command
WESTPAC	Western Pacific
WIA	Wounded in action
XO	Executive officer

Endnotes

Chapter 1

1. Philip Lewin, MD, FACS. *Infantile Paralysis: Anterior Poliomyelitis.* Philadelphia: Saunders, 1941.
2. Sister Elizabeth Kenny (1880–1952) was an Australian nurse trained in pediatric care. After treating polio patients, she devised her own treatment method. While controversial, she travelled throughout Australia, Europe and the United States treating polio patients and opening clinics, including the Sister Kenny Institute in Minneapolis. She trained many others in this method.
3. Elizabeth Kenny, *The Treatment of Infantile Paralysis in the Acute Stage.* Minneapolis, MN: Bruce Publishing Company, 1941.
4. Dr. John Aseff, Medical Director of the Post-Polio Clinic at the National Rehabilitation Hospital in Washington, DC, explanation of re-innervation of motor neurons in recovering polio patients.
5. Prostigmin is used to treat the reversal of Nondepolarizing Neuromuscular Blockade. Prostigmin may be used alone or with other medications. Wikipedia.
6. Daniel J. Wilson, "A Crippling Fear: Experiencing Polio in the Era of FDR." *Bulletin of the History of Medicine*, Johns Hopkins University Press (Fall 1988), p. 487.
7. Ibid. p. 491.

Chapter 2

1. The Hoover Commission was a group appointed by President Harry S. Truman in 1947 to recommend organizational and administrative changes in the federal government. It was named for President Herbert Hoover.
2. General Lyman Louis Lemnitzer CBE (August 29, 1899–November 12, 1988) served as the fourth chairman of the Joint Chiefs of Staff from 1960 to 1962. He is one of only four officers in the history of the United States Army to have actively served as a general during three major wars (World War II, Korea and Vietnam). Wikipedia.

Chapter 3

1. See John Spencer, "The Challenges of Ranger School and How to Overcome Them," (USMA West Point, New York: Modern War Institute, 2016). https://mwi.usma.edu/challenge-ranger-school-can/.
2. Paul J. Scheips and David P. Griffin, "The Role of the Army in the Oxford, Mississippi Incident 1962–1963," OCMH Monograph No. 73M (June 24, 1965).
3. Ibid.
4. Ibid.
5. Ibid.

6. See comments by General Curtis E. LeMay, the Air Force chief of staff who attended all security briefings. See also an interview with Colin D. Heaton, *Above the Reich*. (New York: Dutton Caliber, June 8, 2021), pp. 340–3.
7. Dino A. Brugioni, "The Invasion of Cuba," *The Quarterly Journal of Military History* (Winter 1992).
8. Ibid.
9. The Grange is the National Farmers' Association that advances agriculture and animal husbandry.
10. One of these was Sergeant Charles Robert Jenkins who deserted for North Korea in January 1965. See Jenkins's memoir with Jim Frederick, *The Reluctant Communist: My Desertion, Court-Martial, and Forty-Year Imprisonment in North Korea*. University of California Press, 2008.

Chapter 4

1. Captain Antonio Michael Mavroudis (June 6, 1939–October 28, 1967) from New York, NY, was posthumously promoted to major.
2. According to various sources, including Max Hastings, "between 1965 and 1968, 310,011 served there, along with 346 advisers." See Max Hastings, *Vietnam: An Epic Tragedy*. New York: Harper Collins, 2018, p. 371. On the 1967 NVA number of 232,000 in the south by the end of 1967, see p. 389.
3. Many books mention the Kit Carson Scouts. See James E. Livingston, Colin D. Heaton and Anne Marie Lewis, *Noble Warrior: The Story of Maj. Gen. James E. Livingston USMC (Ret.) Medal of Honor*. Minneapolis, MN: Zenith Press, 2010, pp. 12–13.
4. See Major General William B. Fulton, *Vietnam Studies—Riverine Operations 1966–1969*, St. John's Press, 1972, pp. 17–19.
5. Ibid.
6. Ibid.
7. See Major General Ira A. Hunt, Jr., *The 9th Infantry Division in Vietnam—Unparalleled and Unequaled* (University Press of Kentucky, 2010), p. 7.
8. Fulton, p. 20.
9. Ibid, p. 172.
10. Ibid, pp. 21–2.
11. Dong Tam was the home ashore for the Mobile Riverine Force including the Army's 9th ID and the Navy's Riverine Assault Force.
12. Fulton, p. 23.
13. Henry Cabot Lodge, Jr. (July 5, 1902–February 27, 1985), American diplomat and Republican United States senator from Massachusetts; United States ambassador to South Vietnam from August 25, 1965 to April 25, 1967. See Lodge opinions in Hastings, p. 380. See also Fulton, pp. 24–5.
14. On media bias and disinformation see Hastings, pp. 379–80.
15. See also Fulton, p. 45.
16. Ibid, p 27. For detailed information on Westmoreland taking command of MACV and Major General DePuy, his operations officer, see Hastings, p. 209.
17. For discussion of approval of concept, see Fulton, pp. 26–31.
18. The USS *Colleton* (APB-36) was a Benewah-class barracks ship.
19. This was one of many such operations in late 1967 and early 1968. Later, on January 10, 1968, the 2nd Brigade, 9th ID accounted for 47 enemy KIA after bitter fighting in the same area of Dinh Tuong Province.
20. Fulton, pp. 138–9. See also Neil Sheehan, *A Bright Shining Lie: John Paul Vann and America in Vietnam*. (Vintage, 1989), which details specifics related to this chapter.

21. Fulton, p. 139.
22. The 3/47 Infantry Battalion Combat After Action Report (AAR), Battle of 4 Dec 1967, dated January 11, 1968.
23. Ibid.
24. Ibid.
25. Again, on media reporting, see Hastings, pp. 379–80.
26. Fulton, p. 142.
27. Covered under Articles 19 and 53 of the Geneva Convention of 1949, ratified in 1950, and The Hague Conventions.
28. This includes whether a hospital, church, mosque, etc.
29. See 3/47th Infantry Battalion AAR for Rach Ruong Canal Battle.
30. *The Old Reliable* (newspaper) February 21, 1968.
31. Fulton, p. 149.
32. Ibid, pp. 155–6.
33. Ibid.
34. See 3/47 Infantry Battalion Combat AAR, Can Tho Battle, dated March 24, 1968.
35. General Creighton W. Abrams, Jr. (September 15, 1914–September 4, 1974) was one of the most effective tank commanders during World War II, becoming famous during the battle of the Bulge. He commanded U.S. forces in Vietnam from 1968–72. He was U.S. Army Chief of Staff (1972–74). Due to German intelligence reports following the failed Ardennes Offensive Adolf Hitler issued the order to "get that Jew Abrams." Ironically, Abrams was a Methodist.
36. Hastings states that "in 1967 ... one in five soldiers were already using marijuana, a proposition that three years later would rise to half. The in-country percentage of known pot smokers grew from one quarter in 1967 to two-thirds in 1971." Hastings, p. 382. These figures have been hotly contested, and don't reflect accurately the use of marijuana in my unit in 1967–8.
37. President Ronald Reagan in his address to the VFW National Convention, August 18, 1980.

Chapter 5

1. 2nd Lieutenant William Laws Calley was the only man charged involving C Company, 1st Battalion, 20th Infantry Regiment and B Company, 4th Battalion, 3rd Infantry Regiment, 23rd ID. The death toll was 347 according to the U.S. Army, others estimate that more than 500 people died. Calley was convicted and sentenced to life in prison, which was reduced to 10 years.
2. An article entitled "The Memory Hangs on ..." by Sophia McDermott-Hughes reprinted in *The Middlebury Campus* (April 28, 2022).
3. Ibid.
4. Ibid.
5. An article by Robert Keren entitled "Campus Unrest in the '60s and '70s," January 24, 2014. https://www.middlebury.edu/announcements/2014/01/campus-unrest-60s-and-70s.
6. McDermott-Hughes.
7. Ibid.
8. Ray Stewart, ed. "Marine Corps' Amphibious Operations in the Vietnam War"; also Peter Brush, "Marine Corps Armor in the Special Landing Force (SLF) Vietnam War 1965–1970." Both articles are found in Marine Corps Vietnam Tankers Historical Foundation document (Annex G-11—Marine Corps' Amphibious Operations in the Vietnam War (https://mcvthf.org/Book/ANNEX%20G-11.html).
9. Ibid.
10. Ibid.

11. See *JCS and the War in Vietnam, 1960–1968*, Office of Joint History, (OJCS, 2009), p. 162.

12. Ibid. pp. 162–3.

13. Paul Henry Nitze (January 16, 1907–October 19, 2004) served as U.S. Deputy SECDEF, SECNAV, and director of policy planning for the U.S. State Department.

14. Paul Culliton Warnke (January 31, 1920–October 21, 2001) was opposed to expanded involvement and told Lyndon Johnson after the Tet Offensive the war was "unwinnable."

15. *JCS and the War in Vietnam*, p. 163. For an excellent discussion of the Clifford Committee's review of Operation *Durango City* see pp. 161–8.

16. Ibid. p. 163.

17. David Dean Rusk (February 9, 1909–December 20, 1994) was the United States secretary of state from 1961 to 1969 under presidents John F. Kennedy and Lyndon B. Johnson, the second-longest serving secretary of state after Cordell Hull from the Franklin Roosevelt administration.

18. Wikipedia article on Clark Clifford's life cites a *Washington Post* article interview with Dean Rusk.

19. Admiral David M. Rubel earned the Navy Cross on May 3, 1945, for shooting down 19 of 24 kamikazes attacking his ship, the USS *Aaron Ward* while a gunnery officer. He later served during Korea and Vietnam.

20. USS *Paul Revere* Plan of the Day—Shellback Initiation Program, dated September 9, 1970.

21. Ibid.

22. The USS *Indianapolis* delivered the first atomic bomb to Tinian on July 26, 1945, and was sunk by submarine I-58 on July 30, 1945. Only 316 of the 1,100 crew survived to be rescued. The captain of the "*Indy*" was court-martialed for losing his ship. Decades later he was exonerated.

23. General Tran Thien Khiam (1925–2001) was a South Vietnamese soldier and politician. During the 1960s, he was involved in several coups. He helped President Ngo Dinh Diem put down a November 1960 coup attempt and was rewarded with a promotion. Later in 1963 he was involved in the coup that assassinated Diem. Wikipedia.

24. Nguyen Van Thieu (April 5, 1923–September 29, 2001) was a South Vietnamese military officer and politician who was the president of South Vietnam from 1967 to 1975.

25. Amphibious Group Three All Hands Letter From Keelung, Taiwan, November 15, 1970, signed by RADM David M. Rubel.

26. Ibid.

27. Ibid.

28. Ibid.

29. In World War II, Typhoon Cobra also known as "Halsey's Typhoon" struck the United States Pacific Fleet in December 1944. Three destroyers sank and 790 sailors were killed.

Chapter 6

1. Elihu Root (February 15, 1845–February 7, 1937) was a lawyer and Republican politician, who served as secretary of state and secretary of war in the early 20th century. He was also a U.S. senator from New York and received the 1912 Nobel Peace Prize. Wikipedia.

2. William E. Farrell article in *The New York Times*, "800 US Paratroopers Open War Games in Egypt," (November 15, 1981).

3. Muhammad Anwar el-Sadat (December 25, 1918–October 6, 1981) was a military officer, and third president of Egypt, from October 15, 1970 until his assassination by fundamentalist army officers on October 6, 1981. Wikipedia.

4. Loren Jenkins article in *The Washington Post*, "U.S. FORCES Airlifted to Egypt for Joint Maneuvers" (November 9, 1981).

5. Military Airlift Command (MAC) abstract summarizing MAC's role in *Bright Star*, date and author unknown.
6. As recorded in my 82nd Airborne Division Parachute Jump Log.
7. Farrell, *New York Times*.
8. Ibid.
9. 82nd Airborne Division Parachute Jump Log.
10. These statistics were reported in Lieutenant Colonel Harold L. Timboe, "Mass Casualty Situation: GALLANT EAGLE 82 Airborne Operations: A Case Report," MC USA, *Military Medicine Journal*, Vol. 153, April 1988.
11. Ibid.
12. Robert Lindsey, "Winds Suspected in Chute Accidents," *The New York Times* (April 1, 1982).
13. 82nd Airborne Division All American Review Program—28 May 1982.

Chapter 7

1. Dr. Garretson was best known as a compassionate and caring physician and teacher. His practice was primarily at Norton Hospital and the University of Louisville Hospital. He was the neurosurgical division chief and chairman of the Department of Neurological Surgery from 1971–97, as well as president of the Neurosurgical Institute of Kentucky.
2. Casper W. Weinberger, *Fighting for Peace: Seven Critical Years in the Pentagon* (Warner Books, 1990), p. 7.
3. Ibid, p. 381.
4. Ibid, p. 383.
5. See Peter Grier, "Congress Seeks Greater Control Over U.S. Actions in Persian Gulf," *Christian Science Monitor* (24 September 1987).
6. Weinberger, p. 401.
7. Ibid, p. 403. Also https://en.wikipedia.org/wiki/USS_Stark_incident.
8. USS *Vincennes* (CG-49) was a Ticonderoga-class guided missile cruiser that shot down Iran Air Flight 655 on July 3, 1988, during an engagement with Iranian gunboats.
9. Weinberger, p. 402.
10. Ibid, p. 403.
11. Molly Moore, "Pentagon Tips Its Cap to Weinberger," *The Washington Post* (November 18, 1987).

Chapter 8

1. Lieutenant Colonel Timothy A. Wray, "The Army's Light Infantry Divisions: An Analysis of Advocacy and Opposition," essay written for the National War College (January 1, 2005).
2. Ibid.
3. Omar Efraín Torrijos Herrera (February 13, 1929–July 31, 1981) was the commander of the Panamanian National Guard and military leader of Panama from 1968 to his death in 1981.
4. See the 1990 book *7th Infantry Division (Light) JUST CAUSE 20 Dec 89–30 Jan 90* (listed in WorldCat Online Global Catalog); also, "Operation *Just Cause*: The Planning and Execution of Joint Operations in Panama February 1988–January 1990" a monograph by Dr. Ronald H. Cole published by the Joint History Office, Office of the Chairman of the Joint Chiefs of Staff (November 1995), p. 6.
5. See John T. Correll, "A Small War in Panama," *Air Force Magazine* (December 1, 2009).
6. John Dinges, *Our Man in Panama: How General Noriega Used the United States And Made Millions in Drugs and Arms* (New York: Random House, 1990).

7. Lieutenant Colonel James H. Embrey, "Operation JUST CAUSE: Concepts for Shaping Future Rapid Decisive Operations," in *Transformation Concepts for National Security in the 21st Century*, edited by Williamson Murray (U.S. Army War College Press, 2002, 2002), p. 200.

8. See *7th Infantry Division (Light) JUST CAUSE*, WorldCat Online Global Catalog.

9. Ibid. See also Cole, "Operation JUST CAUSE," pp. 6–8.

10. Embrey, pp. 200–1.

11. Ibid.

12. Ibid.

13. Ibid, pp. 201–2.

14. Ibid, p. 201.

15. Cole, p. 17.

16. Embrey, pp. 203–4.

17. Ibid, p. 202.

18. Ibid, p. 202.

19. Cole, pp. 10–11.

20. On the Dignity Battalions see https://en.wikipedia.org/wiki/Dignity_Battalions.

21. Ibid.

22. See Keith Kellogg, *War By Other Means: A General in the Trump White House* (Regnery Publishing, 2021) for his brigade's role in *Nimrod Dancer* and *Just Cause*.

23. Cole, p. 27.

24. Ibid.

25. Ibid, pp. 29–30.

26. Ibid, p. 30.

27. Ibid, p. 34.

28. Ibid. p. 42.

29. Cole, p. 56.

30. Cole, p. 67.

31. Embrey, p. 220.

32. Cole, pp. 51, 53.

33. The invasion of Grenada, Operation *Urgent Fury*, launched on October 25, 1983. The U.S. with six Caribbean nations invaded the small island nation after the murder of the prime minister by the Cuban and Soviet friendly People's Revolutionary Government, endangering American students on the island.

34. See *Army Times* article covering General Stiner's candid assessment of *Just Cause* (February 26, 1990).

35. In addition to the cited references, much of the data and statistics in this chapter are taken directly from various official documents including unit/agency after action reports, historical summaries, briefings, and actual aircraft deployment schedules published by the following sources: Office of the Joint Chiefs of Staff, Department of the Army, 7th Infantry Division and the Air Force 60th and 62nd Air Lift Wings. *Operation JUST CAUSE Lessons Learned, Volumes I–III*, published by the Center for Army Lessons Learned (CALL) in October 1990, were also used as corroborating sources to help ensure the accuracy of the data.

Chapter 9

1. For the history of the JACADS project see "Johnston Atoll Chemical Agent Disposal System (JACADS) Final Environmental Impact Statement" prepared by U.S. Army Corps of Engineers (November 1, 1983).

2. Part of the Nunn-Warner agreement included removal of chemical weapons from West Germany in 1986. The German government, as did all Europeans, wanted the chemical munitions removed from their continent.

3. The Army's Capstone Program aligned all Army reserve and Army national guard units under active gaining commands for peacetime and wartime purposes.

Chapter 10

1. Mutaib bin Abdullah Al Saud (born March 6, 1952) is a Saudi military officer and member of the royal family who served as Saudi Arabia's minister of the National Guard from 2013 to 2017. Previously he was commander of the National Guard from 2010 to 2013. Wikipedia.

2. Abdullah bin Abdulaziz Al Saud (August 1, 1924–January 23, 2015) was the king of Saudi Arabia from 2005 to 2015. During his reign, tensions increased between conservative Salafi (Wahhabi) forces and liberal reformers.

3. The Khobar Towers was a terrorist attack using a truck bomb at Building #131, part of a housing complex, near the King Abdulaziz Air Base on June 25, 1996. It was the living quarters for coalition forces who were assigned to Operation *Southern Watch*, enforcing the Iraqi no-fly zones. Wikipedia.

4. As reported in the May 13, 2003 edition of the Kuwait News Agency (KUNA).

5. Fahd bin Abdulaziz Al Saud (1921 or 1923–August 1, 2005) was King of Saudi Arabia from June 13, 1982 until his death in 2005. Wikipedia.

6. Saudi government statement, "Riyadh Names 12 Perpetrators," dated June 7, 2003.

7. Stephen R. Weisman, "Toll in Saudi Arabia Rises to at Least 20," *New York Times* (May 13, 2003).

8. *The Daily Telegraph* (May 16, 2003).

Chapter 11

1. USPHS is one of the nation's uniformed services committed to the service of health. Its 6,000 officers serve in agencies across the government.

2. NOAA corps is a uniformed service of the government that monitors oceanic and atmospheric conditions and supports major waterways. Its 320 officers serve aboard ships, fly aircraft and conduct marine and aviation operations.

3. Many of the dates/times and subjects of the visits to Congress, RADs and NAUS events cited in this chapter were taken directly from my 2005–11 personal desk calendar diaries and from articles in the 2005–11 editions of the bi-monthly NAUS *Uniformed Services Journal*.

4. *Flat Stanley* is a character in a children's book series written by New York City born author Jeff Brown. The first *Flat Stanley* book was published in 1964. The book series is popular among children of all nationalities.

5. The John Abbott House and Museum, added to the National Register of Historic Places 1n 1976. Wikipedia.

6. The dates and times of congressional visits and other events are taken directly from my NAUS desk calendar diaries and NAUS journals.

Chapter 12

1. See the December 2020 edition of ABMC "Commemorative Sites Booklet" for descriptions of commemorative sites.

2. U.S. Coast Guard Cutter *Tampa* was sunk by Kapitänleutnant Hans Wolf Hertwig commanding UB-91 on September 26, 1918. Hertwig surfaced to recover survivors and found none among the 131-man crew.

3. At dawn on October 31, 1941, the USS *Reuben James* was torpedoed and sunk near Iceland by U-552 commanded by Kapitänleutnant Erich Topp. Of the 144-man crew only 44 enlisted men survived. Wikipedia.

4. Due to the incredible success of the 5th and 6th Marine Regiments they were awarded the French Fourragère, that nation's senior unit award, akin to a Presidential Unit Citation.

5. Jeffrey Goldberg, ed. *The Atlantic* (September, 2020).

Bibliography

7th Infantry Division Light JUST CAUSE 20 Dec 89–30 Jan 90. WorldCat Online Global Catalog.

"9th Infantry Division in Vietnam." *OCTOFOIL Magazine* editions 1 and 2 for period January–June 1968.

3/47th Infantry Combat After Action Reports during the Vietnam War for the period November 1967–April 1968.

3/47th Infantry OPORD 3-68 (TIGER Coronado X), dated 21 January 1968 during the Vietnam War.

82nd Airborne Division. *All American Review* Program. May 28, 1982.

82nd Airborne Division. G3 End of Tour Report, dated May 25, 1982.

American Heritage interview at https://www.americanheritage.com/exclusive-interview-clark-clifford April 1977, Volume 28, Issue 3.

Army Times. "Assessment of JUST CAUSE." February 26, 1990.

Brugioni, Dino A. "The Invasion of Cuba," *The Quarterly Journal of Military History*, Winter 1992.

Brush, Peter. "Marine Corps Armor in the Special Landing Force (SLF) Vietnam War 1965–1970." Annex G-11, https://mcvthf.org/Book/ANNEX%20G-11.html.

Bush, George H. W. and Brent Scowcroft. *A World Transformed.* New York: Knopf, 1998.

Cole, Ronald H. "Operation Just Cause—The Planning and Execution of Joint Operations in Panama, February 1988–January 1990." Joint History Office, Office of the Chairman of the Joint Chiefs of Staff, November 1995.

Company C, 3/47th Infantry Company Battle Rosters, during Vietnam War, dated November 1 and December 1, 1967.

Correll, John T. "A Small War in Panama." *Air Force Magazine*, December 1, 2009.

The Daily Telegraph, May 16, 2003.

Dinges, John. *Our Man in Panama: How General Noriega Used the United States—And Made Millions in Drugs and Arms.* New York: Random House, 1990.

Embrey, James H. and Williamson Murray, eds. "Operation JUST CAUSE: Concepts for Shaping Future Rapid Decisive Operations," in *Transformation Concepts for National Security in the 21st Century*, U.S. Army War College Press, 2002.

Encyclopedia Britannica 2020 online.

Fulton, William B. "Vietnam Studies—Riverine Operations 1966–1969." New York: St. John's Press, 1972.

Garner, Godfrey. *Brothers in the Mekong Delta: A Memoir of PBR Section 513 in the Vietnam War.* Jefferson, NC: McFarland & Company, Inc., Publishers, 2020.

Gettysburg College Freshman Fraternity Date Book, 1957.

Goldberg, Jeffrey, ed. *The Atlantic*, September 3, 2020.

Grier, Peter. "Congress Seeks Greater Control Over U.S. Actions in Persian Gulf." *Christian Science Monitor*, September 24, 1987.

Hastings, Max. *Vietnam: An Epic Tragedy.* New York: Harper Collins, 2018.

Heaton, Colin D. and Anne Marie Lewis. *Above the Reich: Deadly Dogfights, Blistering Bombing Raids, and Other War Stories from the Greatest American Air Heroes of World War II, in Their Own Words.* New York: Dutton Caliber, 2021.

Heaton, Colin D. and Anne Marie Lewis. "Belgian Volunteer in the Waffen SS," interview with Waffen SS *Oberfuehrer* Leon Degrelle, *Military History* ISSN-0889-7328. Weider History Group, Leesburg, VA, November 2006.

Heaton, Colin D. and Anne Marie Lewis. *Occupation and Insurgency: A Selective Examination of The Hague and Geneva Conventions on the Eastern Front 1939–1945.* New York: Algora Publishing, 2008.

Home of the Merciful Savior (HMS) For Crippled Children. Medical records/doctor's notes for William "Billy" Matz, 1944–1953.

Hunt, Ira A. Jr., *The 9th Infantry Division in Vietnam—Unparalleled and Unequaled.* University Press of Kentucky, 2010.

JCS and the War in Vietnam, 1960–1968. Office of Joint History, OJCS, 2009.

Jenkins, Charles Robert and Jim Frederick. *The Reluctant Communist: My Desertion, Court-Martial, and Forty-Year Imprisonment in North Korea.* University of California Press, 2008.

Joint Task Force South in Operation *Just Cause* Oral History Interview with MG William Matz, 7th Infantry Division, conducted by Dr. Robert K. Wright of the U.S. Army Center of Military History, April 30, 1992.

Jordan, Robert W. *Desert Diplomat: Inside Saudi Arabia Following 9/11.* Potomac Books, 2015.

Kellogg, Keith. *War by Other Means: A General in the Trump White House.* Regnery Publishing, 2021.

Kenny, Elizabeth. The *Treatment of Infantile Paralysis in the Acute Stage.* Minneapolis, MN: Bruce Publishing Company, 1941.

Lewin, Philip, MD, FACS. *Infantile Paralysis: Anterior Poliomyelitis.* Philadelphia: Saunders, 1941.

Lindsey, Robert. "Winds Suspected in Chute Accidents." *The New York Times*, April 1, 1982.

Livingston, James E., Colin D. Heaton, and Anne Marie Lewis. *Noble Warrior: The Story of Major General James E. Livingston USMC (Ret.), Medal of Honor.* Minneapolis, MN: Zenith Press, 2010.

Matz family collection of documents and photographs.

Matz, William. 82nd Airborne Division Parachute Jump Log, 1962–82.

McIntosh, Terry. *The Youngest Green Beret: A True Story Set in the Mekong Delta, Vietnam 1968–1969.* Independently published, 2019.

Military Airlift Command (MAC) abstract summarizing MAC's role in *Bright Star*, date and author unknown.

Military Medicine Journal, Vol. 162, December 1997.

Murray, Williamson. "Transformation Concepts for National Security in the 21st Century," *Strategic Studies Institute*. U.S. Army War College Press, 2002.

National Rehabilitation Hospital. Post-Polio Clinic medical records for William Matz, 2014–2023.

NAUS desk diary calendars and NAUS journals, 2005–11.

Office of the Joint Chiefs of Staff, Department of the Army. "7th Infantry Division and the Air Force 60th and 62nd Air Lift Wings." *Operation JUST CAUSE Lessons Learned*, Volumes I–III. Center for Army Lessons Learned (CALL), October 1990.

The Old Reliable, February 21, 1968.

Pennsylvania Polio Survivors Monthly Newsletters, 2021–4.

Pompeo, Michael. *Never Give an Inch: Fighting for the America I Love.* Broadside Books, 2023.

Powell, Colin. *It Worked for Me: In Life and Leadership.* New York: Harper Audio, 2012.

Powell, Colin. *My American Journey.* New York: Ballantine Books, Updated edition, 2003.

Scheips, Paul J. and David P. Griffin. "The Role of the Army in the Oxford, Mississippi Incident 1962–1963." OCMH Monograph No 73M, June 24, 1965.

Schwarzkopf, Norman H. *It Doesn't Take a Hero: The Autobiography.* New York: Bantam Books, 1992.

Sheehan, Neil. *A Bright Shining Lie: John Paul Vann and America in Vietnam*. Vintage, 1989.

Spencer, John. "The Challenges of Ranger School and How to Overcome Them." USMA West Point, New York: Modern War Institute, 2016. https://mwi.usma.edu/challenge-ranger-school-can/.

Stewart, Ray, ed. "Marine Corps' Amphibious Operations in the Vietnam War." Also Peter Brush (author), "Marine Corps Armor in the Special Landing Force (SLF)" Annex G-11, https://mcvthf.org/Book/ANNEX%20G-11.html.

Thayer, Jim. *Tango 1-1: 9th Infantry Division LRPs in the Vietnam Delta*. Pen & Sword Military, 2020.

Timboe, Harold L. "Mass Casualty Situation: GALLANT EAGLE 82 Airborne Operations: A Case Report." *Military Medicine Journal*, Vol. 153, April 1988.

Uniformed Services Journals, 2005–11. NAUS, Springfield, VA.

Union, Tod. *Vietnam Infantry Tactics: 9th Infantry Division in The Vietnam Delta: Vietnam War Memoirs*. Independently published, 2021.

Vietnam War Panel Discussion, Gettysburg College, October 19, 2021.

Weinberger, Casper W. *Fighting for Peace: Seven Critical Years in the Pentagon*. Warner Books, Inc., 1990.

Weisman, Stephen R. "Toll in Saudi Arabia Rises to at Least 20." *New York Times*, May 13, 2003.

Wilson, Daniel J. "A Crippling Fear: Experiencing Polio in the Era of FDR." *Bulletin of the History of Medicine*, Johns Hopkins University Press, Fall 1988.

Wray, Timothy A. "The Army's Light Infantry Divisions: An Analysis of Advocacy and Opposition." Essay written for the National War College, January 1, 2005.

Index